T0173722

GENESIS
of the
COSMOS

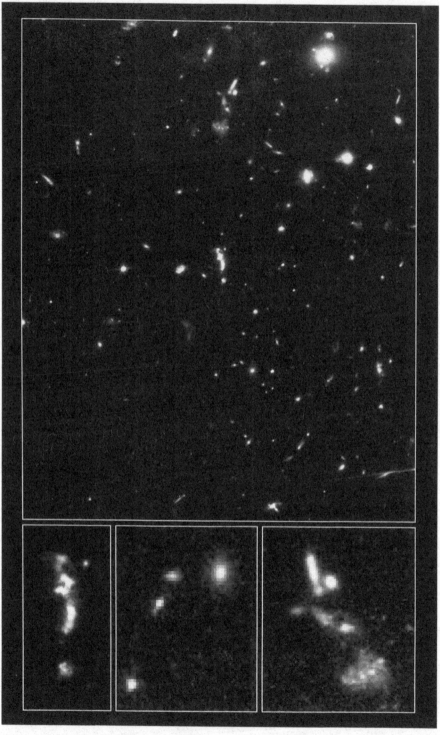

Distant cluster of galaxies as observed by the Hubble Space Telescope. Photo courtesy of NASA's Space Telescope Science Institute.

GENESIS
of the
COSMOS

The Ancient Science
of Continuous Creation

Paul A. LaViolette, Ph.D.

Bear & Company
Rochester, Vermont

Bear & Company
One Park Street
Rochester, Vermont 05767
www.InnerTraditions.com
Bear & Company is a division of Inner Traditions International

Copyright © 1995, 2004 by Paul A. LaViolette

Originally published in hardcover in 1995 by Park Street Press under the title *Beyond the Big Bang*

All rights reserved. No part of this book may be reproduced or utilized in any form or by any means, electronic or mechanical, including photocopying, recording, or by any information storage and retrieval system, without permission in writing from the publisher.

THE LIBRARY OF CONGRESS HAS CATALOGED A PREVIOUS EDITION OF THE TITLE AS FOLLOWS:

LaViolette, Paul A.
Beyond the big bang: ancient myth and the science of continuous creation / Paul A. LaViolette
p. cm.
Includes bibliographical references and index.
ISBN 978-1-59143-034-6 — ISBN 978-1-59143-839-7 (e-book)
1. Cosmology. 2. Creation. 3. System theory. 4. Microphysics. 5. Science–History. I. Title.
QB981.L327 1995
113—dc20   94 – 29978

ISBN of current title *Genesis of the Cosmos:* ISBN 978-1-59143-034-6

Printed and bound in the United States

10   9

Text design by Virginia Scott Bowman
This book was typeset in Minion with Futura as the display typeface

CONTENTS

PART 3

CHANGING THE PARADIGM

✳

PREFACE

NOT ALL SCIENTIFIC DISCOVERIES come about as a result of a long and arduous transcendental journey. However, this one did. It began one night in the spring of 1968 when a series of insights began flooding into my mind, concepts at once simple yet of considerable advancement. The ideas themselves were as amazing as the manner in which they were coming to me, as if sensed from some other level. Superimposed on the background of the music I had been listening to in my Johns Hopkins dorm room came notions of flux, balance, and dynamic equilibrium. I was shown these principles harmonizing together, forming the very essence of existence. Like an attentive pupil, I absorbed them.

There seemed to be an urgency about the whole affair. I was given to understand that, sometime back during the course of its development, Western science, or more specifically physics, had mistakenly taken the wrong turn. It was not that its experiments were improperly construed or that its observations were improperly made; it had to do with the theoretical framework that had been set up to interpret them. The errors were at a very basic assumptive level. I was shown that the classical physicist's view that physical reality is comprised of inert structures at its most basic level was wrong. I saw that nature at its most fundamental level is instead in perpetual balanced change, like life itself.

This new ecological metaphysics filled me in the months that followed. I began seeing how, through recurrence, natural processes formed the enduring systems around us, atoms, living organisms, solar systems, galaxies, and so on. I understood that there was hierarchical structure to this vast chain of being, with systems nested within systems in a repeating manner. And so began the formulation of "my theory of existence," a theory of systems founded on the principle of *process*.

Almost five years later, I discovered, with some relief, that others had followed this same path. Seminal thinkers such as Ludwig von Bertalanffy, Kenneth Boulding, and others had banded together seventeen years earlier and formed the Society for General Systems Research, which now had a membership of over a thousand. Here was not one individual, but a whole society of scientists who, like myself, were studying the fabric of nature and finding it to be governed by specific laws of organic process. And there were others who had also blazed the way, such as

Pierre Teilhard de Chardin and Alfred North Whitehead. The burden of responsibility to communicate these basic truths to others was lifted from my shoulders.

But my journey was far from finished. At about that same time, I had become engrossed in learning about a new branch of thermodynamics that studied how certain systems whose constituents abide in a state of incessant flux spontaneously generate orderly patterns from their internal chaos. Writings by Prigogine, Nicolis, and others; a picture of exotic chemical waves spiraling on the front cover of *Science* magazine; a paper by Einstein speculating on the fundamental nature of matter— all these combined together one night to spawn in an insightful flash a new theory of microphysics. A new dimension had opened up for me, and I was for some days sensing the flux in all things, even in rocks—things that my former academic physics training had taught should be lifeless.

When it first came forth, the theory was crudely formed. It would take several years to hone and refine its concepts. This new approach to subatomic physics, which I later came to call *subquantum kinetics,* was based not on mechanics but on chemistry, not on static structure but on process. This new microphysics followed an avenue that had been overlooked by eighteenth- and nineteenth-century ether theorists for the simple reason that chemical waves and "dissipative structures" had not yet been discovered at that time. The only models then available for understanding wave and particle phenomena were mechanical in nature. Subquantum kinetics was not just pleasing from a philosophical standpoint. It surpassed conventional physics in a number of respects. It explained the origin of matter and energy, the structure of matter, how force fields arose, and how they induced movement. It could also solve many of the problems that plagued standard physics.

As the theory took shape, I realized that a new mission once again lay ahead of me, a responsibility to communicate a discovery of great importance. As before, I was to find that others had followed this path. But in this case, these former theorists lived in an era much further back in time, before the dawn of recorded civilization.

This shocking discovery happened in 1975 while I was working as a consultant in Boston. A friend, after hearing me describe subquantum kinetics, commented that the basic idea sounded a lot like the Tarot metaphysics and urged me to take a class on the subject. Several weeks later I noticed that a weekly adult eduction class on the Tarot was scheduled in the Harvard Square area, so I decided to attend.

After the first few lectures it became apparent that there was indeed a strong similarity between the symbolic meaning of the major Tarot arcana and concepts basic to my process physics. I recognized that the Tarot was describing the science of system genesis, the process by which ordered forms (systems) spontaneously spring into being. More specifically, it was presenting a theory of how our universe came into being!

I was stunned. Here was a physics based on systems principles discovered mostly in the latter half of the twentieth century that were known to people many thousands of years ago. All that I had learned about Western civilization's rise from a primitive past lay hopelessly shattered.

The next year I moved to Oregon to begin work at Portland State University on a doctorate in general system theory. This was one of the few universities in the world offering a degree in this fascinating field. While there, I continued refining subquantum kinetics and made a major theoretical breakthrough in the summer of 1978. 1 also continued investigating the ancient roots of this esoteric physics. I had found that this science of creation was not unique to the Tarot. The same principles also appeared in the symbology of astrology, in the I Ching, and in certain ancient myths describing the world's creation.

After years of hard work, my theory of subquantum kinetics was finally published in 1985 in a special issue of the *International Journal of General Systems.* Subsequently, many of its astronomical and cosmological predictions were published in a number of other scientific journals. In 1994, I published a book on the theory entitled *Subquantum Kinetics* (see www.etheric.com). Now it is time that the full story about its ancient origins is known. This book reveals for the first time advanced scientific wisdom that has remained hidden for so long in our ancient myths and esoteric lore. Consider this legacy left by our predecessors and wonder— who were they?

I would like to thank my father and mother, Fred and Irene, for the long hours they spent helping me edit this manuscript. I would also like to thank my sister Mary, Larry Svart, Ann Richards, Marilyn Ferguson, Rosemary Loeine, Tom Abshier, Carolyn Halsey, Rosi Goldsmith, and others for their editorial assistance.

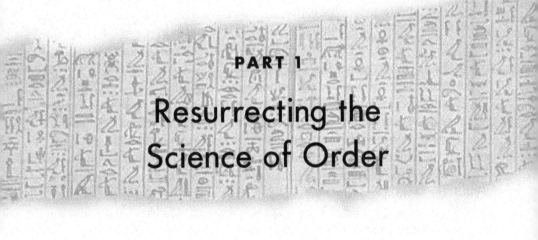

PART 1

Resurrecting the Science of Order

1

A LOST SCIENCE REDISCOVERED

OUR ANCIENT LEGACY

From ancient times to modern, people have been fascinated by the idea that Earth was once peopled by a civilization that had developed a highly advanced science. In his *Dialogues,* Plato writes about how the early ancestors of the Greeks and Egyptians had developed a technically advanced culture, but their intellectual achievements had been later lost due to humanity's endurance of a global conflagration and deluge that culminated around 11,600 years ago.[1]

The notion prevalent among many paleontologists that human civilization began for the first time only 6,000 years ago is challenged by ancient texts that trace the culture and science of China, India, and Egypt to a civilized prehistory of considerably greater antiquity. For example, Manetho, an Egyptian priest who lived in the third century B.C.E., reported that the divine and mortal dynasties that reigned prior to the beginning of the historical Egyptian empire spanned a total of 24,927 years. The historical period of ancient Egypt is recognized to have begun following the unification of Upper and Lower Egypt under Menes, an event that is variously dated between 4240 B.C.E. and 3200 B.C.E. Consequently, Manetho maintained that Egyptian prehistoric civilization began more than 30,000 years ago.

The Royal Papyrus of Turin gives an even earlier date. It lists the kings who ruled over Upper and Lower Egypt prior to Menes and the total length of their reigns as 36,620 years, thereby placing the origin of Egyptian prehistoric civilization at around 42,000 years before the present.[2] A passage from the writings of Herodotus may corroborate this early date. After stating that 340 generations of kings and priests had ruled over Egypt, Herodotus said that he had been told that during this long succession of centuries "the sun moved from his wonted course, twice rising where he now sets, and twice setting where he now rises." The Egyptologist R. A. Schwaller de Lubicz has suggested that this seemingly puzzling passage be taken in an astronomical sense, that during this long period of rulership the Earth's axis had precessed through one and a half Great Cycles, slowly translating the vernal equinox so that it passed one and a half times through the zodiac sign sequence. As a result, on two previous occasions the spring sun would have risen in the opposed constellation, where it today sets.[3] Given that it takes about 26,000 years for the poles to make a complete Great Cycle revolution, Schwaller de Lubicz estimates a passage of 39,000 years up to the time of Herodotus, placing the beginnings of Egyptian civilization at around 41,500 years before the present.

Mysterious megalithic structures of possible prehistoric origin have been found at various sites throughout the world, such as the Temple of the Sphinx adjacent to the Giza Sphinx; the cyclopean fortress at Baalbek in Lebanon; the strange statues on Easter Island; and the pre-Incan fortresses, temples, and walls found at Ollantaytambo, Tiahuanaco, and Sacsahuaman in the Peruvian Andes. The immense size of the blocks and the way they are interlocked to form some of these monuments suggest that their construction utilized technologies far in advance of those available to known ancient civilizations. Could these be relics of a technically advanced race that once inhabited the globe and met its decline during a period of climatic turmoil?

Geological evidence indicates that from about 14,000 B.C.E. until about 9600 B.C.E., the Earth's climate experienced abrupt changes that brought it out of the last ice age. This harsh transition was marked by periods of rapid ice sheet melting and withdrawal, continental flooding, and a prolonged episode of mass animal extinction. If civilized antediluvian empires collapsed during this interval, as legend appears to suggest, it is conceivable that certain of their scholars would have had the foresight to attempt to preserve some of their science for transmission to future generations in the hope that it would one day be understood.

Such transmission of knowledge could explain why certain of the ancient cultures were able to spring into existence fully developed from the very start of their recorded histories. In the case of ancient Egypt, Schwaller de Lubicz notes that from the dawning of the historical period with the appearance of the first known pharaohs there existed a complete writing, a carefully established calendar, a social order, a census, and a perfectly ordered myth and cult, all of which point to the existence of a long civilized epoch preceding the historical period.[4] In particular, he points out that the Egyptian Sothic calendar, which was in use as early as 4240 B.C.E., was quite advanced by most standards. Based on observations of the heliacal rising of the star Sirius, known to the ancient Egyptians as Sothis, it yields a highly accurate solar year of $365^{1}/_{4}$ days. This implies that the Egyptians had an exceptional knowledge of astronomy inherited from prehistoric times.

The Egyptologist John Anthony West agrees with Schwaller de Lubicz's assessment of ancient Egyptian culture. He remarks:

> Every aspect of Egyptian knowledge seems to have been complete at the very beginning. The sciences, artistic and architectural techniques and the hieroglyphic system show virtually no signs of a period of "development"; indeed, many of the

achievements of the earliest dynasties were never surpassed, or even equaled later on. This astonishing fact is readily admitted by orthodox Egyptologists, but the magnitude of the mystery it poses is skillfully understated, while its many implications go unmentioned.

How does a complex civilization spring full-blown into being? . . . The answer to the mystery is of course obvious, but because it is repellent to the prevailing cast of modern thinking, it is seldom seriously considered. *Egyptian civilization was not a "development," it was a legacy.*[5]

DO MYTHS ENCODE SCIENTIFIC TRUTHS?

If advanced knowledge from prehistoric times had been transmitted orally across the intervening "dark age," it is quite possible that it would have been conveyed in the form of symbol-laden myths and lore. By concealing abstract scientific concepts in entertaining stories that were meaningful to tribal cultures, this science would have survived through the generations even though its carriers might not understand its significance.

Tradition holds that fragments of such an advanced science have survived in astrology, in the Tarot, and in the symbolism of certain ancient myths. The specifics of this hidden knowledge, however, have long eluded Hermetic scholars. It is as though they had been left to fathom the depths of cryptic physics texts containing just equations with no accompanying words of explanation. Nevertheless, with the scientific understandings that have emerged over the past several decades, it is now possible to fill in the missing background material and reconstruct this lost science.

Modern scientists have only recently begun to understand and formulate concepts that explain how living systems function and to generalize these ideas to form the interdisciplinary field of study called *general system theory*. Key to this new understanding is that natural systems maintain their physical form or state of order through the continuous operation of energy-expending, order-building processes. By understanding that process is fundamentally important to structure, systems theorists have also been able to explain how natural systems first came into being. Applying these order-genesis concepts to a variety of scientific disciplines, they have been able to describe how tornadoes form in a turbulent atmosphere, how chemical concentration patterns self-organize in reacting chemical solutions, how proto-organisms may have first developed out of nutrients in the Earth's primeval oceans, how new social orders emerge, and how creative thoughts spontaneously come into being. As recently as 1973 these ideas were applied to the microphysical realm to develop a theory of how matter and energy first came into being in our universe. So it is indeed quite surprising to find that specific myths and lore from the dawn of recorded history use similar concepts to construct a highly sophisticated cosmological science.

When certain creation myths are closely examined, the various actions or attributes of their main characters are found to portray, through metaphor, a sequence of fundamental natural principles that describe how the physical world first came into being out of a primordial vital flux. This same scientific theory of creation is found in the Egyptian story of Atum, the Sumerian myth of Anu and Apsu, the *Enuma Elish* myth of ancient Babylon, as well as in various creation myths from ancient Greece, India, China, and Polynesia. It is also encoded both in the Tarot and in the lore of astrology. Moreover, scattered fragments of it appear in the process-oriented metaphysical writings of Heraclitus, Lao Tzu, and Confucius and in ancient Hindu and Buddhist mystical teachings.

A common theme that runs through many of the ancient traditions is their depiction of earth and sky separating from an initially unitary state at the dawn of physical creation. For example, the ancient Egyptians had a myth explaining that the physical universe was created precisely at the moment when the air god Shu separated his two children, Geb (earth) and Nut (sky), from their initially unitary love embrace. Sumerian myth relates that physical creation began when the heaven god An and earth goddess Ki uncoupled from their initially commingled state. According to Akaddian myth, the universe began to form when the warrior hero Marduk split the body of the defeated ocean goddess, Tiamat, thereby forming the sky and earth from her two halves. We encounter the same separation metaphor in the book of Genesis, where the moment of creation is marked by God's division of the primordial waters into heaven and earth. Even in the Southern Hemisphere, the Maori of New Zealand have a myth stating that the physical world became created when Tane, the god of the forests, separated his parents Rangi (sky) and Pappa (earth) from their initial love embrace. The ancient Chinese P'an-ku creation myth also depicts this bipolar emergence. It relates that in the beginning there existed chaos, which was contained in a single cosmic egg. Chaos then separated into earth (yin) and sky (yang), the two being pushed apart by the dwarf P'an-ku, who for 18,000 years grew in size at the rate of ten feet per day.* It is said that out of his body grew the cardinal mountains, stars, planets, thunder, lightning, rain, rivers, seas, soil, and rocks.

Could these myths be more than just entertaining stories? Consider, for example, the myth of P'an-ku. According to the ancient Chinese, yin and yang signified feminine and masculine aspects of an all-pervading ether, the subtle substance that forms all material things. This ancient ether, however, was substantially different from the mechanical ether of the classical era of modern physics. Whereas the ether of classical physics was inert and unchanging, its more ancient predecessor was

---

*After 18,000 years of growth, P'an-ku would have reached a height of about 21,000 kilometers, which happens to be about 60 percent larger than the actual diameter of the Earth.

conceived to be an active substance, one that transmuted and reacted much like a reacting chemical solution. Its yin and yang polarities were conceived to continuously transmute into one another in a never-ending cyclical process, yin alchemically transforming into yang, and yang alchemically transforming into yin. In their "separated," postcreation state, yin and yang were said to reciprocally alternate in dominance in repeating cycles, a predominantly yang condition always evolving into a predominantly yin condition, and back again in cyclic fashion. In effect, the P'an-ku myth portrays the creation of the physical world as the primordial self-emergence of an etheric wave pattern that proliferated to generate all physical form. Only during the early twentieth century did modern science discover that subatomic particles, the building blocks of our physical world, were formed of fields having wavelike characteristics.

Since the yin-yang wave was conceived to arise in an ether that was continually transforming, it is better likened to a *chemical wave* than a mechanical wave. Chemical waves are fundamentally very different from mechanical waves. They form only in solutions that are in the process of reacting, consuming energy-rich reactants and releasing energy-depleted reaction products. More specifically, they arise only from reactions that proceed in a closed loop much like the yin-yang transformation loop of ancient Chinese metaphysics. Such waves consist of cyclic variations in the concentrations of chemical species engaged in this ongoing, looping reaction. Provided that their underlying chemical reaction continues to function, they are able to arise spontaneously and persist, despite the ever-present tendency of molecular diffusion to disperse them. Because molecular diffusion also plays an important role in their formation, chemical waves are often termed *reaction-diffusion waves*.

The Belousov-Zhabotinskii reaction offers a good example of the chemical wave phenomenon (see figure 1.1).[6] This reaction was first observed in laboratory experiments carried out in 1958 by the Soviet chemist B. P. Belousov, and it was this discovery that first attracted scientists' attention to the existence of chemical waves. The phenomenon became more widely known ten years later when the chemists A. N. Zaikin and A. M. Zhabotinskii perfected the reaction by adding a dye indicator to make its waves visible.

Whereas mechanical waves were known to physicists of the Renaissance period and ancient world, reaction-diffusion waves became known for the first time to modern science only in the late twentieth century. This is understandable since, unlike mechanical waves, reaction-diffusion waves are not readily apparent in nature. Although the B-Z reaction can be easily reproduced in a high school classroom with the proper chemical reagents, its discovery was made possible only because chemistry had progressed to a considerably advanced level. So it comes somewhat as a surprise to find ancient myths modeling the process of physical creation after this phenomenon.

Reaction-diffusion processes are a fundamental part of cell metabolism and,

Figure 1.1. Chemical waves produced by the Belousov-Zhabotinskii reaction in a solution covering the bottom of a petri dish. The waves are made visible by a dye indicator that alternates from red to blue. The patterns form naturally as bull's-eyes but can also be made to adopt spiral configurations. Photos courtesy of Arthur Winfree and Fritz Goro.

broadly speaking, serve as the underpinnings of all ecological and social interactions. So, by suggesting a reaction-diffusion type of ether, the ancient mythmakers were developing an organic, living-system approach to physics and cosmology. Actually, all of us experience reaction-diffusion waves in our everyday life without being aware of them. The nerve impulses and wave forms that form the basis for all thought, feeling, and sensation are actually chemical reaction waves. However, scientists have become aware of this only recently, following the completion of many years of carefully controlled physiological studies. Broadly interpreted, "reaction-diffusion" waves are also commonly present in nature in the form of predator-prey population waves, domains of high or low animal population density that migrate through an ecosystem environment over the course of many years. Yet the wave-like movement of such population fronts normally passes unnoticed unless one has the foresight and interest to conduct careful demographic surveys. So, in view of the elusiveness of the reaction-diffusion phenomenon, it is intriguing to find that, well before the time of the Greek atomists, ancient philosopher-scientists had chosen a reaction-diffusion medium as the primordial substrate for giving birth to physical form.

The ancient creation science conceives all physical form, animate or inanimate, to be sustained by an undercurrent of process, a flux of vital energy that is present in all regions of space. Such ongoing metabolic activity might even be regarded as manifesting a kind of vital consciousness. This view has much in common with the animistic worldview of aboriginal tribal cultures, which infers the presence of life-like consciences or spirits in all things, even in inanimate objects such as rocks and rivers or the Earth itself. Myths and esoteric lore conveying this ancient science similarly endow the natural world with sacred qualities. The ether, conceived to serve as the substrate for physical form, is portrayed on the other hand as the domain of the spirit. This view of a vast, living beyond contrasts sharply with the sanitized mechanistic paradigm of modern physics, which has denied the existence of an unseen supernatural realm and forged a wedge between science and religion.

It is reasonable to expect that early attempts to reason about nature's elusive subatomic realm would have been inspired by mechanical rather than chemical phenomena. Almost everything in our everyday experience is perceived in terms of the relationships and relative motions of solids, liquids, and gases making up our environment. So it is not particularly surprising to find the ancient Greek atomist philosophers Leucippus and Democritus proposing that all matter is composed of mechanical agglomerations of *atoma*, inert and lifeless particles suspended in a void. Aristotle perpetuated this mechanical view when he taught that all physical things are formed from *hylê* (ele), an unobservable lifeless prime matter whose substance could become mechanically formed like moldable clay. His concept survived into the classical era of physics (seventeenth to nineteenth centuries), where it became reconceived as the *luminiferous ether*, a frictionless, elastic, solid substance that was supposed to have served as the carrier of light waves. Light waves were conceived to be mechanical waves that moved through this ether substance in much the same way that vibrational stresses move forward along the length of a pipe when one end is hit with a hammer.

Unlike chemical waves, mechanical waves cannot arise spontaneously from the medium itself, but must be created by some outside disturbance. Thus the mechanical ether of classical physics was conceived to have a passive rather than an active role in wave transmission; it served merely as a wave carrier, not a wave creator, light waves being simply bulk movements of its inert medium. This mechanical ether concept prevailed during the formative years for classical electromagnetism when empirical discoveries began to reveal the laws of electricity and magnetism. Electric and magnetic force fields were imagined to consist of mechanical stresses communicated through the elastic ether substance to distant locations. As for their notion of matter, some classical physicists adhered to Aristotle's view and considered material particles to be mechanically formed out of the ether itself, while others leaned more toward the Democritean view and considered particles to be entities distinct from the ether, much like stones immersed in a pool of water. Although physicists abandoned the ether concept near the beginning of the twentieth century, this mechanical view of particles and their fields was retained and still haunts physical theory today.

The history of science has shown that mechanical, machinelike models tend to predominate when a field of investigation is in its infancy, later to be replaced by more lifelike, organismic concepts. Whether it is the field of biology, business administration, sociology, or psychology, early theories that leaned heavily on mechanistic thinking were later replaced by more sophisticated theories formed within the organismic paradigm. Phenomena once described in a reductionistic, piecemeal fashion later became viewed in holistic contexts, and linear cause-effect descriptions were replaced by more sophisticated nonlinear multi-interactive models. Physicists, though, have been slow to relinquish the mechanical paradigm that has encompassed them for so long, perhaps because microphysics deals with a realm that evades direct observation. Nevertheless it is quite striking that in an era well before the time of Democritus scholars had adopted an organic physics worldview, one that far surpasses in sophistication the simple animistic ideas of primitive tribal cultures.

As we will find in considering the material that follows, certain ancient creation myths are products of a much greater level of intellectual achievement than myth historians have led us to believe. Apparently, the scientific content of these works had been overlooked for these many centuries because the historians studying them did not have the proper scientific grounding to comprehend their contained scientific metaphors. Only as recently as the late twentieth century have scientific developments advanced our understanding of nature to the point that now, for the first time, we are able to perceive in myth the outlines of a physical science remarkably similar to what theorists are just now beginning to develop. Far from being crude tales devised by primitives, certain ancient myths record a sophisticated cosmological science of exceeding brilliance, one that rivals the contemporary big bang theory in its predictive accuracy and is in many ways more aesthetically pleasing.

This ancient reaction-kinetic science leads to a theory of physical creation that is quite different from that espoused by the big bang theory of contemporary science. The big bang cosmology proclaims space, time, and existence to be an improbable transient phenomenon, of relatively finite extent, that emerged essentially out of nowhere together with physical form. The ancient science, on the other hand, theorizes that physical creation has come into being from a preexisting prime substance, or ether, and that space, time, and this ether are infinite and virtually immortal. Whereas the big bang theory postulates that material existence emerged all at once in the briefest fraction of a second from an infinitesimally small point, the ancient science describes matter and energy creation as a continuing process taking place over a period of many billions of years. The big bang theory proposes that the universe and the space in which it exists emerged explosively, leaving distant galaxies to rush away from our own at incredible speeds. The ancient science, on the other hand, makes no such claim; it assumes that space is static, neither expanding nor contracting, and that it has been so for all time. Finally, the big bang theory postulates the emergence of a quantum of energy of such great power that this single burst produced all the matter and energy in the universe, yet it offers no clear explanation as to how this enormous amount of energy came into being. The ancient science, on the other hand, proposes the emergence of an incredibly small energy blip, far smaller than the energy required for creating a single subatomic particle. Not only does it explain how this seed arose from the primordial ether, it also explains how this seed grew to produce a material particle, a process which repeated ad infinitum would generate an entire star-studded cosmos.

MODERN THEORY MIRRORING ANCIENT IDEAS

The reaction-diffusion ether that these ancient creation myths and lore appear to be portraying closely resembles that proposed in the novel physics methodology called *subquantum kinetics,* developed in the mid-1970s.[7] Subquantum kinetics conceives all material subatomic particles and energy quanta to be wavelike concentration patterns that self-organize in the continuously transforming ether whose reaction kinetics are specified by five reaction-diffusion equations collectively called Model G. This theoretical approach was itself inspired by work done on the Brusselator, a computer-simulated reaction-diffusion system developed in the mid-1960s by scientists at the Free University of Brussels.[8] The Brusselator reaction system differs from the Belousov-Zhabotinskii reaction and from natural predator-prey systems in that, in addition to producing propagating waves, it also generates stationary wave patterns. The Model G ether displays a similar ability to form both stationary and propagating wave patterns, a property that allows it to generate analogs of both material subatomic particles and propagating radiant energy waves.

It comes as somewhat of a surprise to find that certain ancient creation myths and esoteric lore describe through metaphor a Model G–like ether because the idea for this kind of ether could not have come about simply by observing laboratory chemical reactions. As far as we know, there are no chemical reaction systems that are able to spawn stationary wave patterns. The wave-ordering behavior of Model G and the Brusselator can be discovered only by simulating the equations on an electronic computer.

In describing how physical form came about, these prehistoric scientific traditions usually begin by describing, in a metaphorical fashion, the primordial ether, the prime substance out of which all physical form later took shape. The ancient Egyptians, Sumerians, Akaddians, and Hindus personified this medium as a life-giving primordial sea or ocean. The Tarot, astrology, and the ancient Egyptian myths of Atum and Osiris give particularly detailed accounts of the reactive, transmutative quality of this ether and even specify the manner in which its different fractions combine and mutually interact.

Almost all of these science-laden creation myths and lore describe the ether's initial state as one of omnipresent chaos. Some myths portray this condition of disorder as being collectively produced by warring gods who randomly battle one another for control of the cosmic empire. These stories then relate how an etheric mother goddess one day seeks out a place protected from this surrounding chaos and gives birth to a unique son whom she subsequently nurtures to maturity. This youth eventually develops into a powerful warrior who ventures out of his fertile "womb" to successfully resist the order-destroying effects of his ferocious adversaries. Variously named Horus, Zeus, or Marduk, he organizes his allies into a coherent resistance force that ultimately prevails over chaos to establish the warrior's ordered dominion over the ether. As he does so, physical order is brought into being.

The newborn god-child metaphorically depicts a localized concentration fluctuation in the ether that arises spontaneously like a tiny cry in the dark. This blip striving to express itself would consist of a localized bunching of the ether's subtle constituents. In physical terms, it would correspond to an extremely feeble energy pulse spontaneously arising in space. The drama of the child's birth, growth, and eventual emergence to rulership described in these myths accurately depicts how this energy pulse grows in size, resists the eroding effects of competing fluctuations born in its chaotic environment, and finally develops into a full-size stationary-wave pattern of subatomic dimensions (see figure 1.2). In physical terms, this wave would be a self-created subatomic particle.

Although concepts such as childbirth, mother-child nurturing, growth, combat, and territorial authority are part of the immediate life experience of primitive cultures, this does not necessarily imply that the cultures that originated this lore were themselves primitive. If ancient scientists had, in fact, designed mythic time capsules for oral transmittal, they would have chosen to frame their physics in terms of concepts that were entertaining and had meaning in the context of everyday tribal life, otherwise their message would not have been perpetuated through very many generations. Early prehistoric cultures depended on accurate memorization rather than a written language to pass on important ideas and concepts, so the idea of conveying encoded stories seems quite plausible.

The story of P'an-ku, related earlier, does not describe a drama of nurturing nor tell of an ultimate battle between order and chaos, as do some of the other creation myths. Nevertheless, it expresses in a succinct manner a number of the key features involved in the modern science of order genesis. It portrays how a concentration pulse, born from chaos—the fluctuating ether—grows in size, driving two interrelated etheric elements, yin and yang (X and Y), to evolve into a condition of disparity, the concentration of yang (sky) rising to arch over the falling concentration of yin (earth). This initial condition of low X/high Y concentration, then expands outward through the ether to form a regular wave pattern of alternating yin and yang polarity, generating an X-Y concentration wave similar to that produced by the Brusselator reaction system. The modern science of subquantum kinetics uses similar concepts to describe how a quantum of matter such as a proton spontaneously emerges from the chaotic reactive ether.

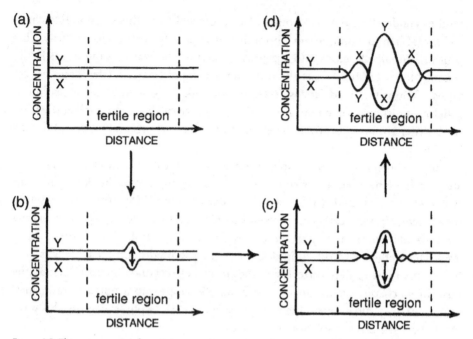

Figure 1.2. The emergence of a stationary wave pattern in a two-variable reaction system such as the Brusselator or Model G: (a) initially, the X and Y reactants maintain uniform concentrations; (b) a tiny concentration fluctuation emerges simultaneously in the interlinked X and Y reactants; (c) this X-Y fluctuation grows in size and intensity; and (d) finally, the X and Y reactant concentrations evolve into a stable stationary wave pattern.

One thing that stands out is the diversity of cultures having myths that portray the moment of creation as the emergence of a condition of disparity between two interrelated species, earth and sky. Although these various myths and lore employ different characters and symbols, the underlying physical emergence theme they convey is essentially the same. In all cases, the concept they convey effectively portrays how an initially uniform and featureless ether "self-divides" to produce a bipolar *X-Y* wave pattern.

The notion that ancient esoteric lore such as the I Ching, astrology, and the Tarot may have been purposely designed to encode a highly sophisticated science of cosmogenesis does not negate the possibility that the same lore might embody additional levels of meaning. Just as the concepts of nonequilibrium thermodynamics have been applied in modern times to a variety of natural phenomena ranging from an explanation of the origin of unicellular life to how the mind creates ideas, so too the basic principles encoded in this ancient lore may be broadly interpreted for application to various natural phenomena. Indeed, this ancient knowledge could open the way to a truly general unified science. By conceiving the universe to function on principles very much like those that operate in biological, sociological, economic, or psychological systems, the ancient science provides the conceptual umbrella that finally joins physics with the life sciences.

The necessary theoretical foundation for understanding this creation science may be found in the recently formulated field of general system theory and in the new physics and chemistry subdiscipline known as *nonequilibrium thermodynamics*. Lacking the unique perspective afforded by these new sciences, mythologists have tended to interpret these myths literally. For instance, in interpreting the metaphor of Geb separating from Nut in the Atum creation myth, mainstream Egyptologists have concluded that the ancient authors of the myth actually meant that the physical sky had physically separated from the physical earth at the dawn of creation. Such a crude interpretation presumes that these myths were created by people who possessed a level of knowledge substantially lower than the physics metaphor interpretation would imply. Given the common contemporary belief that prehistoric humans existed at a relatively primitive level of intellectual development, mythologists would have had little motive to question the validity of their assessment. Yet could it be that the primitiveness inferred from these myths is a characteristic not of the originators of the myths but of their interpreters?

Of the many known ancient creation myths and lore, only a few are found to encode this advanced cosmogenic physics. For example, out of a total of forty-five stories presented in one published collection of creation myths, only three were found to portray this science of creation.

In determining whether a given myth or lore shows a strong correlation with a modern creation physics paradigm, we should look for the following criteria. First, there should be some indication that the ancient myth or esoteric lore attempts to describe the origin of the universe or the process of physical creation. Second, the myth or lore must metaphorically express, through the described personality traits or actions of its gods or characters, specific concepts or principles corresponding to concepts or principles encountered in the modern science of system genesis. Third, the myth or lore should present these metaphorical traits or actions in a sequential order that corresponds to the logical order of events inherent in the systems genesis paradigm.

The ancient traditions that do show evidence of this science do not all present the science with the same degree of detail. There are some that give particularly thorough accounts, such as the Tarot, astrology, and the ancient Egyptian creation myths of Atum and Osiris. Ancient creation myths such as the Sumerian story of An and Ki and the Polynesian story of Rangi and Pappa express this physics in lesser detail. Nevertheless, even these less detailed accounts present a sufficient number of properly sequenced concepts that they too should be regarded as carriers of this cosmogenic knowledge.

Such variation in the level of detail is to be expected. For example, if some single prehistoric civilization or group of sages had redundantly encrypted this ancient science into a number of myths and lore for communication to the future, these messages would undoubtedly have undergone varying degrees of deletion or modification during the thousands of years of oral transmission. As a result, some would preserve their original scientific message better than others. Using a gene analogy, it is as if some of the surviving myths and lore preserve a more complete set of gene fragments inherited from the original creation physics "gene." Although certain gene fragments might appear in one myth and not in others, enough would be present in any one myth or lore that a conversant scientist could recognize the encoded creation science. Moreover, when these mythic time-capsule messages are studied collectively, a maximum number of "gene" information carriers are viewed, allowing the ciphered science to be seen with greatest clarity, in the same way that a holographic image becomes clearer as more parts of the underlying hologram are illuminated.

C. G. Jung has suggested that the striking similarities evident in myths from different cultures reflect the presence of universal motifs or archetypes embedded deep in the human psyche and arising from an impersonal collective unconscious that each person individually inherits.[9] Such archetypes are unconsciously expressed in myth through the language of metaphor. However, the specific sequences of images and events seen in the creation myths and esoteric lore examined here, metaphors sequenced in a certain logical order to portray complex scientific concepts, is more appropriately attributed to the conscious than the unconscious mind. Did ancient mythmakers purposely make use of universal archetypes to encode the concepts they were trying to express—metaphors whose meaning would have been understood by cultures many thousands of years in the future? If so, how did the originators of these ciphers come to conceive this prehistoric wisdom? Are we to attribute all of this to the workings of highly intuitive minds? Or was it the ultimate distillation of a science that was carefully wrought through the interplay of theory and experimentation? Could modern scientists just now be rediscovering sophisticated systems concepts once known to these ancient magi?

Or have human unconscious minds for generations been telepathically receiving an advanced science of otherworldly origin? Experiments have shown that telepathy transcends immediate spatial boundaries. Remote viewers are known to be able to mentally "see" events taking place on the other side of the planet. Spiritual masters claim to access information by transcending the bounds of time and space. Could the sensitive unconsciouses of ancient mythmakers have been unknowingly receiving this advanced physics knowledge and expressing its unfamiliar and abstruse concepts in the form of metaphorical stories and metaphysical teachings of varying detail, stories whose underlying meaning would one day be recognized and appreciated by future scientists?

2
PROCESS AND ORDER

According to my whole biological outlook, I am rather committed to the ancient Heraclitean concept that what is permanent is only the law and order of change.

LUDWIG VON BERTALANFFY, *GENERAL SYSTEM THEORY*

THE IMPLICIT AND THE EXPLICIT

Western science has sought to discover the ultimate makeup of the physical world by searching for nature's ultimate "building blocks." Atoms were believed to be fundamental until it was discovered that they were made up of subatomic particles. Later high-energy physicists theorized about the existence of even smaller subdivisions of matter called quarks. Throughout this atomistic crusade, there has been a tendency to regard *structure* as the primary basis for physical reality and to relegate *process* to second place, being thought to arise from changes in the arrangement of these more fundamental structures.

The ancient science of creation takes a very different approach by instead positing *process* as the basis for the physical world, actively maintaining even the most basic of physical structures. Although they produce observable physical form, these processes are themselves not "physical" but "meta"-physical, that is, "beyond" the physical. They occur in a subtle ether that is inherently unobservable.

Whereas modern physics recognizes the existence of only observable physical phenomena and concerns itself exclusively with studying those phenomena, the ancient science recognizes also the existence of the metaphysical realm and focuses primarily on describing that realm as a way of understanding the physical. This more ancient view portrays the metaphysical realm, the domain of the spirit and soul, as being more fundamental and primordial, the physical world being a more transitory form of existence generated from the metaphysical. The physical is not portrayed as separate from the metaphysical, nor any less sacred; rather, it is a more visible aspect of the metaphysical. Expressed another way, whereas modern physics traditionally recognizes the existence of only the *explicit order*, the more outwardly apparent structural-physical order of matter and energy, the prehistoric science recognizes that this explicit order is an expression of a more fundamental and less obvious *implicit order*, the functional order inherent in the arrangement and interplay of the underlying ether processes.

This notion that the physical universe emerged from an underlying spiritual realm hidden from the senses is a central teaching of most religions. Hindu and Buddhist teachings, for example, state that the perceived physical world, termed *maya*, is not the real essence of things, but arises from a more fundamental MAYA that represents existence in all its dimensions of being. MAYA is said to be the all-generating, self-transforming divine substance. The Hindus personify the dynamic aspect of the divine substance by the mother goddess, Shakti, whose name translates as "cosmic energy." This supreme power is conceived to generate and animate the illusory display, *maya*.

Often this relation of *maya* to MAYA is portrayed with the metaphor of mind and thought, the physical world being compared to a thought that has formed within the universal mind. According to Eastern teachings, those who transcend the world of appearances to perceive their true identity with this inner realm of mind free themselves from *samsara*, the continual cycle of soul reincarnation, and enter the state of *nirvana*, oneness with the universal.

The ancient science begins its explanation of the origin of physical things by describing the divine primordial ether. It characterizes this ethereal substance as actively transmuting and shows how the interplay of its processes results in the fetal emergence of a particle of matter, in effect an ordered wave pattern in the ether. By starting at a metaphysical etheric level to describe how these generative processes operate, these myths and lore tell their story of creation from a point of view removed from the immediate physical world. This clearly indicates that these early theoreticians were capable of a very sophisticated level of abstraction, certainly far greater than paleontologists traditionally ascribe to Stone Age people.

In the ancient Egyptian story of Osiris (see chapter 6), process is personified by the god Osiris, whose death and resurrection form the central theme of the myth. Osiris was associated with flowing water, the life force in plants, the reproductive power in animals, and more generally with the vital activity that maintains and propagates the order of life, all metaphors that collectively imply the notion of order-creating activity. Coffin Text 330 relates Osiris with both process and "the Order," Mayet, the natural order of the world:

> *Whether I live or die I am Osiris,*
> *I enter in and reappear through you,*
> *I decay in you, I grow in you,*
> *I fall down in you, I fall upon my side.*

The gods are living in me for I live and grow in the corn
    that sustains the Honored Ones.
I cover the earth,
whether I live or die I am Barley,
I am not destroyed.
I have entered the Order,
I rely upon the Order.
I become Master of the Order,
I emerge in the Order.[1]

Especially during the Old Kingdom, the earliest period of dynastic Egypt, these mythical characters were not regarded as personal "gods" or divine personalities requiring spiritual worship, although cults later emerged that did practice such idolatry. Rather, gods such as Mum and Osiris were symbols of specific functional principles or aspects of nature. For this reason, some Egyptologists, such as Schwaller de Lubicz, refer to them as *Neters* rather than gods; *Neter* is synonymous with the Hindu concept of *deva*.

Curiously, the story of Osiris does not begin with the creation of order, but rather with its downfall, symbolized by the murder and dismemberment of Osiris. Although these destructive actions may sound pointless and crude, they actually present a brilliantly crafted scientific discourse that explains the natural tendency of ordered states to become disordered. This fundamental concept holds an important status in modern physics and is set forth in mathematical form in the second law of thermodynamics. This natural tendency is quite masterfully expressed in the poem "Under" by Carl Sandburg:

I am the undertow
Washing tides of power
Battering the pillars
Under your things of high law.

I am a sleepless
Slowfaring eater
Maker of rust and rot
In your bastioned fastenings,
Caissons deep.

I am the Law
Older than you
And your builders proud.

I am deaf
In all days

Whether you
Say "Yes" or "No."
I am the crumbler:
Tomorrow.

(Chicago Poems)

Some might find it odd that an ancient myth about the creation of order would begin instead with the destruction of order. Yet to a modern systems theorist, it is not all that unusual. For it is well known that in nature the processes that create ordered forms are themselves generated as a result of the *destruction* of order. By its nature, process stems from the ever-present tendency for states of order to consistently deteriorate, dissolve, and disperse.

THE LAW OF CHANGE

Process and its counterpart, the destruction of order, are both associated with the tendency for nonequilibrium states to attain a condition of equilibrium. An example is the phenomenon of heat flow, in which a hot object placed in a cold environment, or alternatively a cold object placed in a hot environment, will tend to approach a state in which it is at thermal equilibrium with its surroundings. Similarly, the pattern of alternating hot and cold temperature shown in figure 2.1a will ultimately tend toward a condition in which temperature is uniform everywhere. The initial nonuniform temperature distribution constitutes what thermodynamicists would call a state of spatial order, characterized by some hot areas and some cold. The final state of temperature uniformity, on the other hand, is considered a more disordered state, because the former spatial pattern has vanished. With the passage of time, thermal order evolves toward a more probable condition of thermal disorder.

This approach to equilibrium is what constitutes process, in this case the process of heat flow. More specifically, it is *irreversible* process, for nature never performs its equilibration feat in reverse: objects initially at thermal equilibrium with their environment never spontaneously heat up or cool off. In other words, a system that is initially in a low state of order will not become more ordered.

Some processes may seem at first to be exceptions to this rule. For example, plant photosynthesis tends to build up order as less ordered raw materials such as carbon dioxide and water are taken in and combined into more ordered energy-rich sugar molecules. Yet we find that such order-building processes are ultimately powered by processes that destroy order. Photosynthesis, for example, is ultimately driven by the Sun's outgoing flux of heat and light, which is an example of energy flowing from a hot to a cold region, from the Sun's surface at 5,800° Celsius to outer space at a few degrees above absolute zero, with absolute zero at -273° Celsius (-459° Fahrenheit).

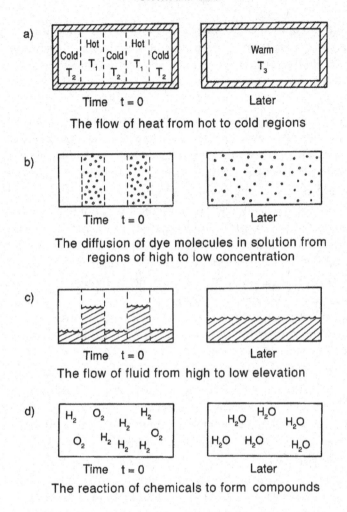

Figure 2.1. Several examples illustrating nature's tendency to equilibrate toward states of maximum disorder.

Thermodynamics has mathematically formalized this important disordering principle of nature as its "second law," which briefly states that if a system is isolated from its environment over time, it will tend to become increasingly disordered, and never the reverse. This law is usually phrased in terms of *entropy*, a thermodynamic quantity that physicists and chemists use to measure order, with lower entropy corresponding to greater order. The second law states that an isolated system always tends to evolve toward a state of increased entropy.

The *isolated* (or *closed)* system is one that is isolated from its environment and may be defined as unable to exchange matter or energy with its surroundings. It is necessary to specify a state of isolation when stating this general rule because some systems are able to counteract the increase of entropy by importing energy or energy-rich raw materials from their environment and using those resources to continuously rebuild their decaying order at a suitably fast rate. So, physicists specify that the second law is universally true only for processes taking place within closed boundaries.

Figure 2.1 shows various examples of entropy increase operating in nature. In each case, an initially less probable, nonequilibrium state (greater order) proceeds toward a more probable state of equilibrium (lesser order). The increase of entropy is more evident in the first three examples, where the initial state appears as some form of spatial order, a patterning of either temperature, molecular concentration, or fluid elevation, which becomes homogenized as a result of heat conduction or mechanical mixing. In the fourth example, the system's constituents are already uniformly distributed to begin with. Here, entropy increase occurs primarily because chemical reactions transform the system's constituents to a lower state of potential energy, with the amount of potential energy signifying the amount of stored-up energy that is available to do work.

Actually, higher potential energy is closely related to greater order. In each example in figure 2.1, the destruction of order is accompanied by the conversion of potential energy, thermal (2.1a), osmotic (2.1b), mechanical (2.1c), or chemical (2.1d), into *heat* (molecular kinetic energy). With equilibration, each system becomes more disordered as its molecules become more randomly arranged and more kinetically energized than when they started.

Those who originated the Osirian creation myth were apparently cognizant of this thermodynamic principle of entropy increase, for the machinations of the evil god Set brilliantly portray the operation of the second law. Inscriptions from the third millennium B.C.E. tell us that Set represents the negative traits of destruction, unregulated violence, disorder, darkness, death, and decay of the body, all of which illustrate the principle of order destruction. He was also associated with thunderstorms, earthquakes, desert windstorms, sickness, confusion, and quarreling, all of which represent instances of either physical, biological, or mental disarray. Even Set's birth was said to be chaotic. Section 205 of the Pyramid Texts describes him as one who at birth "broke out in violence."[2]

Hieroglyphs represent Set in the form of a jackal-like beast, sometimes having the body of a man (figure 2.2). Just as jackals in a pack can destroy prey many times their own size, in accordance with the second law, natural processes tear down prevailing states of order with their little nips and bites. This law of dissipation is also metaphorically portrayed in Set's plot to murder his brother Osiris. Osiris portrayed the process of order creation; his death at the hands of Set and Set's confederates metaphorically depicts the universal principle of order destruction.

The myth of Osiris, which is presented in chapter 6, relates how the evil Set tricked his brother Osiris to lie down in a coffin. Set's seventy-two confederates

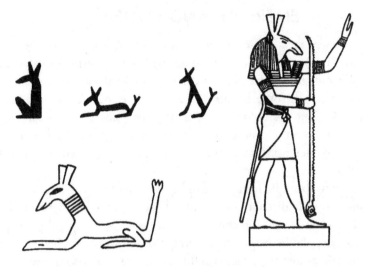

Figure 2.2. Hieroglyphs of Set, the Egyptian god of destruction.

then slammed the lid shut, bound the coffin, and threw it into the Nile, where it was carried out to sea, eventually washing up on the shores of Byblos in Phoenecia. After an intense search, Isis, the wife of Osiris, finally retrieves the body of the deceased Osiris from Byblos and conveys it on a boat to Buto in the Nile Delta. However, Set discovers the body and in a great rage rips it into fourteen pieces and scatters them about in the Nile swampland. The entrapment of Osiris within the coffin symbolizes a state of order isolated within a closed system. Just as a living system inevitably dies when cut off from its life-sustaining supply of food and oxygen, Osiris begins to die as he is shut in his coffin. It is also significant that the death of Osiris comes about as his coffin floats out to sea; the flowing Nile symbolizes the passage of time and the accompanying tendency for ordered states to evolve toward equilibrium. Osiris's final dismemberment by Set and the subsequent dispersion of his body parts by the currents of the Nile depict the final state of thermodynamic equilibrium, exemplified by a state of complete disintegration. Just as Set is assisted by a mob of six dozen conspirators, so too does the entropy process realize its objective of order destruction through myriads of disorganized actions.

This principle of entropy increase is also found in the Babylonian creation epic, where it is personified by the mother goddess Tiamat and her raucous, quarreling children. Tiamat attempts to subdue her rebellious son Marduk, the champion of order, by giving birth to various monsters and dragons. This principle of order destruction is also presented in the ancient Greek Olympian creation myth, personified in the destructive behavior of the Titans, Cyclopes, and Hundred-Handed Ones, all of whom are children of the earth goddess Gaea.

ENTROPY AND ORDER

Although the ancient creation science acknowledges the existence of the entropy principle, it does not regard decay as an inescapable fact governing the fate of the universe. This is quite reasonable, for if order always relentlessly decayed and never sprang into existence, the ordered universe would never have come into being in the first place. Ancient cosmologists apparently recognized that order creation could occur under certain conditions, namely, in circumstances where order-destroying entropy-increasing processes were properly harnessed to create local "pockets" of order.

In the water flow example in figure 2.1c, as water rushes downhill, it proceeds toward a more probable distribution, a state of lower potential energy, and in so doing obeys the second law. Yet if this flow is properly harnessed so that it passes through the turbine of a hydroelectric generator, useful work may be produced in the form of electric power. If we focus on the turbine generator to the exclusion of the surrounding environment, we will find that order has indeed increased within the boundaries of that system, as electric power is a form of order.

But it must be admitted that the turbine generator is not an isolated system and hence does not fall under the dictates of the second law. It imports water at a high potential energy (low entropy) and exports this water at a low potential energy along with a large quantity of waste heat (high entropy). Systems such as this, which function with matter and energy throughputs, are instead termed *open systems.**
The waterfall produces order locally within this open system without violating the second law because it simultaneously generates a proportionately greater amount of disorder (heat and kinetic energy) in the environment. So the total entropy change, or the sum of the entropy decrease produced in the open system plus the entropy increase produced in the environment, is such that on the whole entropy has increased; order has on the whole tended toward disorder.

Chemical oxidation (figure 2.1d) can also create order in open systems. We are all very much aware of oxidation's order-destroying tendencies when the metal parts of our automobiles begin to rust or when a forest catches fire. But, when properly harnessed within the cylinders of our engine to combust fuel (and destroy its chemical order), oxidation propels our car, thereby creating a useful form of ordered activity. Like the hydroelectric turbine, an automobile engine is an open system; it imports oxygen and fuel at high potential energy and exports waste heat and combustion products at low potential energy. Again, this order-creating activity does not violate the second law. The ordered work produced by the automobile

---

*Open systems that are closed to matter transport have traditionally been called "closed systems" by thermodynamicists. Following the convention used by most systems theorists, however, we classify such systems as "open" systems and use the term "closed system" as a synonym for isolated system.

engine, the open system, is won at the expense of producing a greater amount of waste-heat disorder in the environment.

In addition, there are many kinds of naturally occurring open systems that create order in an analogous fashion, one example being living organisms. They counter entropy by using energy-dissipating processes to continuously rebuild themselves. For example, animal organisms take in food, material rich in energy and order and low in entropy; break down and oxidize this ingested material; and finally release the reaction products as chemical wastes and low-grade heat. As they do so, the energy released from this digestion and oxidation drives the biochemical cellular processes that maintain and repair the organism's cell structures. In this way, the organism counteracts the effects of entropy-increasing processes that continuously destroy its organic order. Thus, by continuously breathing, eating, eliminating, importing, and exporting, organisms are able to stay alive and even to evolve toward states of *increased* order. Plants function in a similar way, except that they synthesize their food internally through the conversion of low-entropy light energy. In either case, the survival of the living organism is critically dependent on its open-system character, its ability to exchange matter and energy with its environment.

Cosmology seeks to explain the genesis of the most abundant of all forms of order, namely, the material universe. Our observations of the physical world teach us that only open systems spontaneously produce order. Could it thus be true that the matter and energy composing our universe is also an example of order arising in an open system? The ancient philosophers, as we shall soon see, imagined the physical universe to be just such an open system, built up from wave patterns arising in an ether whose substance was continuously transmuting and reacting much like the chemical reactions that sustain a living organism. If the ether's vital substance is imagined as executing a continuous sequence of transmutations through a consecutive series of ether states, eventually transmuting into states that form the basis for our physical universe and then passing on to others, the ether states composing our physical universe would constitute an open system. Although ether substance would be leaving those states by entropically transmuting to "lower" states, these states would be replenished with fresh substance transmuting from "higher" up. Thus by conceiving of the physical universe as an open system ordering phenomenon, the ancient creation science lays the groundwork for a feasible theory of cosmic creation.

Besides living organisms and the chemical-wave-producing Belousov-Zhabotinskii reaction mentioned in chapter 1, several other examples of order-generating open systems could be cited, such as a candle flame, a convection pattern, or a city. In each case, the fluxes that pass through the system's boundaries are vital to the system's survival. For example, without the continuous flow of oxygen and vaporized paraffin, a candle flame would quickly become extinguished. Again, the honeycomb cellular convection pattern produced in a pan of heated oil persists only as long as heat is applied (see figure 2.3). If the heat source is removed, convection stops and the liquid becomes uniformly calm.

A city must continuously import raw materials, energy, labor, and money and must export products and/or services. If it does not, it will disintegrate through the loss of its inhabitants and the collapse of its social order. In 1973, when the Arab oil embargo interrupted the otherwise continuous flow of energy imports that fueled the world economy, people became painfully aware of the importance of the flow of resources to the effective operation of their economic systems. The world recession that was precipitated in the early 1980s after the United States began to restrict its money supply to control inflation also illustrates the considerable extent to which national economies depend on a smooth flow of resources.

The human mind is also an example of an open system. Here we broaden the conventional definition of open system to include systems that exchange information with their environment. This is quite permissible because matter and energy may be categorized as kinds of information. Like a living organism, the mind is always active, continually structuring and processing sensory and emotional data, even during states of sleep or meditation. Conceptual systems, such as a body of scientific knowledge, a system of beliefs, or a language, are also characterized by a state of flux. For example, if young minds do not continuously educate themselves with the concepts that make up a branch of science, and if they do not communicate and share these ideas with one another, as the population turns over and sages pass

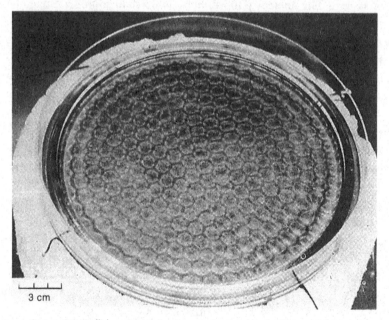

Figure 2.3. Cellular convection patterns in a pan of oil heated from below.
Photo courtesy of E. Koschmieder.

away, such a knowledge system would eventually die with them and be forgotten. The ancient science of cosmic creation may have faded away in such a fashion.

Open systems are characterized by a continuous turnover of their components. Living systems, for example, renew their structures at an astounding rate by regulating metabolic processes that simultaneously build up and break down their intricate physical forms. Their structures remain essentially unchanged, despite the continuous turnover of the molecules that compose them. The turnover times for protein in humans ranges from as little as half an hour for adenosine triphosphate (ATP), which powers all cell processes, to 175 days for hemoglobin in red blood cells (table 2.1).[3] Individual cells making up body tissues also turn over exceedingly fast, created through mitosis (cell division) and destroyed through desquamation (peeling off) or autolysis (self-destruction). The average residence time for cells in rat tissue ranges from more than one hundred days for bone and muscle to as little as a day or so for cells lining the small intestine. During the organism's early years, anabolic processes prevail over catabolic processes, and this results in physical growth, order creation. At maturity, however, these countervailing processes approach a balance, thereby maintaining the organism's tissue mass in a more or less constant condition called a *steady state*.

Categories of Open Systems

1. **Hydrodynamic systems:**
   Tornadoes, hurricanes, cellular convection patterns
2. **Hydromagnetic systems:**
   Sunspots, solar convection cells, ball lightning, electric arcs
3. **Reaction–diffusion systems (broadly defined):**
   Predator-prey oscillations, security price oscillations, chemical clock oscillations, chemical waves, nuclear reaction oscillations, subatomic particles, and waves
4. **Synthesizer systems:**
   Primitive single-celled organisms
5. **Emotional/cognitive systems:**
   Creative thoughts, fantasies, dreams
6. **Hybrid systems (combinations of above categories):**
   Candle flame (categories 1 and 3), biological organisms (categories 3 and 4), socioeconomic systems (categories 3, 4, and 5)

TABLE 2.1
Residence Times of Components in a Cell

SUBSTANCE	SPECIES	TURNOVER TIME
Red blood cell hemoglobin	human	175 days
Proteins of musculature	human	10 days
Protein (body average)	human	5 days
Proteins of liver and serum	human	14 hours
Red blood cell ATP glycolosis	human	0.4 hours
Liver mitochondria	mouse	16 days
Total protein	rat	1 day
Protein of liver, serum, and internal organs	rat	8 hours
Glucose (body average)	rat	1 hour
Cytochrome-a in wing muscle	grasshopper	0.01 seconds

Ludwig von Bertalanffy, an early pioneer in the theory of open systems, recognized that living systems are not just in a state of flux, they *are* flux. To describe their dynamic nature, in 1942 he coined the German word *Fliessgleichgewicht*, which means "patterned flow equilibrium." He noted that one should look beyond the facade of structure to the processes underlying structure to understand the true essence of living systems:

> From a more general viewpoint, we begin to understand that besides visible morphologic organization, as observed by the electron microscope, light microscope and macroscopically, there is another, invisible, organization resulting from interplay of processes determined by rates of reaction and transport.[4]

Bertalanffy taught that the structural form of a living thing is an outwardly visible expression of its ordered biochemical processes, hence, that the primary order of organisms is in their processes, not in their structures. In other words, he taught that an organism's explicit order is the visible manifestation of its less apparent implicit order. In a similar fashion, the ancients viewed the physical universe as an open system, a visible *Fliessgleichgewicht* produced by underlying ether transmutation processes. They recognized that the order of the universe was not found in the observable world, but in the unobservable ether realm, contained in the orderly arrangement and interaction of ether reaction processes.

An open system must maintain itself in a nonequilibrium state, otherwise the irreversible processes that sustain its order will reach a state of rest at equilibrium and leave the system's structure to rapidly deteriorate. For this reason, open systems are often called *nonequilibrium systems.* The Russian-born Belgian theorist and Nobel laureate Ilya Prigogine, known for his accomplishments in understanding the thermodynamic behavior of such systems, has often spoken of the importance of the nonequilibrium condition to the creation of order. In *Order out of Chaos*, he and Isabelle Stengers state, "We come to one of our main conclusions: At all levels, be it the level of macroscopic physics . . . or the microscopic level, *disequilibrium is the source of order. Disequilibrium brings 'order out of chaos.'*"[5]

Ironically, open systems sustain their own order by continuously feeding upon order in their environment and dissipating that foreign order. For example, a living system sustains its order-creating biochemical reactions by continuously ingesting food (environmental order), breaking down its complex molecules, and dissipating, or dispersing, to the environment the waste heat calories that result from this breakdown process. Hence system scientists sometimes characterize living organisms as *dissipative structures,* a term that emphasizes their need to continuously increase entropy in their environment to stay alive. The term may be equally well applied to chemical waves, social organizations, ecosystems, mental thoughts, and, as we shall see, even to subatomic particles (subatomic wave patterns) arising in an alchemic ether. In all cases, the system's form arises because of the operation of underlying dissipative processes.

THE SECRET OF LIFE

The ancients went beyond merely portraying the process concept. In addition, they sought to encode a fundamental subtle truth about process that only recently has come to be appreciated by modern science, namely, that the entropic tendency toward disorder and order-creating irreversible process are two sides of the same coin. Process can be either the mainstay of order or the enemy of order. It is both creator and destroyer; it creates one kind of order by taking advantage of the decay of some other kind of order. The prehistoric mythmakers were aware that process can have both a light side and a dark side, for in the Osirian creation myth, Osiris (order-creating process) and Set (the order-destroying entropy-increase principle) are depicted as brothers. A similar duality of nature is evident in Hindu mythology. The Hindu mother goddess is sometimes depicted in the form of Shakti, who signifies the feminine creative principle, and other times in the form of Kali, who signifies the principle of destruction.

The Osiris mysteries cleverly illustrate how the dark side of process, the "Set" decay principle, can be made to serve the light side, the Osirian order-creation principle. Set, who portrays the increase of entropy through his murder of Osiris, is challenged to battle by Osiris's son Horus, the god of light. Horus eventually defeats Set and brings the kingdom once again under the rule of order. Horus does not kill Set, however. For Set (dissipation) is a power that cannot be annihilated, only temporarily restrained or redirected. Horus instead *utilizes* Set by making him subservient to Osiris and thereby redirecting him toward creative ends.

Details of this part of the Osirian passion are given in the Pyramid Texts, which date from approximately 2480 to 2140 B.C.E. These inscriptions relate that Horus wished to tell his deceased father the good news of his victory over Set. So, leaving the world of light, he traveled down into the *Dwat*, the underworld or spirit world, where the unconscious soul of Osiris rested in a lifeless state. Osiris here appears to signify the etheric transmutation process functioning at a minimal level very close to equilibrium. Upon finding him, Horus tells Osiris that Set has been defeated. Through this verbal communion, Horus places Set "under" Osiris and Set "lifts" Osiris up, reviving him to consciousness. The soul of Osiris becomes the spirit of life and growth and bestows upon Horus its *ka*, life force or vital essence. Osiris now becomes the vital power behind Horus, who signifies the newly emerged ordered state. Even in the resurrected state, Osiris does not return to the realm of the living. He is destined to remain in the underworld ruling over the souls of the dead. Nevertheless, he derives fulfillment through his son, who now reigns in the realm of light and whom he supports through his *ka*.

Without the help of Set (entropic process), Osiris (creative process) is powerless. This subtle connection between entropy and creative flux is described in a hymn inscribed in the pyramid of Wenis. Beginning in section 258 of the Pyramid Texts, Horus speaks to his resurrected father as follows:

> This is Horus [speaking], he has ordained action
> for his father,
> he has shown himself master of the storm,
> he has countered the blustering of Set,
> so that he [Set] must bear you—
> for it is he that must carry him [Osiris] who is [again] complete.[6]

Rituals enacting this part of the Osirian pageant portrayed Set in the form of a boat bearing the resurrected body of Osiris in the festal procession on the Nile. The metaphor of the subordinated Set "bearing" the recumbent Osiris suggests that order prevails in the etheric realm, that entropic process has been harnessed in an ordered fashion. Moreover, because Osiris (process) is depicted as nourishing Horus (the newly emerged structural form) with his vital essence, this metaphor portrays the profound thermodynamic insight that all living things, and more generally all ordered forms spawned by open systems, depend for their existence on the continual operation of the second law. It explains that such states of order persist through the control of dissipative process.

But the subjugated Set not only represents nonequilibrium processes that sustain open systems found in the physical world, he also represents the irreversible flux that drives the vast primordial ether and sustains the physical universe itself. In this context, the resurrected Osiris borne aloft on the surface of the flowing Nile portrays the insight that the entire physical universe is maintained through the proper harnessing of etheric process.

In the course of his resurrection, Osiris gains the power of "knowing," which includes awareness of Set's real nature. Beginning in section 581 of the Pyramid Texts, a chorus addresses Osiris, saying:

> *Horus has seized Set, he has put him beneath you so that*
> *he can lift you up. He will groan beneath you as an earthquake.*
>
> *Horus has made you recognize him [Set] in his real nature,*
> *let him not escape you;*
>
> *he has made you hold him by your hand, let him not get*
> *away from you.*[7]

The metaphor of Osiris holding on to Set expresses the concept of entropic process being controlled or ordered. At the beginning of the myth, the unwary Osiris had made the fatal mistake of following Set's lead. In the above passage, the resurrected Osiris is warned to always maintain an upper hand and keep process tamed, otherwise the fate of death would be repeated. Osiris (order) holding or controlling Set (entropic process) metaphorically portrays the ordering of the transmutative processes operating in the etheric underworld realm. That is, etheric process must be kept properly ordered, otherwise the physical order it creates would once again succumb to entropy. In this way, the Osiris myth discloses a fundamental secret of physical creation. Namely, physical form (the ether's explicit order) is able to come into being despite the order-destroying effects of entropy because of the preexisting order inherent in the underlying ether processes (the ether's implicit order). If the underlying implicit order were to disappear, the outwardly apparent explicit order would dissolve.

This metaphor of Osiris keeping hold of Set is depicted in the fresco of Osiris shown in figure 2.4. The wooden staff that Osiris is holding and not "letting get away" is, in fact, Set. Looking closely, one can even make out the scoundrel's features. The crown of the scepter displays Set's head with his two ears and long snout, and its bottom end displays Set's distinctive forked tail. Just as a staff is normally used to support oneself in walking, Osiris uses Set as a support. Without Set (dissipation), Osiris (ordered process) could not exist.

This inference that the scepter might signify the concept of process finds support in Egyptian lore. Its symbol, commonly called the "key of the Nile," was known to the Egyptians as the *uas*. It was placed in the hands of all the *Neters* and played a significant role. According to Schwaller de Lubicz, the *uas* represents a living branch

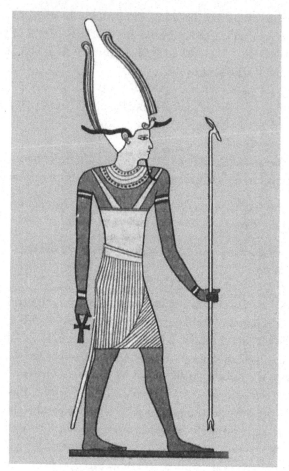

Figure 2.4. The Egyptian god Osiris. From Budge,
The Gods of the Egyptians, pl. 25.

that conducts nourishing, vivifying sap, fluid that ascends without again descending.[8] It denotes the flux of the Word, the primordial ether, the active creative function. An examination of canes and scepters found in pharaonic tombs shows that the *uas* was made from the living branch of a tree that had been cut so as to include a section of the lower source branch as well as two offshoots coming from its upper end (figure 2.5). Sometimes it was left in this natural form and covered entirely with gold, like the one found in the treasure trove of King Tutankhamen. More commonly its appendages were bent to form an angled head and forked tail, perhaps by forcefully molding the wood while it was still wet. Often a face, reminiscent of the evil Set, was formed on this head.

The linear form of the *uas* scepter, arranged with Set's head at one end and tail at the opposite end, suggests the idea of irreversible one-way process, a concept also

Figure 2.5. Variations of the *uas* scepter.
From Schwaller de Lubicz, *Sacred Science*, fig. 29.

suggested by the metaphor of upward-flowing sap. Its vertical orientation, with head pointing up, indicates that process is operating in its ascending, vivifying mode and that Osiris is directing the etheric process toward constructive purposes. Significantly, Osiris holds the staff in his left hand. The left side of the body traditionally represents the less obvious, hidden, submissive aspect, while the right side represents the more obvious, outwardly manifest, dominant aspect. Osiris is thus indicating that the creative process he represents works "behind the scenes" in the metaphysical realm of the ether.

A more comprehensive analysis of the Osirian myth in chapters 6 and 7 provides several additional systems concepts that will aid our understanding. Nevertheless, from the analysis presented to this point we may conclude that the ancient cosmologists who formulated the Osirian creation myth apparently had a very sophisticated understanding of open systems and of the subtle interrelation between entropy and order creation. It was not until recently that modern science began to grasp the subtle concepts understood by the ancients thousands of years ago. Thermodynamics was developed for the first time in the late nineteenth century as a result of the seminal work of Ludwig Boltzmann. At that time its principles dealt primarily with processes taking place within *closed* systems, the decay of ordered states with the final attainment of thermodynamic equilibrium. It was not until the 1920s that Bertalanffy and Defay independently proposed that the discipline be expanded so that it could also be applied to the description of open systems.[9] This new branch of thermodynamics that deals specifically with open systems is termed *nonequilibrium thermodynamics*, or sometimes, *irreversible thermodynamics*.

Modern science did not begin to ponder the "Set principle" seriously until the nineteenth century, and only in the last half of the twentieth century did thermodynamics advance to the point where it also has been able to embrace the "Osiris principle." Only now are we in a position to appreciate adequately the kernels of wisdom hidden in the Osiris myth: life-sustaining irreversible process and order-destroying entropy-increase are one and the same, and living systems maintain their order by mastering entropic processes and using them to their advantage.

The story of Osiris and the myth of Atum were not just folklore tales handed down through oral tradition; these stories were also incorporated into the rulership customs of ancient Egypt. Royal funerary rites and rituals concerned with the transference of power from one reign to the next reenacted the sacred Osiris creation mystery. The king, or pharaoh, was regarded as a living deity and mediator between the world of the living and the world of the dead. Through him, the vital energies of the cosmos were bestowed upon his country, ensuring its power and welfare. During his reign, the king was looked upon as the incarnation of Horus. At death, he became the incarnation of Osiris, and the successor to the throne became the living Horus. Just as Osiris's soul journeyed into the underworld to repose in an unconscious state, the king's soul at death was believed to journey into the underworld. The deceased king was placed in his burial chamber in a horizontal position to represent the recumbent Osiris in his underworld cave. Just as Osiris was preserved from decay by being wrapped in the coils of the primordial serpent Nehaher, so was the king's body embalmed in preservative mummy wrappings. After ascending to the throne, the new king traditionally entered his father's tomb to give him the news that order had been reestablished, paralleling the Osirian myth in which Horus revived Osiris to consciousness by telling him the good news of his victory over Set. Upon hearing these tidings, the late king's underworld soul was believed to return to life to confer his *ka* (vital essence) upon his son, thereby ensuring that the powers of life and growth would again flow to nature.[10] In this way, the new king was thought to derive his power from his deceased father.

Thus the subtle thermodynamic concepts evident in the Osiris mysteries were in this way reenacted in terms of real-life events every time a king died. This ritual may have persisted even during Egyptian prehistory, which could explain why the Egyptians kept records of the reigns of those past kings. Was this royal tradition consciously designed in prehistoric times as a way of transmitting an advanced science to the distant future? If so, it was quite a clever method, for it ensured that this knowledge would be perpetuated. The successor to the throne was bound to participate in the ceremony, because only in this manner could he assume the power of rulership. Moreover, the tradition could only be destroyed if the royal lineage itself was unseated by force. Although the pharaonic succession eventually came to an end, the royal practice of inscribing mythical texts and hymns on the walls of pyramids, temples, and coffins has left us valuable records of this lore that have survived through time.

The ceremonial transference of *ka* from father to son had a practical advantage of ensuring that the tradition would be perpetuated. But in addition, it had an esoteric significance of portraying a scientific concept fundamental to the ancient open-system science, namely that the ordered material world (Horus, the new king) is sustained through the operation of ordered etheric processes (Osiris resurrected in the underworld). Explicit order is derived from and sustained by a preexisting implicit order.

For most of the duration of the Old Kingdom, the passion of Osiris was concerned not with the salvation of human souls, but with the resurrection of divine souls, those of the pharaonic kings. The resurrection ceremony was performed not out of a personal fear of death on the part of the ruling class but rather to ensure the continued prosperity of the empire and to celebrate the miracle of the creation of the cosmic order. Beginning around 2500 B.C.E., however, as the Old Kingdom began to wane, the Osirian ceremonies took on a more personal significance for the masses. A cult of Osiris developed and gained in popularity. With the end of the sixth dynasty around 2250 B.C.E., the old social order collapsed and a chaotic period ensued. Intellectual inquiry flourished during this unsettled period, probing matters such as the relation of God and man and the fate of the soul. The passion of Osiris became reinterpreted in this new context. It was thought that, like the soul of Osiris, the human soul journeyed at death into the underworld. It was thought that if the deceased was given a proper burial with funerary rites, his soul too would become resurrected into a conscious afterlife. Because the spiritually inclined did not have the resources to build pyramids and elaborate tombs, painted and inscribed coffins took their place during this period.[11]

CLOSED- VERSUS OPEN-SYSTEM PARADIGMS

Almost every field of science has passed through a stage in its development in which it viewed natural systems as closed systems, that is, as inanimate mechanisms. For example, medical texts in medieval western Europe described human physiology in mechanical terms, comparing the organs of the body to the parts of a complex mechanism consisting of levers, pipes, and pumps. Descartes also expounded this view in his writings during the seventeenth century. This approach sought to describe the body's physical structure with little reference to the underlying order-creating processes responsible for its formation. Later, with the invention of the steam engine and the development of thermodynamics, the body became viewed as a "heat engine," a *chemodynamic machine* transforming food energy into useful work, for example, muscle action. But still, this view paid little attention to the processes that sustained the body. Physiologists and physicians today are aware that the human body is an open system, a structure maintained through the continuous action of biochemical processes.

The machine analogy at one time also dominated the business profession. Organizations generally managed their internal affairs in a military-like fashion, with orders flowing down through a chain of command to workers who were automatons in a machinelike system. Managers trained with this model tended to ignore or even suppress social processes naturally present in the workforce that otherwise could have helped transform their company into a more profitable institution. In recent decades, however, the open-system approach has come to the fore.

Organization theorists today agree that a business firm functions much like a living organism.[12] Increasingly, managers are recognizing that healthy firms are those in which workers and management intercommunicate to form an integrally functioning whole. They are turning away from the view of the firm as a rigid entity, responsible only to its own objectives, and are embracing a more open-system concept that views the firm as an evolving unit functioning within a larger social whole and critically dependent upon and responsible to its environment.

A similar progression from closed- to open-system views is evident in the fields of psychology and neurophysiology. Although computerlike machine models of the human mind still command considerable attention today, open-system models are increasingly making inroads. The new view conceives of the mind as a kind of evolving life form, with thoughts as ordered patterns that emerge into awareness from fluxes of sensory data and feeling tones recursively circulating in the brain at a subconscious level.[13]

The field of microphysics still has a way to go before it entirely escapes from the closed-system paradigm. This should not come as a surprise. Compared to these other disciplines, microphysics deals with phenomena that are quite far removed from everyday experience. Even when scrutinized with specialized measuring instruments, nature is found to be particularly obstinate about revealing its secrets at the subatomic level. While it is quite an easy matter to measure precisely both the position and momentum of a moving body at the macroscopic level of everyday experience, a precise measurement at the quantum level is impossible. The objects of study are so small that the measuring probe has a major effect on the particle being probed. Consequently, as an observer obtains increasingly more precise knowledge about the position of a particle, at the same time he must be satisfied to settle for increasingly less precise knowledge about the particle's momentum. In the optimum situation, the true values of a particle's position and momentum can be known only within certain ranges. Formally stated, this condition of observational uncertainty is known as the Heisenberg uncertainty principle.

Below the quantum level, direct observation is impossible. Since the physicist's probes are necessarily composed of matter or energy, by their nature they can only sense the presence of other matter and energy quanta. Consequently, there is no direct way to verify whether or not a subatomic particle or energy wave is composed of smaller structures or to determine whether these component structures may be inert and unchanging or actively transmuting and interacting. A subatomic particle could actually be a pattern composed of ever-so-tiny and undetectable etheric units maintained in a state of transmutative flux, and physicists would have no way of knowing it.

Common sense tells us that nature should operate in a consistent fashion. Since all ordered structures in nature that are easily accessible to direct observation are open systems, it is reasonable to suspect that subatomic particles and energy waves may also be open systems. If force fields are also a physical manifestation resulting from the operation of some underlying subquantum flux, then higher-order structures bound together by force fields (for example, atoms, molecules, crystals, planets, solar systems, and so on) should also be viewed as dissipative structures. As discussed in later chapters, ancient creation myths and lore present precisely such an open-system theory of particles and fields, as does the contemporary subquantum kinetics methodology.

Meanwhile, during most of the twentieth century, physicists continued to view subatomic particles and energy waves as closed-system phenomena, as self-sufficient entities whose only relation to space is one of occupancy. The twentieth-century philosopher Alfred North Whitehead has referred to this as the doctrine of simple location. Accordingly, contemporary particle physicists expect that a subatomic particle should continue to exist even if ideally placed within a closed box that prevented components of any kind from entering or leaving. Modern physics inherited this mechanistic model of self-sufficient matter from Newton, who in turn took the concept from the ancient Greek atomists. Physicists later discovered that atoms were composed of smaller structures, such as protons, neutrons, and electrons, and they theorized that protons and neutrons, in turn, were composed of an amalgam of still smaller structures called quarks and gluons. Like the atom, however, they also viewed these finer subdivisions of matter as isolated systems.*

This view of matter and energy has led contemporary physicists to presume that the universe as a whole is a closed system and that its macroscopic evolution is governed by the dictates of the second law of thermodynamics. They state that the universe achieved its greatest state of order at the moment of the big bang explosion and that things have been tending toward greater disorder ever since, as space has continued to expand and energy has continued to disperse. Asked about the distant future, they tell us that all the stars in all of the galaxies will eventually burn out and everything will grow very cold. When this ultimate "heat death" is reached, all that will supposedly remain will be a bunch of cold burned-out stellar cores wandering through the dark reaches of space together with a scattering of black holes, all with a temperature of no more than a few degrees above absolute zero. So modern cosmology appears to be firmly in the grip of the "Set principle."

Paul Davies, a contemporary science writer and cosmologist, presents such a tragic outlook in his book *The Runaway Universe*.

---

*Based on the outcome of certain key experiments performed during the 1970s and early 1980s, physicists have begun to abandon this localized, closed-system view of individual quantum structures in favor of a more nonlocalized view. According to this new outlook, discussed further in chapter 13, some quantum properties, such as a subatomic particle's orientation, can depend on the quantum state of its partner particle, even if the two are separated by a considerable distance. Consequently, two such particles are thought to maintain nonlocal connections that allow them to exchange information mutually through space. Still, in many other respects, the closed-system model dominates physicists' thinking.

The unpalatable truth appears to be that the inexorable disintegration of the universe as we know it seems assured, the organization which sustains all ordered activity, from men to galaxies, is slowly but inevitably running down, and may even be overtaken by total gravitational collapse into oblivion.[14]

The final gravitational collapse he mentions here concerns the belief that attractive gravitational forces acting between galaxies will one day stop space from expanding and then cause it to begin contracting. This is theorized to lead to a final event, nicknamed the "Big Crunch," in which the contents of the entire universe are smashed together to disappear into the depths of a black hole, never to return. Sociology once earned the title "the dismal science," after Thomas Malthus called people's attention to the inevitable catastrophes that await the world if population growth goes unchecked. It appears that modern cosmology should also share this title.

The closed-system approach to modern cosmology also has difficulty accounting for the origin of the universe. The second law proclaims that if the universe is a closed system, material order (matter and energy) should not have come into being of its own accord. The first law of thermodynamics also rules out the possibility of physical creation. This energy conservation assumption, which claims that energy can be neither created nor destroyed, only converted from one form into another, has worked fairly well for the kinds of laboratory experiments thermodynamicists conduct, but it encounters quite an embarrassing situation when taken into the cosmological arena. Namely, the very existence of the physical universe proves that at some time in the past this law must have been violated in a very significant way.

Cosmologists have attempted to minimize the apparent contradiction by suggesting that the universe emerged in a single big bang explosion and that the energy quantum producing this explosion was created in an extremely short instant of time, an instant so short that anything might be possible, even a tremendous violation of the energy conservation law. Consequently, twentieth-century cosmology has proposed that the act of physical creation was a highly improbable and perplexing event and that the more usual circumstance, which has prevailed for the vast majority of the time, has instead been the destruction of order. It pictures the universe continually running downhill, somewhat like a windup clock gradually spending its mainspring. Fortunately, the open-system ether physics encoded in ancient myths and lore offers a much more optimistic cosmology that is scientifically feasible. Since thermodynamics permits order genesis to occur in a universe that functions as an open system, matter and energy creation become the norm, rather than a primordial abnormality.

3

THE NEW ALCHEMY

Nature has no system? It has, it is life, that streams from an unknown center to an unknowable periphery. The study of Nature may therefore be carried on unceasingly; it may be taken in detail or studied as a whole in length and depth.

GOETHE

HOLISTIC SCIENCE: AN ANCIENT IDEA

Modern science has tended to become increasingly complex and specialized as its base of knowledge has grown. Due to the sheer magnitude of the accumulated information, no one person can expect to become proficient in all of it. For the sake of practicality, science has specialized itself into numerous disciplines. Unfortunately, these disciplines have tended to develop in isolation from one another, each with its own particular jargon that encumbers interdisciplinary communication.

Nevertheless, many have hoped that humankind might one day develop a unified science, one that could allow us to penetrate beyond the artificial boundaries of the disciplines and see nature as a whole. In fact, through the pioneering work of certain visionaries, significant progress has been made in this direction during the past century. One of the most significant advances toward a synthesis of the sciences has come with the development of general system theory (GST) and related disciplines such as nonequilibrium thermodynamics. Interestingly, these theoretical tools have led to the development of a generalized theory of creation very similar to the creation science encoded in ancient myths and lore.

Nature is made up of a variety of systems, which despite their differences share certain characteristics. General system theory seeks to describe these shared characteristics by means of generalized principles. Just as physics has derived general laws to describe physical phenomena, such as the laws of gravitation and electrostatics, general system theory has discerned certain general principles that describe systems regardless of their type. In effect, it is a science of organization. It concerns itself with how systems come into being, how they grow and develop, how they function to maintain their ordered forms, and how they eventually die.

General system theory offers a holistic way of understanding our world. It has an interdisciplinary scope, potentially encompassing all the sciences and professions, from physics and chemistry to psychology and business administration. It does not replace the existing arts and sciences; rather, it provides a means for organizing in a more efficient manner what is already known. If the various disciplines of knowledge are viewed as separate countries, general system theory might be thought of as a kind of Esperanto that allows these disciplinary countries to communicate with one another and discover that they are all living in the same world, that the phenomena they have been separately studying are incredibly similar. This similarity provides a framework upon which a unified science might be built.

The Austro-Hungarian-born scientist and philosopher Ludwig von Bertalanffy is credited as being the founding father of general system theory. Bertalanffy made his first public presentation of this approach in 1937 and 1938 while lecturing at the University of Chicago. He was not the only one to think along these lines, however. Others, such as the biologist A. Bogdanov, the economist Kenneth Boulding, the psychologist and biomathematician Anatol Rapoport, the biologist Ralph Gerard, and the philosopher-scientist Ervin Laszlo, had also independently followed a similar path. In many ways, general system theory developed as a "grassroots" movement.

Nevertheless, Bertalanffy is generally recognized as the originator of general system theory owing to his early thinking and extensive publication on this subject. His first insights came from the field of biology. In 1928 he published a book on developmental biology,[1] in which he advanced the idea that living organisms are organized systems and are best studied as whole entities. He understood that the parts and processes of a living system are organized in accordance with certain principles and proclaimed that the fundamental task of the biologist was to discover these laws. After years of work, he developed his theory of open systems, which recognized that a living organism's uniqueness lies in its ability to utilize imported food energy to sustain metabolic processes that continuously maintain and rebuild its form and that such processes were necessary to the organism's survival.

Bertalanffy later generalized the principles he was discovering to include systems studied in other disciplines. He eventually came to realize that systems at all levels of nature's hierarchy are governed by a similar set of natural laws and that the interdisciplinary search for such laws constituted a valid branch of scientific investigation that he later called "general system theory."[2] In 1954 Bertalanffy and his colleagues Boulding, Rapoport, and Gerard founded the Society for General Systems Research. This international organization, since renamed the International Society for the Systems Sciences, is an affiliate of the American Association for the Advancement of Science and holds annual meetings attended by scientists, educators, and practitioners from all walks of life.

The movement to develop general system theory was largely a response to the spread of scientific reductionism during the early and middle part of the twentieth century. Reductionists sought to explain all natural phenomena, including biological, social, and psychological phenomena, solely in terms of the mechanistic laws of physics and chemistry. This trend had its origins in the scientific revolution of the sixteenth and seventeenth centuries, when the Aristotelian intuitive, holistic, organic approach to understanding nature was replaced by the "resolutive" or analytical method of the philosopher René Descartes.

Certainly, the efficient application of the analytical method facilitated the rapid increase in the modern world's reservoir of scientific knowledge. This approach advocates investigating a system by subdividing it into its elementary parts and processes and focusing on each of these separately to simplify the data-gathering process. By overemphasizing analysis, however, modern science has neglected the holistic approach. The universe, once viewed as an organic whole, became dissected into a meaningless mass of disconnected mechanisms. Through efficient-minded specialization, Western society created a host of social and environmental problems that it might otherwise have avoided if it had adopted a more holistic perspective. Modern society found itself in an era that, to use the Hopi Indian term, may rightfully be called *koyaanisqatsi*, "life out of balance." To deal effectively with problems arising in our increasingly complex society, analysis must be balanced with holistic synthesis. As Bertalanffy has aptly put it, analysis without synthesis is like "trying to walk with one leg."

To explain how ordered forms spontaneously emerge in nature, general system theorists have arrived at a general theory of system genesis that is very similar to the creation physics encoded in ancient myths and lore. Its development may be traced partly to the writings of Bertalanffy and Defay in the late 1920s and early 1930s and to more recent developments in the field of nonequilibrium thermodynamics that were spearheaded by George Nicolis, Ilya Prigogine, and others. As suggested diagrammatically in figure 3.1, this theory of order genesis is applicable to a wide variety of disciplines, including quantum physics. Using a theory essentially identical to that described in ancient creation myths and esoteric lore, modern systems theorists have been able to explain how ordered wave patterns emerge in certain nonlinear chemical reaction systems, how cellular convection patterns emerge in a heated fluid, how the first single-celled organisms might have emerged out of the early ocean, how termite societies build their immense dwellings, how new technologies emerge to replace old ones, how thoughts arise in our minds, and how material particles and energy quanta might emerge from a preexisting ether.

The development of subquantum kinetics in the mid-1970s marked the first time that general system theory concepts had been applied to microphysics, with the result that microphysics, which traditionally had been grounded in the closed-system mechanistic paradigm, became incorporated into the organic, open-system

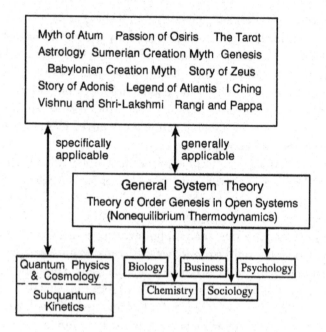

Figure 3.1. The correspondence between concepts in ancient myths and lore and modern scientific thought.

paradigm. The ancient cosmogenic science finds its closest correspondence with modern scientific thought in this recent extension of general systems concepts into physics. From a historical perspective, this ancient open-system cosmology was deciphered in modern times only after science had developed the framework for a similar unified science based around the open-system paradigm. Indeed, these ancient creation myths and lore provide evidence of a prehistoric cosmological science of exceeding brilliance.

THE HIERARCHY OF BEING

The universe is found to be a vast hierarchy of systems nested within systems. A given system participates in forming the system lying directly above it in the hierarchy, its *supersystem*, and at the same time it is composed of systems residing at a level immediately below it in the hierarchy, its *subsystems*. This evolved hierarchy subdivides into three primary vectors of system evolution: the *material evolution vector* of "nonliving" systems; the *life evolution vector* of biological organisms and social organizations; and the *mental evolution vector* of systems born from the mind (see figure 3.2). Any organized form known to science should potentially be a member of one of these three evolution vectors. Other philosophers have similarly segmented natural evolution into three distinct stages. For example, Ludwig von Bertalanffy called these the

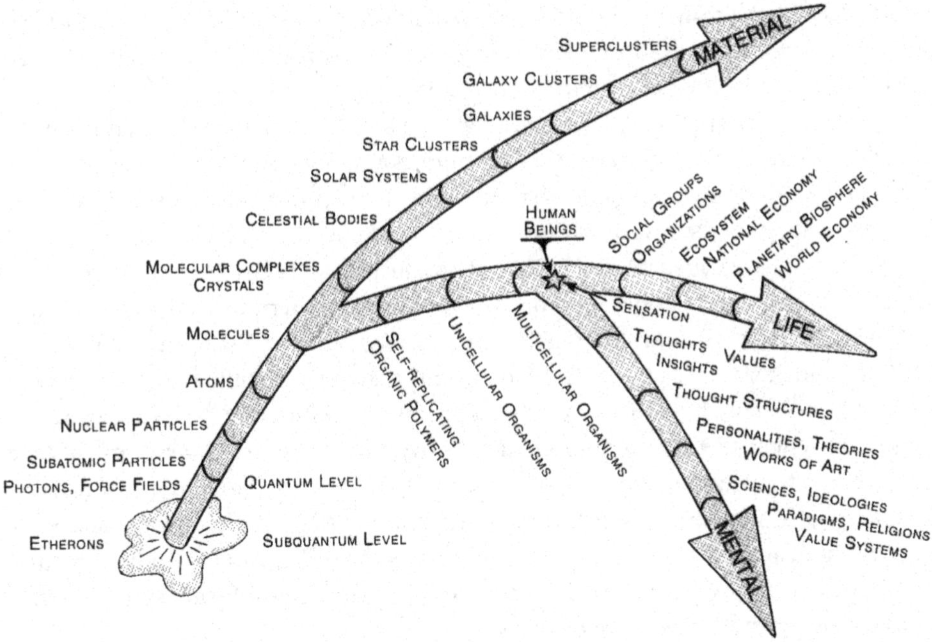

Figure 3.2. Vectors of system evolution: the material, life, and mental evolution vectors. The direction of each arrow denotes the order in which the systems on a given branch came into being.

inanimate, living, and *symbolic* levels, whereas the philosopher Pierre Teilhard de Chardin termed them *prebiological, biological,* and *social* stages.

A natural system at any level of this hierarchy might be defined as any kind of enduring organized entity or form that is continuously repeated or re-created as a result of the dynamics or mutual interactions of more basic components. An atom, a candle flame, a living organism, a society, a galaxy, a belief system, or a body of knowledge are examples of natural systems. Most systems scientists agree that natural systems belonging to the life and mental evolution vectors require ongoing processes to sustain their forms and hence are best categorized as open systems. All structures along these vectors are examples of explicit order that has emerged as a result of a pre-existing implicit order or organization inherent in the processes that create them.

Regarding the various force field–organized systems that compose the material evolution vector, we have no direct observational knowledge as to whether the force or energy fields that organize these systems are products of closed systems, since we have no way of directly sensing what makes up these fields. Our instruments can sense fields, but the underlying etheric processes that may or may not be producing them elude direct observation. Physicists have traditionally assumed that force fields and subatomic particles, both real particles and virtual particles, are forms that exist in their own right as closed systems requiring no "hidden processes" to sustain them. As a result, modern cosmology has had to tolerate the chicken-egg problem that this closed-system view creates.

A general system theorist, however, taking a broader perspective of nature's hierarchy, would seriously question this arbitrary assumption. Applying the principle of parsimony, he or she would infer that structures at the elusive microphysical level, like those making up the rest of nature's system hierarchy, are dependent on an underlying flux. Viewed as such, all natural systems, including those of the material vector, would be classified as open systems. Consequently, modern general system theory shows that the entire universe is a living thing, as was believed in ancient times.

By studying various kinds of open systems that are accessible to detailed observation and seeing how their underlying processes produce ordered wave patterns, general system theorists have been able to hypothesize what processes might take place in a transmuting ether to generate wave patterns that are physically realistic analogues of subatomic particles and energy waves.* Ancient philosopher-scientists could have arrived at their open-system physics through a similar study of nature. So let us now consider how various kinds of open reaction-diffusion systems found in nature produce their wave patterns.

PREDATOR-PREY CYCLES

Although we may not think of it as such, an ecosystem, with its numerous predator and prey animal species, is a kind of reaction-diffusion system. Ecologists use the term "predation" in describing the consumptive transformations that take place among animal species and "migration" to denote the movement of animals through the ecosystem environment; generally speaking, however, predation and migration are reaction and diffusion processes.

Predator-prey systems typically involve large numbers of animal species interacting with one another. But for simplicity let us consider a predator-prey system that consists of a single predator species and a single prey species: foxes eating rabbits. On the whole, this operates as an open system as shown in figure 3.3. Biomass enters the system in the first transformation (a) as food eaten by the rabbits and leaves the system in the third transformation (c) as foxes that die of old age or starvation and presumably serve as the food input for other organisms. These three transformations describe the implicit order (or functional order) of the predator-prey system. They specify the behavior of the predator and prey animals, the *way* in

---

*Such inductive reasoning is commonly used in theoretical physics. For example, the seventeenth-century physicist and mathematician Sir Isaac Newton reasoned by induction when he inferred that the laws of mechanics he had derived from observations of the macroscopic world also applied to the microscopic world of atomic matter.

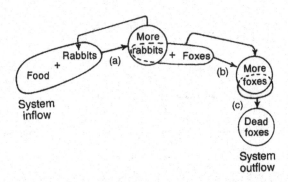

Figure 3.3. Positive feedback and coupling in the fox-rabbit predator-prey system. The arrows directed to the right indicate a transformation of biomass from one species to another: (a) lettuce greens into rabbits; (b) rabbits into foxes; and (c) live foxes into dead foxes.

which encounters take place between the individual animals that make up the system—the manner in which the system processes transform.

The system's implicit order displays two salient features, positive feedback and coupling. Positive feedback refers to a situation in which the output of a transformation process subsequently becomes input to that same process. The first and second transformations (a and b) are examples of this. In the first example, the first generation of rabbits (input rabbits) eats lettuce and doubles the size of its population as new rabbits are born (output rabbits). But this high second-generation population "feeds back" to convert more lettuce into even more rabbits. With an unlimited supply of lettuce and no predators, this vicious cycle would cause the rabbit population to grow exponentially over time. Such growth is called *nonlinear* because a given amount of the input variable (for example, rabbits) results in a much larger output of that same variable. Similarly, foxes eat rabbits and produce even more foxes, which cycle back to consume even more rabbits and produce even more foxes. Given the ideal situation of a very large supply of rabbits, the fox population would also tend to soar in a nonlinear manner.

But since these two positive-feedback loops are mutually *coupled*, the output of the rabbit loop serving as the food input for the fox loop, the two growth trends end up limiting each other. Instead of each species population growing indefinitely, each cycles up and down around a certain long-term average value. Positive feedback and coupling must both be present for the population system to oscillate. Whereas positive feedback tends to disrupt the equilibrium of the system by driving the two populations toward extreme values, coupling tends to balance the system by bringing its populations closer to the average.

Regardless of the type of interacting species involved, foxes and rabbits or wolves and deer, the oscillatory pattern is generally the same. Each species exhibits population maxima that are peaked and short-lived and population minima that are

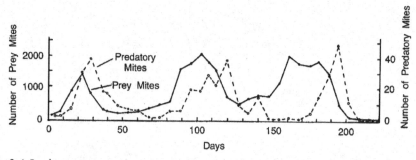

Figure 3.4. Predator-prey population cycles for the six-spotted mite and its predator mite, Typhlodromus, observed in a controlled environment. Data from Huffaker, "Experimental studies on predation," p. 343.

dish-shaped and more drawn out in time. Also, the predator population peak always lags slightly behind the prey population peak. A typical oscillatory pattern is shown in figure 3.4 for a system consisting of two species of mites, one preying on the other. While the magnitude and length of these cycles may vary erratically owing to statistical variations in the predator-prey interaction, the alternation of population sizes between maximum and minimum values occurs in a relatively orderly manner.

This cyclic oscillation constitutes the predator-prey system's *explicit order*. It arises as a result of unique features present in the system's preexisting implicit order, namely the attributes of positive feedback and coupling. Whereas implicit order is *behavioral*, explicit order is *quantitative*, showing up as a difference in *amount*, more rabbits versus fewer rabbits or more foxes versus fewer foxes.

The explicit order displayed in figure 3.4 is best described as timewise or temporal ordering. A predator-prey system can also produce population oscillations, however, that are ordered spatially as well as temporally. These cycles have the form of population waves that advance through the ecosystem environment. They arise because animals foraging for food naturally tend to migrate (or diffuse) from regions of high to low population density. As a result, a wave of high prey population density will advance through the environment, followed by a wave of high predator population density. When the prey population is the variable X and the predator population is the variable Y, an X-wave front (region of increasing X) moves forward followed by a Y-wave front (region of increasing Y), accompanied by a decreasing trend in X. As X proceeds toward its minimum, Y also begins dropping to a minimum, thereby completing one cycle of the advancing predator-prey wave. With the predator population at a minimum, the conditions are now ripe for the advance of another X-Y predator-prey wave.

Predator-prey waves occur as *nerve impulses* in virtually all the higher forms of animal life. They form the physical basis for our mind's consciousness. In this case, sodium ions behave as the prey and potassium ions behave as the predator. The propagating neuroelectrical impulse first allows sodium ions to enter the nerve fiber through the membrane wall. The rising sodium ion concentration (prey population) reduces the membrane's permeability to sodium ions and increases its permeability to potassium ions. The resulting rise in potassium ion concentration (predator population) eventually reduces permeability to both ions, resetting the membrane to its initial impermeable state. Through a domino effect, these predator-prey ion interactions are able to sustain the propagation of a solitary electrical pulse along the length of the nerve fiber. In effect, of all reaction-diffusion wave phenomena, these predator-prey waves come closest to our own direct experience. Yet, despite this subjective familiarity, it took years of biochemical research to discover their objective nature.

Certain ancient creation myths and lore describe a primordial transmuting ether having two reaction intermediates, X and Y, whose wavelike patterns form the material particles of our tangible universe. As in the predator-prey system, this explicit order emerges from an implicit order that includes aspects of both nonlinear growth and coupling, characteristics essential to wave production. The ancient ether physics, however, appears to describe a primordial ether that spawns a periodic stationary wave pattern of precise wavelength as its first emergent waveform. Such a waveform bears a closer similarity to the stationary wave patterns produced by the Brusselator reaction-diffusion system than to predator-prey waves, because the latter are inherently propagative and variable in length from one cycle to the next. Before discussing the Brusselator, however, it is useful to examine the Belousov-Zhabotinskii reaction, a wave-producing chemical concentration that illustrates some of the concepts portrayed in the ancient science of creation.

CHEMICAL WAVES

The reaction-diffusion wave phenomenon is not commonly observed in chemical reactions carried out in the laboratory. There is one chemical solution, however, that has been found to exhibit the phenomenon, the Belousov-Zhabotinskii reaction, or "B-Z reaction." Belousov, the Russian chemist who originally discovered the reaction, worked with a mixture of sulfuric acid, citric acid, potassium bromate, and ceric sulfate. He noted that as these ingredients reacted, the cerium metal ion concentration in his solution oscillated between two alternate ionic valence states, the *cerous* (valence 3) and the *ceric* (valence 4). As this occurred, the solution changed from yellow to colorless and back again to yellow with clocklike precision. Such behavior had never before been observed in an inorganic chemical reaction. Work on this interesting reagent was continued by Dr. Zhabotinskii at the Institute of Biological Physics near Moscow. It was hoped that by investigating this easily reproducible laboratory reaction something could be learned about biological clock phenomena such as the

Krebs cycle, the circadian rhythm, and the heartbeat cycle.

Zhabotinskii found that these chemical oscillations could be made more visible as red-blue color changes by adding to Belousov's reagent an iron sulfate dye indicator called ferroin. This colorful version of the B-Z reaction has become popular as a science class demonstration, fascinating both science and nonscience majors alike. Initially, when the chemical ingredients are mixed together, the solution has a red color, the ferroin dye indicating that the cerium ion is in its chemically reduced cerous state. After an incubation period lasting ten to thirty minutes, however, the solution's color changes, as if magically, to a blue hue, with the ferroin dye indicating that the cerium ion has become oxidized predominantly to its ceric state. As the reaction proceeds and the chemical conditions in the solution continue to change, the dye once again reverts to its red chemically reduced state.

If a beaker full of this solution is kept stirred to ensure homogeneity of the reactants, these color changes will take place simultaneously throughout the entire solution. A complete cycle can take five to ten seconds to complete, depending upon the recipe used to prepare the solution. By sensing the solution's hue with a light meter sensitive to blue light, it is possible to record these oscillations with a chart recorder, as shown in figure 3.5. These oscillations are extremely regular, the wave amplitude, period, and shape being the same from one wave to the next. Because of the regularity of its oscillations, the B-Z reaction has been termed a "chemical clock."

Whereas the previously discussed predator-prey system involves just two variables (prey population and predator population), the far more complex B-Z reaction system has a total of three variables (ceric ion, bromous acid, and bromine ion). Whereas the predator-prey system is described by just three reaction processes, or kinetic equations, the B-Z reaction system requires a total of ten

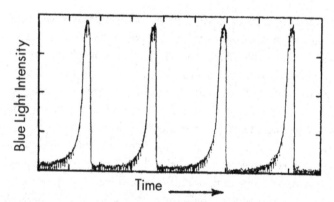

Figure 3.5. A chart recording of periodic color changes in the B-Z reaction. The vertical scale plots the intensity of blue light coming from the solution. When the blue light intensity is low, the solution is in its red state.

kinetic equations for its description. A simplified rendition of the B-Z reaction network, which reduces the ten reactions to just five, is mapped out in figure 3.6. Each letter represents a different state of chemical transformation: A and B signify the initial "fuel" reactants; Ω and Z the final waste products; and W, X, and Y the reaction intermediates (ceric ion, bromous acid, and bromine ion) that engage in oscillation. Like the predator-prey system, the B-Z reaction has the necessary prerequisites that allow it to produce explicit order; it has positive feedback in the form of closed-loop reactions and it has coupling—two distinct reaction pathways that interact with one another (that is, $A \longrightarrow \Omega$ and $B \longrightarrow Z$).

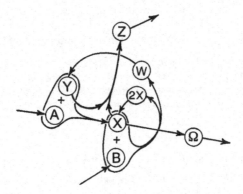

Figure 3.6. A simplified representation of the B-Z reaction pathways. This reaction model, known as the Oregonator, was developed by chemists at the University of Oregon.

The precision of these clocklike oscillations is apparent in figure 3.7, where the simultaneous concentration values of two of the reaction's variables, bromine ion (Y) and ceric ion (W), are plotted against each other. As the reaction oscillates, the (Y, W) coordinate point cycles around the graph repeatedly following the same closed-loop path. The unique orbital trajectory that it traces out is called a *limit cycle*, with the implication that the variables limit themselves to a specific orbit on the graph. If the reaction system concentrations are forced to start at discrepant values, such that the (Y, W) coordinate point initially begins at a spot away from this orbital path (for example, points 1, 2, or 3 in figure 3.7), the reaction system concentrations will change in such a manner that the point rapidly attracts back toward its habitual limit cycle. For this reason, a limit cycle is often called an *attractor*. The predator-prey system has no such unique attractor, which is why its oscillations are inherently erratic.

In addition to chemical clock oscillations, the B-Z reaction is also able to generate propagating chemical waves, advancing fronts consisting of successive regions of high and low concentrations of certain chemical species (cerous and ceric ion). These waves are most readily seen when the solution is poured into a laboratory

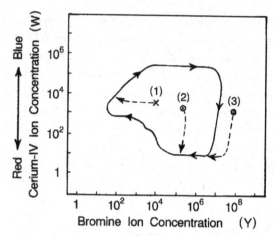

Figure 3.7. An example of a B-Z reaction limit cycle. The coordinates plot concentrations (in arbitrary units) for two of the system's three variables as they evolve over time. Point I marks the location of the system's unstable steady state. Adapted from Field and Noyes, "Oscillations in chemical systems," p. 1881.

culture dish and forms a shallow layer a few millimeters thick. If it is left unstirred, a tiny blue circular patch eventually nucleates in the initially red-orange solution. This blue color then slowly migrates into the surrounding medium as a blue wave front while the central *pacemaker center* turns from blue back to red. A complete cycle of color change may take twenty to sixty seconds, depending upon the characteristics of the particular solution. As the pacemaker center continues to oscillate in regular fashion, it sends out radially expanding rings of red and blue hue, which move forward at a speed of several millimeters per minute. It eventually forms a target pattern consisting of a series of concentric rings similar to the patterns seen in figure 1.1. The ceric ion graph displayed in figure 3.8 presents a cross-sectional view of one such wave train. Systems scientists term such an energy-dissipating wave pattern a *dissipative structure*.

It is also possible to produce several other types of chemical wave behavior. If the culture dish is briefly tilted to produce a slight mechanical disturbance, ring patterns can sometimes be reshaped into spirals that slowly rotate as the wave fronts propagate outward (figure 1.1). Also, by properly adjusting the concentrations of the B-Z reaction ingredients so that the solution remains red throughout and by momentarily touching one spot with a heated needle, it is possible to induce a solitary blue wave to form and propagate outward through the surrounding red solution. In some ways this expanding ring resembles the kind of wave produced when a stone is dropped into a still pond. In this case, however, the disturbance is chemical in nature rather than mechanical. In fact, mechanical agitation of the solution destroys the delicate chemical wave.

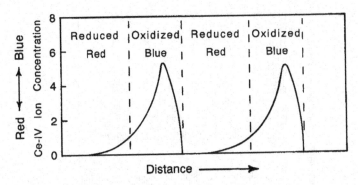

Figure 3.8. A profile of a typical chemical wave train moving from left to right. The graph shows how ceric ion concentration varies with increasing distance from a pacemaker center.

To produce its explicit order, an open system such as the B-Z reaction must have reactions that maintain a state of flux and that flux must proceed at a sufficiently high rate. If the fuel chemicals are allowed to become used up to an extent that the reaction rate falls below a certain threshold value, the concentration oscillations will rapidly decrease in amplitude and eventually cease altogether. In the case of the B-Z reaction, this would be marked by the disappearance of all chemical wave activity. The ferroin dye would revert to its reduced state, and the solution would become uniformly red throughout.

THE BRUSSELATOR REACTION SYSTEM

The ancient myths and esoteric lore that convey the science of cosmic creation describe an ether reaction system that has two variable components, metaphorically portrayed in some myths as earth and sky or in the I Ching as yin and yang. These two ether components, X and Y, are initially uniformly distributed throughout space, but their concentrations randomly fluctuate in magnitude. Furthermore, the myths describe how a particularly large fluctuation (the hero), emerging in a very fertile region of space, grows in size and eventually drives the X and Y concentrations apart to create the core of an ordered wave pattern.

What kind of a reaction system would have an implicit order capable of generating a two-variable ordered stationary wave pattern that these traditions seem to be describing? The Brusselator system appears to be a good choice (see figure 3.9).[3] In fact, it has earned special recognition in the field of chemistry and systems science as being the simplest two-variable reaction-diffusion system capable of spawning waves of regular periodicity.[4] It enjoys a prototype status in the field of chemical kinetics similar to that of the simple harmonic oscillator (spring-and-mass system) in mechanics, the most fundamental generator of mechanical waves.

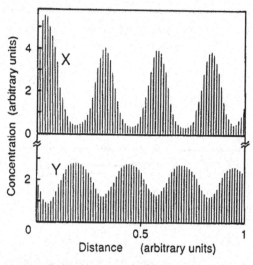

Figure 3.9. Computer simulation of a nonlocalized steady-state dissipative structure generated by the Brusselator in a one-dimensional reaction volume. Data from Lefever, "Dissipative structures in chemical systems," pp. 4977–78.

The Brusselator is specified by just four kinetic equations, which are diagrammed in figure 3.10. A and B are the initial reactants (inputs), X and Y are the reaction intermediates (system variables), and Ω and Z are the final products (outputs). Like the predator-prey system and the B-Z reaction, the Brusselator system includes aspects of both positive feedback and coupling, the two necessary prerequisites for the generation of explicit order. Positive feedback, or what chemists call *autocatalysis,* is present in the third reaction step wherein the output X cycles back to become an input to the same reaction. This gives the Brusselator its necessary nonlinear property.

The second feature, coupling, is evident in the linking of the two distinct reaction pathways $(A \longrightarrow X \longrightarrow \Omega$ and $B + X \longrightarrow Y + Z)$ by the commonly shared intermediate reactant X. In addition, the second and third reaction steps couple by commonly sharing X and Y. Chemists call this kind of double-input sharing *cross-catalysis,* since reaction products must cross over from one reaction to another to form a closed loop with catalytic, reaction-promoting effects. This feature is what gives the Brusselator its ability to generate waves with a precise and regular form. By comparison, the two-variable predator-prey system has just one point of coupling. Foxes remove rabbits without later producing rabbits through some other process. The B-Z reaction does have a cross-catalytic feedback loop; however, its loop is more complex than that of the Brusselator in that it involves the cyclical transformation of three variable reactants $(X, Y,$ and $W)$, rather than just two.

Consequently two features stand out in the Brusselator reaction system that dis-

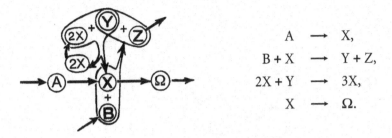

$$A \longrightarrow X,$$
$$B + X \longrightarrow Y + Z,$$
$$2X + Y \longrightarrow 3X,$$
$$X \longrightarrow \Omega.$$

Figure 3.10. A schematic of the Brusselator reaction system.

tinguish its implicit order from that of other nonlinear reaction systems, making it unique as a relatively simple generator of precisely ordered wave patterns: (1) it has only two variable reaction intermediates, X and Y, and (2) its reactions describe a cross-catalysis loop that allows these two variables to cyclically transform into one another, with X changing into Y, then Y changing back into X, and so on.

Interestingly, both of these transformation characteristics are metaphorically portrayed in many of the myths and lore that encode the ancient creation metaphysics. For example, ancient Chinese lore specifies the mutual cyclical transformation of the masculine and feminine ether principles yang (sky) and yin (earth). These two principles may be identified respectively with the second and third reactions; the nonlinear third reaction is feminine by virtue of its nurturing positive feedback properties. Just as the mutually transforming yin and yang are said to constitute an inseparable unity called the *Tao*, which generates all physical form, the Brusselator's mutually transforming X and Y also constitute a unitary self-closing cycle that serves as the generator of its emergent waveforms.

Such simultaneous transmutation of two polar opposite states into one another is a fundamental concept in alchemy and is traditionally called "alchemical union." The word *alchemy* is of Arabic origin and signifies "the science of al-Kemit." Since Kemit is the Islamic name for ancient Egypt, *alchemy* translates as "the science of ancient Egypt."[5] The ancient Egyptian story of Osiris makes indirect reference to such a cross-catalytic process where Isis (the feminine principle) assembles the fragmented body of Osiris (the masculine principle) and bonds with him in nonsexual union to spawn Horus, the warrior son, who ultimately defeats Set and brings physical order into being. The absence of Osiris's generative organ from the reassembled body emphasizes the noncoital nature of this union. Process plays a significant role here, for Osiris is the traditional symbol for generative process, and Isis also plays a catalytic role by marginally reviving him from the death state. Their cooperative union involves the acceleration of process that would occur in a cross-catalytic exchange. Like the yin and yang principles in the ancient Chinese myth, the bonding of Isis and Osiris should be interpreted as an alchemic union involving

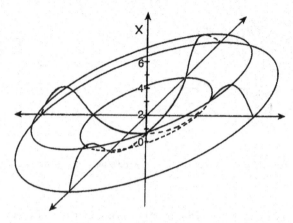

Figure 3.11. Concentration profile of a steady-state dissipative structure generated by the Brusselator in a two-dimensional reaction volume. Data from Nicolis and Prigogine, Self-Organization, p. 182.

cyclical transmutation of polar opposites. Horus, then, signifies the seed of explicit order that arises from this cross-catalytic womb (implicit order) to constitute the first particle of matter.

The Brusselator reaction system cannot be reproduced in the chemist's laboratory because reactions involving the simultaneous combination of three molecules, as specified in its third reaction $2X + Y \longrightarrow 3X$, are rarely observed. Nevertheless, systems scientists have been able to simulate on the computer its reaction processes and study its various ordering phenomena. The concentration profile in figure 3.9 was produced by a computer simulating the Brusselator reactions in a one-dimensional (tube-shaped) volume. Provided that "fertile" conditions prevail through the entire length of the reaction volume, the X and Y concentration profiles build up to the wave patterns shown here, or to their mirror-image wave patterns in which peaks replace valleys and valleys replace peaks. In either case, once the X and Y variables depart from their initially uniform steady-state values, the resulting wave pattern attains its final shape in a seemingly predestined fashion. The X and Y concentrations always deviate in opposite senses owing to the reciprocal interaction of X and Y in the X-Y cross-catalytic loop. Thus "earth" and "sky" (X and Y) always separate from one another.

The ringlike wave pattern shown in figure 3.11 was produced by computer simulation of the Brusselator reaction in two dimensions. A three-dimensional simulation would produce a spherical wave pattern that consists of a central core surrounded by a series of concentric shells. In performing two- and three-dimensional simulations, it is also possible to produce dissipative space structure wave patterns that rotate about a central axis. That is, in addition to the radial wave pattern that produces the concentric series of rings or shells, each such ring or shell would contain

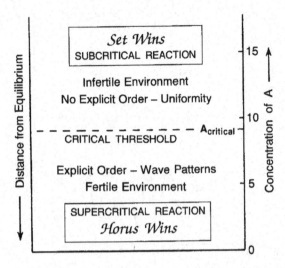

Figure 3.12. How the Brusselator's state of criticality varies with respect to changes in the concentration of its fuel reactant A. In the subcritical state, X and Y remain uniform, whereas in the supercritical state they form stationary wave patterns.

a secondary concentration wave disposed around the circumference of the shell and moving about its geometric center like a horse around a racetrack.

The Brusselator produces wave patterns only if the concentration ratio of the two input reactants $B{:}A$ is greater than a certain critical ratio value $(B{:}A)_{critical}$ that is called the *critical threshold*. When the concentration ratio is below this critical value, the system is *subcritical;* when the ratio is exactly at this threshold, the system is *marginally stable;* and when the ratio is greater than this threshold, the reaction system becomes *supercritical.** Under subcritical conditions, the reaction medium presents an infertile environment for the development of a wave pattern (explicit order). In other words, if Horus were to battle Set under these conditions, Set would win, and the law of entropy would prevail (see figure 3.12). When the Brusselator is in this infertile or subcritical state, X or Y concentration pulses arising as deviations from the mean concentrations tend to regress over time, with the result that the X and Y reactants would maintain constant and uniform steady-state concentrations throughout the reaction volume, forming no ordered spatial pattern.

As A is decreased relative to B (or B is increased relative to A) causing the $B{:}A$ ratio to rise, the Brusselator becomes increasingly less subcritical. When the ratio is equal

---

*Similar terms are used in the nuclear energy field. A nuclear reactor is a kind of nonequilibrium reaction-diffusion system, one that involves the diffusion of subatomic particles (neutrons) and their reaction with nuclear particles (uranium-235). It generates a steady flow of power when the reactions in its core are operating exactly at the critical threshold.

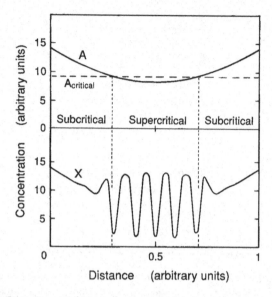

Figure 3.13. Computer simulation of a localized dissipative structure (lower profile) generated by the Brusselator in a one-dimensional reaction volume. Adapted from Prigogine, Nicolis, and Babloyantz, "Thermodynamics of evolution," pp. 38–44.

to the critical value, the reaction medium presents a neutral environment for the development of explicit order in which randomly emerging concentration blips neither increase nor diminish in size. In this case, the battle between Horus and Set results in a standoff. If the $B:A$ ratio rises above the critical threshold so that the Brusselator operates in its supercritical mode, however, the reaction medium then presents a fertile environment for the development of explicit order. Under such circumstances, the cross-catalytic growth-inducing reaction loop formed by the second and third reaction steps would become sufficiently active that a small concentration pulse arising in either the X or Y reactants would be able to grow in size and drive the reactant concentrations away from their former uniform values to eventually form a macroscopic wave pattern.

By assuring that the $B:A$ concentration ratio varies along the length of the Brusselator's reaction volume to produce a fertile (supercritical) region in the center of the volume flanked on either side by infertile (subcritical) regions, it is possible to produce a localized stationary wave pattern, as shown in figure 3.13. This may be done by keeping the concentration of B fixed throughout the reaction volume and allowing A to vary. A is held constant at the ends of the reaction vessel and allowed to progressively decrease in concentration toward the vessel's center. Since A, the "food" reactant, is consumed throughout the reaction volume (due to the reaction $A \longrightarrow X$), its concentration becomes reduced in the volume's midregion and it forms a U-shaped steady-state profile. In regions where A drops below a certain critical threshold value $A_{critical}$ indicated by the horizontal dashed line, the $B{:}A$ ratio becomes sufficiently high that supercritical conditions prevail. As a result, a randomly emerging pulse is able to grow into a stationary wave pattern within this localized supercritical region. Outside of this region the X and Y concentrations dampen out to their subcritical steady-state values. To produce localized stationary wave patterns that can serve as realistic analogs of subatomic particles, the Brusselator must be slightly modified by adding a new variable G to its first reaction $(A \longrightarrow X)$, as described in the next chapter.

CHEMICAL PERSPECTIVES ON COSMOLOGY

The fundamental difference between the cosmology adopted by modern science and that encoded in ancient creation myths and lore can be traced to differences inherent in their underlying paradigms. Modern physics is based on the mechanical model, whereas the ancient cosmology (and subquantum kinetics) is based on the somewhat more complex and more comprehensive reaction-diffusion model, which is applicable to a wide variety of biological and social phenomena. Differences between these approaches can be brought to light by contrasting mechanical and reaction-diffusion wave phenomena observed in nature.

In the case of mechanical waves, the carrier plays a passive role. An impressed force temporarily displaces the molecules of the carrier medium against elastic restoring forces, which soon return the molecules to their original equilibrium positions. Meanwhile, the force of the impulse is passed forward, temporarily displacing adjacent molecules and thereby propagating through the elastic carrier. Mechanical waves can never arise spontaneously; they always form as predictable responses to external causative agents that momentarily disturb the medium's state of equilibrium. In the absence of the wave disturbance, the medium constituents are at rest in a state of mechanical equilibrium.

Both these mechanical waves and the energy waves of modern physical theory require an antecedent energy impulse to bring them into existence. Consequently, in devising a theory for the origin of things, mechanistic physics is confronted with the problem of explaining how the first quantum of energy came into being. Like its mechanical ether predecessor, modern field theory requires an outside agent, a deus ex machina, to impart the initial impulse that emerges as the primordial big bang explosion. Since this twentieth-century mechanistic paradigm does *not* recognize an existence antecedent to the Big Bang, however, it finds itself hemmed into a difficult corner—nothing exists to impart the initial impulse.

The situation is quite different for a cosmology based on the reaction-diffusion paradigm. In the case of reaction-diffusion waves, the underlying medium is inherently active. Not only must the medium's constituents move about (diffuse); they must also react among themselves. To do this, the carrier must maintain a *nonequilibrium* state in which its constituents continually react and change their form.* Moreover, reaction-diffusion dissipative structures may arise either as propagating waves or as stationary waves. Mechanical waves, on the other hand, are inherently dynamic; they can form only stationary disturbances or standing waves when reflected from impenetrable boundaries.

Reaction-diffusion waves come into being quite differently from mechanical waves. Under the proper conditions, they can arise spontaneously (self-organize) without being triggered by an outside disturbance. Their order emerges in a probabilistic fashion and is seeded from the reaction system's own fluctuation "noise." All that is needed is the deviation of a reactant's concentration from its steady-state value by a sufficiently large amount somewhere in the reaction medium. Provided that the reaction system is supercritical, this microscopic *critical fluctuation* is able to grow in size and develop into a macroscopic wave. Ilya Prigogine has coined the phrase *order through fluctuation* to refer to this order-nucleation process.[6] Several differences between mechanical and reaction-diffusion wave phenomena are summarized in table 3.1.

TABLE 3.1
A Comparison of Two Types of Wave Phenomena

MECHANICAL WAVES	REACTION-DIFFUSION WAVES
Carrier is inert, unchanging, nonmetabolic.	Carrier is interactive, undergoing metabolic changes.
Carrier is at a state of thermodynamic equilibrium. It is energy conserving.	Carrier is in a nonequilibrium state. It dissipates energy.
Carrier's entropy is constant.	Carrier's entropy is always increasing.
Carrier plays a passive role in wave transmission; the wave is impressed upon the carrier.	Carrier plays an active role in wave transmission; it sustains the wave.
Waves are never produced spontaneously.	Under certain conditions, waves arise spontaneously.
Waves are a response to an external stimulus.	Waves are an instance of self-organization, an outward expression of the carrier's implicit order.
Wave behavior is represented by a set of linear equations.	Wave behavior is represented by a set of nonlinear rate equations.

---

*Compared with mechanical wave media, reaction-diffusion wave media are more fundamental. Whereas mechanical media just conduct mechanical waves, reaction-diffusion media can conduct both mechanical waves and reaction-diffusion waves. Thus mechanical waves are a special case of the reaction-diffusion wave phenomenon.

The cosmology devised by our ancient predecessors describes a transmuting ether that spawns the material world in the same way that a reacting medium gives rise to dissipative structures. With such a cosmology, there is no need to postulate a creator separate from the creation. The divine may be viewed as one with the creation, encompassing both visible and invisible realms. Rather than conceiving of the act of creation as a specific event long past, the ancient physics suggests that the physical world is sustained on an ongoing basis, with matter and energy being continuously created.

4

THE TRANSMUTING ETHER

Then he who sat on the throne said, "Behold! I am making all things new!"And he said
to me, "Write this down; for these words are trustworthy and true. Indeed," he said,
"they are already fulfilled. For I am the Alpha and the Omega,
the beginning and the end."

REVELATIONS 21

THE PRINCIPLE OF ANTECEDENTS

By carefully considering the blueprint of nature's chain of being, we can conclude that the ancient notion of an ether is quite a sensible concept. The overview in figure 3.2 shows how nature adheres to the following general dictum, called the principle of antecedents: namely, a system always arises from antecedent subsystems. The mental evolution vector arises from the life evolution vector, and that, in turn, arises from the material evolution vector. Unless nature has made an exception for the material evolution vector, this principle applies to subatomic particles and photons as well; such quantum structures must consist of ordered patterns formed in a primordial space-filling ether continuum. This continuum constitutes another vector of existence not directly accessible to our senses.

Through the ages, spiritual masters have testified to the existence of such an invisible nonmaterial realm, which they regard as divine in nature. They say that this divine realm is the true essence of reality and that the observable world is illusory or just one manifestation of the infinite Divine. Modern science takes a very different position in that it recognizes only the observable physical world as real. It subscribes to the doctrine of logical positivism that holds that the "existence" of something is contingent upon our ability to obtain "positive" knowledge of its presence through direct observation. Hence the positivist's domain of reality is restricted to those things perceived directly through the physical senses or indirectly through instrument-aided observation. He refuses to admit the existence of a god, a nonmaterial spiritual realm, or an ether, unless they can be empirically verified. This refusal by many physicists to accept the existence of an unseen realm lies at the heart of the present conflict between science and religion.

Today, logical positivism's last refuge is in the physical sciences. Although it enjoyed some degree of popularity in certain philosophical circles early in the twentieth century, contemporary philosophers regard positivism as outmoded and unworkable. After all, the process of observation is itself a subjective act. What an observer sees depends on what he looks at and the way he looks at it. A scientist's presuppositions inevitably shape the reality he perceives.

The principle of antecedents, deduced from observations of nature's hierarchical structure, leads us beyond the narrow perspective of positivism. The pattern of nature demands that an inherently unobservable realm exists beneath the observable. Just as life emerges from a preexisting sea of water and hydrocarbon molecules, and just as thought surfaces from a preexisting "sea" of sense data, so too must quantum structures, the building blocks of our physical world, arise and take form out of a preexisting etheric sea. Thus the ether would constitute the "nonmaterial antecedent" of the material world. Although modern physics abandoned the ether concept near the beginning of the twentieth century and replaced it with a vacuum, recent laboratory experiments appear to have confirmed its existence (see chapter 12). The ancient ether concept may be coming back into fashion.

ETHER KINETICS

By understanding how certain types of open chemical reaction systems spontaneously self-organize their molecular reactants into macroscopic wave patterns, we are led to infer that the constituents of the all-pervading ether might similarly self-organize to spawn wavelike patterns as the next rung up in nature's hierarchy, these being the subatomic particles that serve as the building blocks of our material world. Such an ether would be made up of a swarming sea of constituent particles, or *etherons,* which diffuse from one place to another and transform from one etheron state to another like molecules in a reacting chemical solution. Etherons would be far smaller in size than the subatomic particles that physicists are normally familiar with. For example, countless quadrillions of them might occupy the space taken up by a single proton.

In addition to being vastly smaller in size, etherons would differ from subatomic particles in many other respects. For example, there is no reason to suspect that they would have mass or charge, as do material particles. Because our detector probes are composed of matter or energy, they can only sense other matter or energy forms, not the individual etherons that might compose those forms. Attempting to use a material probe to detect an etheron would be the equivalent of trying to use water waves to observe the presence of an individual water molecule. Thus we cannot know for sure the true nature of etherons.

For the present, it is not necessary to know the properties of etherons in any great detail, other than that they would be of various types (*A, B, C,* and so on) and would be able to interact with one another in specific ways. Etherons of a specific type would make up a single ether substrate. For example, all of the *A*-ons in space would together compose the *A* ether substrate, all of the *B*-ons, the *B* ether substrate, and so forth. Certain types would be able to transmute through solitary transformations (figure 4.1a), while others would change their form through reactions with other etherons (figure 4.1b). These transformations would occur in an orderly fashion, following well-defined reaction pathways.

Together, these various kinds of etherons collectively compose a reacting and diffusing heterogeneous medium that we call the *transmuting ether.* The transmuting ether closely resembles the organic conception of space proposed by the twentieth-century philosopher Alfred North Whitehead.[1] According to Whitehead, space is a kind of living organism, an integrated system in which the whole of space is more than just the sum of its parts. Just as the millions of cells forming a human being interact in multiple ways to compose a living entity, Whitehead conceived that every volume of space "takes account of" (or interacts with) every other volume, so as to compose an integral structure. This organic conception very appropriately describes the transmuting ether.

Because of diffusion, processes transpiring in one part of the ether depend in a real way on processes that take place elsewhere in the ether. Thus, as in a reacting chemical solution, diffusion binds the reacting ether together into an organic whole. Since there is no way of directly discovering how etherons actually react with one another to produce physical form, we must proceed in an indirect manner through inference. Given that subatomic particles and photons have precise wavelike properties, the Brusselator can serve as a starting point, since it is the simplest system capable of producing regular wave patterns. To produce concentration patterns that might serve as realistic analogs of physical particles and waves, however, it is advisable to modify the Brusselator's first reaction by adding an additional ether state, *G,* which like *X* and *Y* is free to vary both spatially and temporally. Thus

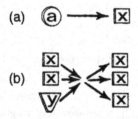

Figure 4.1. Examples of two kinds of
etheron transformations: (a) a solitary transmutation
and (b) a multiunit reaction.

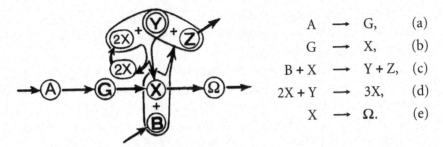

$$
\begin{aligned}
A &\longrightarrow G, & \text{(a)} \\
G &\longrightarrow X, & \text{(b)} \\
B + X &\longrightarrow Y + Z, & \text{(c)} \\
2X + Y &\longrightarrow 3X, & \text{(d)} \\
X &\longrightarrow \Omega. & \text{(e)}
\end{aligned}
$$

Figure 4.2. Model G, a model of the ether reactions that generate the physical universe: (left) mapped out to show its functional order and (right) listed as a series of five reaction steps. The arrows indicate the direction of transformation. Etherons enter the reaction system through states A and B; reside for a time in intermediate states G, X, and Y; and leave the system through states Z and Ω.

the reaction $A \longrightarrow X$ would be replaced by two reactions $A \longrightarrow G$ and $G \longrightarrow X$. The resulting five-step reaction scheme is diagrammed in figure 4.2. Known as Model G, it was originally developed in 1978 as a prospective model for use in the subquantum kinetics physics methodology.[2]

The ordered arrangement of these five reactions in relation to one another would comprise the *implicit order* of the ether. These processes would constitute the soul of the universe, the eternal MAYA that eventually gives birth to *maya*—physical form. The implicit order of Model G, like that of the Brusselator, has aspects of both positive feedback and coupling, features that must be present for a reaction system to be able to produce wave patterns. The positive feedback requirement is satisfied by autocatalytic reaction (d), the coupling requirement by the juncture of the two global reaction pathways $A \longrightarrow \Omega$ and $B \longrightarrow Z$. Coupling is also present in the cross-catalytic loop formed by the sharing of X and Y in reactions (c) and (d), a feature that gives Model G (and the Brusselator) the ability to form waves having precise periodicities. Although Model G involves three variables, G, X, and Y, it may be classified as a two-variable wave generator, since only its X and Y species are members of its cross-catalytic loop. The ancient creation science appears to describe a two-variable wave generator of this particular type.

The various species composing this reacting ether would extend an infinite distance throughout space, their initial and final reactants, A, B, Z, and Ω, maintaining relatively constant and uniform concentration values. The variable ether media, G, X, and Y, would also be uniformly distributed initially, but under the right circumstances their concentrations could depart from their former uniform values and become configured into particulate stationary wave patterns of subatomic size—the subatomic particles that compose the building blocks of all physical form in our universe. Radiant energy, such as light photons and radio waves, would consist of propagating concentration waves. These ether concentration profiles would be equivalent to the energy potential fields that physicists propose as the basis for material particles and energy waves.

We experience the physical world through the forces and impulses that are exerted on our senses or instrument probes. An electrostatic field, material object, or wave of radiant energy makes itself known to us through its ability to exert force. Contemporary physics teaches that forces arise from force fields that are associated with energy potential gradients. Similarly in ether metaphysics terms, forces are produced by sloping ether concentration profiles. Thus electrostatic forces would be produced by gradients in the X and Y ethers, and gravitational forces would be produced by gradients in the G ether. The magnitude of the force would be determined by the steepness of the field's slope. For example, a volume of space having G, X, and Y ether concentrations distributed in a spatially uniform manner as a zero-gradient field (figure 4.3a) would be incapable of generating force. From a physical standpoint, such a region would appear as a vacuum. On the other hand, sloping concentration profiles (figure 4.3b and 4.3c) would be potentially observable, since they could induce sensation-producing forces. The steeper their gradient, the more intense their exerted force. In summary, field potentials, the underpinnings of physical existence, would essentially consist of spatial concentration patterns in the X, Y, and G ether substrates. Table 4.1 summarizes some correspondences that may be made between physical phenomena and their ether equivalents.

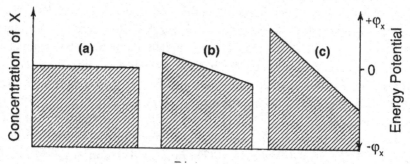

Figure 4.3. The relation between ether concentration and field potential: (a) a "vacuum"; (b) a low-intensity force-exerting field; and (c) a high-intensity force-exerting field. Following physics convention we represent energy potential fields by the Greek letter φ (phi).

TRANSFORMATION AND TIME

To more readily appreciate how the ancient transmuting ether concept leads to the idea of the physical universe as open system, we place an imaginary boundary around the G, X, and Y etheron states as depicted in figure 4.4. This bounded region

TABLE 4.1
Ether Correlates of Physical Fields, Particles, and Waves

PHYSICS PARAMETER*	ETHER CORRELATE
Gravitational potential	G ether concentration
Electrostatic potential	X (or Y) ether concentration
Subatomic particle	Reaction-diffusion wave pattern composed of localized stationary X, Y, and G concentration magnitudes
Light wave	Reaction-diffusion wave composed of localized X, Y, and G concentration magnitudes that propagate forward at the speed of light

*An absolute gravitational or electrostatic energy potential would be the physical equivalent of a G, X, and Y ether concentration. Since physicists have no way of determining the absolute value of a potential, however, they usually refer to potential differences, which may be either positive or negative depending on whether they are measured above or below some arbitrarily set zero potential point. This convention may be accommodated in the ether metaphysics by assigning a specific ether concentration as the point of zero potential.

would circumscribe the domain of the physical universe and its assortment of forms. With the passage of time, etherons initially in precursor states *A* and *B* transmute "into" the etheric domain of our physical universe to temporarily compose ether states *G*, *X*, and *Y*, and then transmute "out of" this domain to compose antecedent states Ω and *Z*. Since etherons would continuously cross this boundary to compose the *G*, *X*, and *Y* ether states underlying our universe, our universe would be a kind of open system.

It would be rather limiting to suppose that the etherons that temporarily compose and sustain our physical universe simply begin their journey in states *A* and *B* and end it in states Ω and *Z*. It seems more probable that they would pass through a vast transmutative sequence extending a considerable way both upstream and downstream of our physical universe. That is, etherons populating state *A* arrive from a preceding state *A'*, which in turn arrive from yet another state *A''*, and so on. The same sort of progression may be hypothesized for substrates *B*, *Z*, and Ω. This consecutive sequence, the *transformation dimension*, is symbolized by the letter *T*.

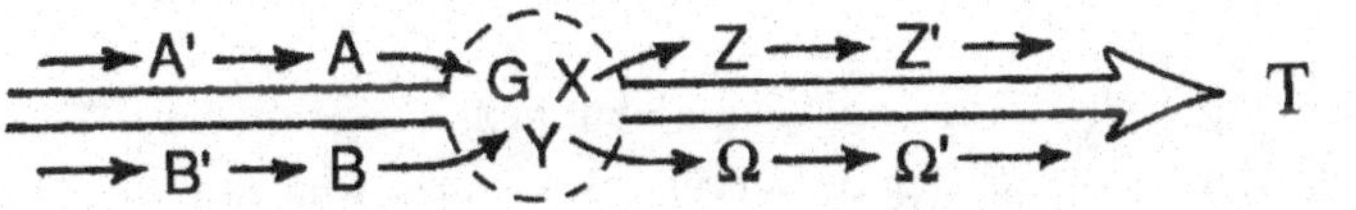

Figure 4.4. A suggested expansion of the ether reaction scheme as it would appear disposed along the transformation dimension. The G, X, and Y ether substrate group would mark the domain of the physical universe.

The transformation dimension, in effect, is generated as a result of the continuous transformation of etherons through their succession of states. Unlike three-dimensional space, which has no preferred direction, the transformation dimension would display a distinct preference, the direction of preferred etheron evolution.

Since this transforming continuum would pass "through" our three-dimensional physical world, the transformation dimension may be thought of as a higher dimension. This is not a fourth dimension of space, however, for etherons in all of these ether states could simultaneously coexist in the same three-dimensional space that contains our physical universe. This entire sequence of etherons could exist along with the G, X, and Y etherons that compose our world, but we would have no way of directly interacting with them or with any explicit order (field patterns) that they might form.

The transmuting ether resembles the *primal flux* of the ancient Greek philosopher Heraclitus (c. 500 B.C.E.). Heraclitus likened the ether to fire. He envisioned the ether to be continuously undergoing change and sustaining a concurrent creation and destruction of its constituents. One surviving fragment of his philosophy states: "This ordered universe *(cosmos)*, which is the same for all . . . always was, is, and shall be an ever-living Fire, kindled in measure and quenched in measure."[3]

The transmuting ether also resembles the *ch'i* of the ancient Chinese and the *prana* of the Hindus. Mahayana Buddhism comments upon this underlying flux in its "doctrine of the impermanence of things." W. Y. Evans-Wentz explains it as follows:

> Just as we discern not the passing of cream into butter, so we fail to comprehend the constant flux of all component objects. The densest aspects of matter, like the most subtle invisible gases, are never exactly the same one second after another; through all alike moves the life-force in its psycho-physical, ever-flowing, ever-structurally modifying pulsation, even as in the body of man.[4]

The way the etheric transmutative flux operates in forming physical creation is brilliantly portrayed by the four-armed Hindu god Shiva Nataraja (figure 4.5), the dancer whose dance is the universe. Shiva, Lord of the Dancers, signifies the ceaseless flow of energy as God (Brahman) manifests the infinite variety of patterns that compose the physical world. Shiva embodies the processes of creation and destruction that continuously operate to sustain the universe. Heinrich Zimmer writes about the dance of Shiva:

> Shiva is Kala, "The Black One," "Time"; but he is also Maha Kala, "Great Time," "Eternity." As Nataraja, King of Dancers, his gestures, wild and full of grace, precipitate the cosmic illusion; his flying arms and legs and the swaying of his torso produce—indeed, they are—the continuous creation-destruction of the universe, death exactly balancing birth, annihilation the end of every coming-forth. The

Figure 4.5. Shiva Nataraja, the Cosmic Dancer.

choreography is the whirligig of time. History and its ruins, the explosions of suns, are flashes from the tireless swinging sequence of the gestures.[5]

Shiva's right dorsal hand represents the creative principle. The tiny hourglass-shaped drum that this hand holds and rhythmically shakes emits the primal sound of cosmic creation (figure 4.6), OM (AUM)—the sound "not made by any two things striking together." Hindus associate sound with the ether element.* Hence Shiva's resounding drum signifies the process of etheric creation, the sound energy etherons produce as they pop into each new etheron state in their sequential journey. The Om vibration, then, would be a metaphor for the fluctuating ether, the zero-point energy that fills all space. Shiva's left dorsal hand represents the destructive principle. It bears a tongue of flame that suggests that dissolution occurs through a process of transmutation. Some statues portray a skull and new moon in the dancer's hair, symbolic of the processes of death and rebirth that simultaneously operate.

Shiva's balanced hand gestures and poised posture suggest that the processes of

---

*Hindu philosophy teaches that there are five elements, of which ether is the first. The other four (fire, air, water, and earth) unfolded from the ether when the universe was created.

Figure 4.6. The sacred syllable AUM, the seed sound of universal creation, as it appears in Sanskrit Devanagari script. The Cosmic Dancer's gestures seem to form this shape.

creation and destruction maintain a dynamic balance. The dancer's two front hands are brought together as if these two antithetical processes worked in unison, birth (upward-pointing right hand) counteracting death (downward-pointing left hand). His upward-pointing hand signs the "fear not" gesture bestowing protection and peace, while his other hand points downward toward his elevated left foot, the sign of release. Verse 36 of the *Unmai Vilakkam* summarizes this transmutation sequence as follows: "Creation arises from the drum: protection proceeds from the hand of hope: from fire proceeds destruction: the foot held aloft gives release."[6]

The Cosmic Dancer signifies the transmuting ether. The processes of creation, preservation, and destruction-through-release that Shiva enacts accurately describe the transmutative journey of etherons as they are born into a particular ether state, temporarily reside in that state, and finally pass from that state into another. This birth-death transmutative sequence is further portrayed by the wreath of flames that encircles Shiva and issues from the mouths of the *makara*, a double-headed mythological water monster upon which Shiva stands.*

Shiva's dance illustrates the secret of how the ether generates physical form, *maya*. His graceful swaying motions would represent the oscillations of the G, X, and Y ethers that compose the physical world. Just as Shiva's movements are a manifestation of the shifting balance between creation and destruction, the concentrations of these ethers vary in magnitude as a result of slight shifts in the rates of etheron creation and destruction. For example, the concentration of G depends on

---

*The series of births and deaths experienced by an etheron as it transmutes through its succession of states resembles in many ways the reincarnational cycle of the soul. According to Hindu mysticism, an individual's soul is born into the physical world, lives a life of experiences, and finally passes away at its appointed time, only to reincarnate once again in a new and different earthly life. To acknowledge this terminal transition, Hindus customarily cremate their dead, that is, subject them to the transmuting element of fire.

the rate at which etherons enter state G through the reaction $A \longrightarrow G$ as well as on the rate at which they are annihilated from that state through the reaction $G \longrightarrow X$.

The symbolism of the four hands is also significant. Shiva bears two hands denoting creation and two denoting destruction; etherons would also enter our universe from two source states A and B and leave our universe through two sink states Ω and Z (see figure 4.2). The two outer arms would together represent one transmutation pathway, $A \longrightarrow G \longrightarrow X \longrightarrow \Omega$, while the two hands gesturing up and down would signify the second intersecting transmutation pathway, $B + X \longrightarrow Y + Z$. The conjoining of these four arms in Shiva's body depicts the intersection and interaction of these two transmutative streams. Without this interaction between separate transmutative pathways, the ether fluxes would be unable to generate a physical universe.

In his manifestation as the Lord of the Dancers, Shiva personifies the concept of etheric transmutation. On the other hand, in his manifestation as the sword-wielding god Kala, he becomes "the Lord of Time," "the Black One," and "the Destroyer." Like Osiris and Set in Egyptian mythology, Shiva and Kala together personify two basic thermodynamic principles of nature—order creation and entropy. In yet another manifestation Shiva becomes the god Maha Kala, "Great Time" or "Eternity." As the "timeless" eternal now, Shiva would signify the entire transformation dimension and perhaps other higher dimensions as well.

As suggested in the symbolism of Shiva, there is a close connection between etheric transformation and time. Our experience of past, present, and future arises from our perception of changes and movements taking place within us and within our external three-dimensional world. This ongoing sequence of events, which gives us the feeling of time's passage, arises because etherons continuously transmute through our universe along the transformation dimension.

Just as time flows forward like a river, etherons transmute forward, forming a four-dimensional etheric river whose unidirectional evolution at every moment re-creates our physical world. When we say that etherons transmute "forward," we mean that they change from their existing ether state into a succeeding one. The same etherons that presently compose our physical universe an instant earlier resided at etheric levels situated "above" our universe and an instant later will reside at etheric levels lying "below" our universe. Our particular sensation of time, from our vantage point as three-dimensional beings, arises because we remain stationary with respect to this higher-dimensional flux.

If through an effort of introspection we were able to project ourselves into this fourth dimension and were to allow ourselves to drift with this transmutative flow, transforming downstream along with the etherons, our sense of time's passage would be quite different. It would be one of ever-present change, of continual becoming. The French philosopher Henri Bergson seems to have undergone an experience similar to this. In his *Introduction to Metaphysics* he writes:

> There is one reality, at least, which we all seize from within, by intuition and not by simple analysis. It is our own personality in its flowing through time—our self which endures.... When I direct my attention inward to contemplate my own self ... I perceive at first, as a crust solidified on the surface, all the perceptions which come to it from the material world.... Next, I notice the memories which more or less adhere to these perceptions and which serve to interpret them.... Lastly, I feel the stir of tendencies and motor habits—a crowd of virtual actions, more or less firmly bound to these perceptions and memories.... But if I draw myself in from the periphery towards the center, if I search in the depth of my being that which is most uniformly, most constantly, and most enduringly myself, I find an altogether different thing.
>
> There is, beneath these sharply cut crystals and this frozen surface, a continuous flux which is not comparable to any flux I have ever seen. There is a succession of states, each of which announces that which follows and contains that which precedes it. They can, properly speaking, only be said to form multiple states when I have already passed them and turn back to observe their track. Whilst I was experiencing them they were so solidly organized, so profoundly animated with a common life that I could not have said where any one of them finished or where another commenced. In reality no one of them begins or ends, but all extend into each other.[7]

The transmutation of etherons through our physical universe is quite appropriately identified with the irreversible flow of time that we experience as physical beings. But does time originate from this flux, or is it more all-encompassing? Etherons necessarily endure over some interval while occupying each of their states, and their transition from one etheron state to the next most likely transpires over a certain period of time. So, to accommodate the existence of etherons transforming along the transformation dimension, we must assume the existence of yet another timelike dimension that is distinct from the transformation dimension and that envelops it together with all its evolving etherons, a fifth dimension passing through the other four.

Present-day physicists view time as a reversible quantity. The equations of both classical mechanics and quantum mechanics are such that the equations would still retain their meaning even if time were assumed to move in reverse, the variable t being replaced by $-t$. Such time-reversal symmetry works well for very simple dynamical systems, such as two colliding particles, where it is difficult to tell if the particles are traveling forward or backward in time. But this ideal type of situation is the exception in nature, rather than the rule. Entropy-increasing processes, such as those summarized in figure 2.1, make sense only if time proceeds in one direction. Prigogine notes that of these two concepts of time, it is the concept of reversible time that should be considered a fiction. He explains that irreversibility is either true on all levels of nature or on none; it is not possible to have time-reversible processes taking place at a microscopic level, and then have irreversibility miraculously emerge at a more macroscopic level.[8] He advocates introducing irreversibility into physics at the most fundamental level. This is exactly what the ancient ether metaphysics accomplishes by proposing the existence of an irreversibly transmuting ether as the basis of physical existence.*

THE REALM BEYOND

SPHERE. But where is this land of four dimensions?

SQUARE. I know not: but doubtless my Teacher knows.

SPHERE. Not I. There is no such land. The very idea of it is utterly inconceivable.

SQUARE. Not inconceivable, my Lord, to me, and therefore still less inconceivable to my Master. Nay, I despair not that, even here, in this region of Three Dimensions, your Lordship's art may make the Fourth Dimension visible to me; just as in the Land of Two Dimensions my Teacher's skill would fain have opened the eyes of his blind servant to the invisible presence of a Third Dimension, though I saw it not.

EDWIN A. ABBOTT[†]

If our consciousness were free to move forward or backward along the transformation dimension, our journey would not take us to a place on the Earth in the past or future. Instead, we would move entirely out of the physical universe; we would leave our physical body behind and venture into another totally different realm of existence composed of totally different etheron states. In fact, on such a

---

*Conventional physical theory does not acknowledge the existence of an ether, so working within this established framework, Prigogine has attempted to introduce the concept of irreversible time at the quantum level, rather than at the subquantum level. He and some of his colleagues have suggested that this could be done by properly revising the mathematical operations that quantum theory uses. Unfortunately, this leaves the closed-system, mechanistic paradigm of quantum mechanics largely unaltered in other respects. To properly rectify the encumbrances of conventional theory, much more radical revisions are needed. The slate of existing theory must be wiped clean, and physics must be constructed anew, adopting premises similar to those presented in the ancient ether physics.

[†]Edwin A. Abbott's entertaining book *Flatland: A Romance of Many Dimensions* (New York: Dover, 1952) relates the humorous story of a two-dimensional being's encounter with an entity from three-dimensional space.

trip we might discover the presence of other universes, even some that had worlds inhabited by other intelligences.

For this to be possible, the transmuting ether would have to contain more than just two intersecting etheron transformation pathways. It would instead be a convoluted realm of numerous dividing and intertwining "streams" that together form a complex ether transmutation network. At some of the many junctions where ether streams happen to cross and interact, the streams would be configured in a way that could potentially spawn a physical universe. Some of these crossings might exhibit a looplike reaction pathway similar to that shown in figure 4.2. For realistic particlelike structures to emerge, however, the etheron transformation rates at these junctures must be of the proper magnitude, a point that is discussed more fully in a later chapter.

Given a large enough ether reaction network, it is quite possible that somewhere in its entanglements other "physical" universes would evolve. Such "parallel universes" could be located upstream or downstream of our own, or might even be situated along a totally different transmutation stream. We could never prove their existence because we would be unable to interact with them physically. Although parallel universes would occupy the same three-dimensional space as our own, they might be far removed from us along the fourth dimension.

To visualize how multiple universes might coexist in the same region of three-dimensional space, let us consider the hologram analogy. A hologram is a photographic plate that contains an interference pattern which, when exposed to a beam of laser light of a specific color, produces a three-dimensional light image of a previously photographed object. Scientists have found that they can record many such interference patterns in the same photographic emulsion simply by using a different color laser light to make each hologram. When this hologram is later exposed to light with one or the other of these specific colors, different holographic images are consecutively elicited. In a similar sense, etherons of different types or "colors" could generate different alternate universes, all of which would occupy the same three-dimensional space.

Unlike a hologram, however, the transmuting ether would not require illumination by light waves of specific color or frequency to "project" its universes from preexisting wave patterns hidden in the ether substance. Rather, the ordered wave patterns composing each of these universes would emerge spontaneously, not from precursor structures, but from precursor processes occurring at their junctures in the ether transmutation network.

What is the nature of these ether reaction processes? Do they proceed from a higher to a lower state of activation, like exothermic cellular respiration reactions that consume some initial store of food energy (figure 4.7a)? Do they maintain a roughly comparable level of activation as they proceed along their reaction pathway by absorbing "energy" from some outside source (figure 4.7b)? Most biochemical reactions taking place in cells are of this second sort. The energy they receive

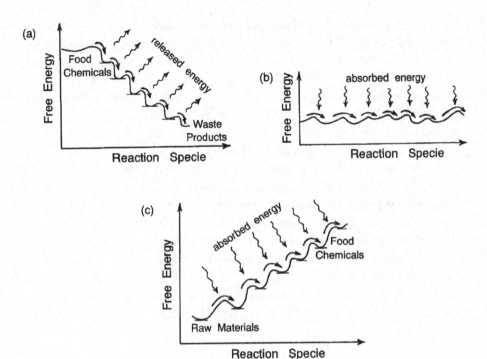

Figure 4.7. Energy diagrams for a variety of biochemical
reaction sequences: (a) respiration; (b) the majority of cellular reactions;
and (c) photosynthesis and protein synthesis.

comes either directly from sunlight or from an auxiliary source of energy stored up in energy-releasing chemicals such as ATP (adenosine triphosphate).* Or, do these ether reactions proceed to higher states of activation like the "energy-banking" reactions of photosynthesis or ATP-powered protein synthesis (figure 4.7c)?

We can only speculate as to which of the above situations best represents the transmuting ether. The third example appears to be the more pleasing alternative, since it depicts a progressive evolution to higher states of order (increasing activation levels). This interpretation is consistent with ancient Egyptian mythologic lore, which portrays etheric process as an ascending regenerative flux.

---

*The organic phosphate ATP is the universal reaction-energizing agent for all living cells. By breaking one of its phosphate bonds, ATP becomes converted into ADP (adenosine diphosphate) plus a phosphate ion and, in so doing, releases a quantity of energy. This released energy is transferred to the host reaction through a process called free energy coupling. In this way a reaction that otherwise might not occur spontaneously would be enabled to proceed. The ADP residue eventually becomes recycled back into ATP by acquiring energy either from sunlight or from the breakdown of fuel molecules such as glucose. Thus the ATP/ADP cycle serves merely as an intermediary for the transfer of energy to recipient reactions.

Like the transformation of base elements into precious metals in alchemy, etherons might ascend to higher states on their journey through the transmutation network. In such an open-system metaphysics, the trend toward increasing order would then become a universal principle, taking place at the etheric level just as it does in open systems at the physical level and presumably as it does at the spiritual level as one's soul proceeds toward a greater degree of perfection.

Still, if our earlier discussion about process is correct, this transmutative process should ultimately be powered by some kind of entropy-increasing flux. What primal energy source or activating agent could provide such a motive flux? It would be most appropriate and in line with spiritual insight to identify this source with a supreme divine power, the "energy" energizing the ether transmutation process perhaps being some form of "divine grace," or "love," bestowed by that power. In such a cosmology, divine love would continually sustain and animate all of creation. The prime mover and the transmuting ether could compose a single multi-dimensional entity. The transmuting ether would constitute this entity's substance, and the divine grace energizing the ether reactions might be an aspect of divine radiation, spontaneously upwelling from within its being.

5

COSMOGENESIS

THE SPONTANEOUS CREATION OF MATTER
The Chaotic "Vacuum" State

Imagine a time before the creation when space, time, and the transmuting ether existed, but no matter, energy, or force fields. During this pre-creation twilight, all of space was dark and devoid of matter. No spark of light or atom of matter appeared in any part of this infinite realm. But this vastness was filled with the ether, a ferment of etheron entities, structures so small that countless billions of them would fit inside the nucleus of an atom. On a rough estimate, a single cubic centimeter of space would have contained well over 10^{65} etherons.*

Let us suppose that these etherons continuously react and transmute from one state to the next following the reaction pathways specified by Model G, diagrammed in figure 4.2. Just as a water reservoir maintains a constant level when water enters at the same rate it leaves, each ether substrate would have maintained a constant, steady-state concentration as etherons entered and left each state at equal rates. Consequently, the primordial ether substrates would have been uniformly distributed throughout all of space at specific steady-state concentration values: A_0, B_0, G_0, X_0, and so on.

If we imagine the ether as an immense sea, its surface would have appeared quite smooth in the eons before the emergence of physical form. Even when viewed on a scale comparable to the diameter of a proton (10^{-12} centimeters), not a single ripple would have been seen. Focusing our view to a much smaller scale, however, a

---

*No attempt is made in subquantum kinetics to characterize the structural nature of etheron entities. Regardless of whether they might be waves or particles, they are assumed to compose the heterogeneous ether and to transform and react according to specified reaction pathways. This minimum estimate of etheron density was arrived at as follows. If a photon could have an energy as high as 10^{17} electron volts, it would have a corresponding wavelength of approximately 10^{-21} centimeters and would occupy a spatial volume of approximately 10^{-63} cubic centimeters. If one hundred etherons were packed together to form the peak of this energy wave (its concentration maximum), these etherons would have a density of 10^{65} etherons per cubic centimeter.

different picture would emerge. These seemingly smooth concentrations would now be seen to exhibit a subtle coarseness composed of tiny fluctuations rising and falling like waves on a choppy sea. These random pulses arise because the reaction and diffusion events that determine the etheron concentrations themselves depend on encounters taking place in a statistical fashion between individual etherons. Just as stock exchange prices vary erratically from day to day as a result of the collective buying and selling behaviors of millions of individual investors, etheron concentrations within a specific volume of space fluctuate randomly about their mean values, rising one moment and falling the next.

Since an etheron concentration fluctuation is a local nonuniform distribution of etherons, each such momentary impulse constitutes a microscopic quantity of explicit order. As discussed in chapter 2, systems scientists equate a nonuniform substrate with a state of low entropy and high potential energy (high potential energy being an accumulation of energy available for performing work). Consequently, these ether substrate fluctuations represent tiny pulses of energy that spontaneously emerge from the ether.* With the appearance of each pulse, the entropy of the ether substrate spontaneously decreases, only to increase once again as the nonuniformity disperses and the fluctuation disappears. The spontaneous decrease of entropy associated with the appearance of each fluctuation does not violate the second law of thermodynamics because the transmuting ether is an open system, not a closed system.

The ether is able to spawn material particles only if it is operating in a fertile supercritical state. Whereas the fertility of the Brusselator may be controlled by varying the concentration of reactant A, in Model G, the G ether takes over the duty of being fertility regulator. Whether or not the ether reactions are supercritical depends on whether the G ether concentration (gravity potential) lies above or below a certain critical threshold value, $G_{critical}$. In very early primordial times G's steady-state value, G_0, lay *above* $G_{critical}$. This would have made the ether transmutation processes proceed smoothly in an infertile *subcritical* mode. Spontaneously emerging fluctuations would have tended to dampen out, leaving the etheron concentrations homogeneously distributed at their steady-state values. Such an initial, pre-creation "vacuum" state is depicted in figure 5.1a.

The Emergence of the Supercritical Womb

Suppose that somewhere in the vast reaches of space, the G substrate concentration dips below its critical threshold value to create a supercritical region, as shown in

---

*Compatible with the "zero-point energy" concept, but very different from the vacuum fluctuation concept of quantum field theory.

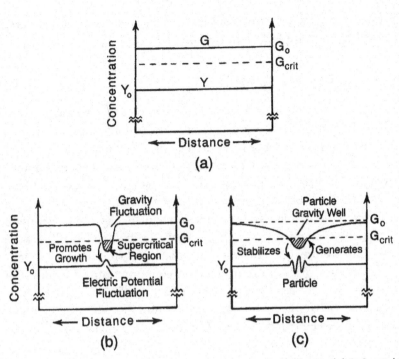

Figure 5.1. The creation of a subatomic material particle as predicted by Model G. Spatial profiles of Y and G ether concentration are displayed for three stages in the materialization process: (a) the subcritical pre-creation state; (b) a localized supercritical region and electric potential impulse; and (c) the formation of a self-stabilized material particle.

figure 5.1b. A gravity well or "*G*-well," of this sort might develop spontaneously as a localized gravity potential fluctuation, or it might come about in the form of a general lowering of the gravity potential over a very large region of space due to long-range variations in G_0. There is a very small but nevertheless finite probability that a large *X* and *Y* electric potential fluctuation would arise within this *G*-well. The greater the size of this fertile region, the more likely the chance of such a fluctuation occurring within its perimeter. The nurturing deviation-amplifying conditions within the *G*-well region help this seed to grow rapidly in size and eventually develop into a primordial subatomic particle (figure 5.1c).

So two ingredients are needed to begin the process of matter creation, a gravitational energy well and a coinciding *X* or *Y* electric potential pulse of critical size. The *G*-well may be thought of as the "mother" or "nourishing womb," while the inseminating electrical energy pulse may be thought of as the "child," the emerging precursor that with sufficient nurturing brings about the ordered state. In the myth of Osiris, this nurtured-growth concept is symbolized by Isis suckling her son Horus in the protection of the Nile Delta swamps. Isis signifies the fertile *G*-well and Horus signifies the growing *X*-*Y* electric potential pulse. A similar theme is apparent in the ancient Greek and Babylonian creation myths that describe how Zeus and Marduk are suckled into adulthood.

The Struggle for Emergence

Should an electric potential *(X-Y)* pulse arise, it must first weather the eroding effect of the omnipresent chaos if it is to eventually grow into a subatomic particle. This confrontation between the newly emerging order and preexisting chaos is a recurring theme in several ancient creation myths as well as in the symbology of the Tarot and astrology. The ancient Egyptian, Babylonian, and Greek creation myths all describe an extended battle between the champion of order (the emergent fluctuation) and the forces of destruction that attempt to maintain the regime of uniform chaos (entropic processes that tend to dissipate the fluctuation). In all these myths, the hero's victory ultimately leads to the creation of an ordered physical universe.

If it is to survive and grow, an emerging concentration pulse must contend with two kinds of order-destroying processes: etheron diffusion and erosion by competing fluctuations. In the first instance, as etherons diffused from higher concentrations in the seed pulse to lower concentrations in the environment (or alternatively from higher concentrations in the environment to lower concentrations in the seed pulse), the pulse would tend to dissipate. In the second instance, the body of the pulse would become eroded if a pulse of opposite polarity were to arise adjacent to it, engulf it, and partially cancel out its concentration hill or well. On the other hand, if this neighboring pulse were to have a similar polarity, the two pulses would instead mutually reinforce each other's growth.

The creation of order in the transmuting ether is somewhat like developing intelligible expression in a crowd of people. If each person in a crowd expresses himself individually in an attempt to be heard above his neighbors, the net result is unintelligible noise. Independent uncorrelated attempts to produce order yield only chaos. On the other hand, if somewhere in this unruly crowd a group of individuals band together to express themselves coherently with a unified voice, their collective expressions become understandable. Their microscopic orders band together to create a macroscopic expressed order.

The will to create order is a double-edged sword. On the one hand, random fluctuations ensure that macroscopic explicit order does not arise. As fluctuations spontaneously arise, combat one another, and dissolve, they collectively generate uniform ether concentrations at the macroscopic level, the same way that blips of noise appearing on an untuned TV set produce a featureless field of "snow." The subquantum chaos that produces this uniformity is responsible at the same time, however, for bringing about the ultimate destruction of this uniform state. For each of these randomly arising fluctuations is actually a miniature form of explicit order. If one such seedling of order happens to grow sufficiently large in size, it can disrupt the prevailing uniformity by forming a macroscopic concentration pattern.

Since etheron pulses emerge in a random manner, there is a certain probability that order-destroying fluctuations will temporarily be absent from a given region of space. If a single concentration fluctuation had the good fortune of arising in the midst of such a favorable environment, it would stand a much better chance of growing in size and becoming large enough to spawn an ordered wave pattern. The ancient stories of battling gods describe a similar protected upbringing for the champion of order. The hero is born and raised in a sheltered area, where he is hidden from his enemies. During his residence there he is able to mature into a strong warrior prepared to confront the forces of disorder. By using this hideaway image, these ancient creation stories are actually giving details about an important aspect of the order-genesis process, one that is important not only to the genesis of physical form but also to the formation of dissipative structures in a variety of open systems.

The Critical Fluctuation

Order-destroying processes such as diffusion and erosion decrease the size of a pulse by eating away at its boundaries. Consequently, they scale according to the pulse's surface area; that is, their destructive effects increase according to the *square* of the pulse's physical diameter. The ether reaction processes that work to increase the size of the pulse instead scale according to the pulse's volume; that is, their order-creating effects increase according to the *cube* of the pulse's physical diameter. So, as the physical size of the pulse increases, the order-enhancing processes increase at a much faster rate, and when the pulse has grown past some critical size, they become more important than the order-diminishing processes. As a result, a pulse smaller than this critical size will tend to shrink in size because erosion and diffusion would have the upper hand. The opposite situation would hold for a pulse larger than this critical size. In this case, the deviation-amplifying reaction processes would dominate over entropic processes and cause the concentration pulse to grow in size and increasingly depart from the steady-state values present in its environment. A pulse that reaches this critical size is called a *critical fluctuation*. A critical fluctuation has essentially won the battle against chaos, for it is able to continue to grow, unhampered by the erosive effects of entropy, and develop into a subatomic particle. In the Osiris myth, Horus symbolically reaches the critical size when he has grown into a mature warrior and is ready to confront Set in battle.

The Breaking of Symmetry

As the fluctuation grows, its X and Y concentrations increasingly depart from their steady-state values but in opposite directions. In other words, "sky" separates from "earth." Depending upon its initial polarity, the fluctuation may proceed toward either of two possible polarities, either positive (low *X*/high *Y*) or negative (high *X*/low *Y*), as in figure 5.2. In making this choice, the ether concentrations proceed from an initial state of symmetry (*X* and *Y* spatially uniform) to a polarized asymmetrical state. Hence in the course of this transition, symmetry is said to become "broken." The technical term for such an ordering event is *spontaneous symmetry breaking*. Symmetry breaking is "spontaneous" because the precise time and place where this event takes place is unpredictable. Whether or not a particle emerges at any given moment depends upon the outcome of the subquantum chaos present in the ether as fluctuations randomly arise and struggle with one another. A similar unpredictability is encountered in the biological procreative process of egg fertilization.

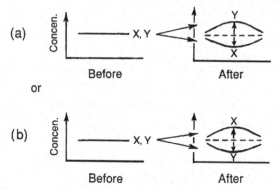

Figure 5.2. Two alternative field polarities that can emerge in the ether at the moment of spontaneous symmetry breaking.

The Emergent Subatomic Particle Wave

Eventually, the deviating *X* and *Y* concentrations will stabilize into a spherically symmetric stationary wave pattern similar to that shown in figure 5.3. This would constitute a subatomic particle. If the particle is of positive polarity, it will have a low *X*/high *Y* core surrounded by shells alternating between high *X*/low *Y* and low *X*/high *Y* concentration. The complementary negative polarity particle would be just the mirror image of this, forming as a wave pattern that proceeds outward from a high *X*/low *Y* core. In either case, the *X* and *Y* components that form the particle's electric potential field would vary with radial distance in a wavelike fashion, with *X* reaching a maximum where *Y* reaches a minimum, and vice versa. Figure 5.4 shows how the *Y* component of the electric potential field, ϕ_y (phi-y), varies with respect to radial distance from the particle's center. The *X* electric field component, ϕ_x (phi-x), which is not shown here, would form a wave pattern that is a mirror image of the *Y* field.

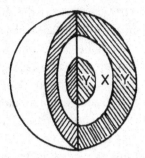

Figure 5.3. A cross-sectional view of the concentric shell-like ether concentration pattern making up a subatomic particle.

Particles with a positive core polarity, the proton and positron (antielectron), would exhibit a positive electrostatic charge and a positively biased *Y* field pattern (figure 5.4, upper left shaded area). Their antiparticles, the antimatter proton and the electron, would exhibit a negative charge and a negatively biased field pattern (figure 5.4, upper right shaded area). These biased field patterns cause the particles to form around themselves extended field gradients by which they are capable of exerting electric, magnetic, and gravitational forces on distant bodies.

These particle-like dissipative structures are unique in that they exhibit both

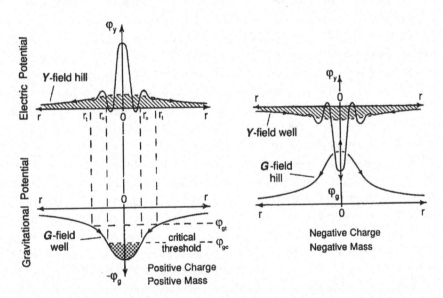

Figure 5.4. Energy potential field profiles for a positively charged particle (left) and its negatively charged complement (right) shown in radial cross section. The vertical axes plot the field's magnitude, and the horizontal axes plot radial distance from the particle's center. Shaded regions indicate the electric potential biases associated with positive or negative charge.

particle-like and wavelike characteristics at the same time. The spacing of their radial shells denotes the subatomic particle's characteristic wavelength, which is comparable to what physicists call the *Compton wavelength*. Modern physics, on the other hand, has failed to come up with a model capable of explaining in a unitary fashion both the wave and particle characteristics of a subatomic particle. Physicists have ignored the problem by assuming that a subatomic particle spontaneously and unexplainably changes into either a point-mass particle or a traveling wave depending upon whether the observer happens to experimentally observe one or the other of its two aspects. They call this two-sided doctrine the *wave-particle dualism*. This positivist conception leads to the preposterous conclusion that matter has no existence independent of observation. The unitary wavelike particles predicted by the ancient ether metaphysics do not require such a contrived view of nature.

Parthenogenesis (Virgin Birth)

The transmuting ether matter-creation process can also be expressed in terms of the anthropomorphic imagery of ancient mythology. For matter creation to begin, a fertile womb must first be present; that is, a supercritical *G*-well must develop in the ether. Next, a child must be allowed to arise within this nurturing region—an embryonic energy fluctuation must be permitted to nucleate. The spontaneous impulse fertilizing the supercritical region comes from chaos naturally present in the ether and is not imparted by an "outside" agent. Within the nourishing protective environment of the mother, the child grows to maturity and develops into a critical fluctuation. In so doing, his large size makes him invulnerable to the surrounding chaos. As an autonomous power, he battles the forces of destruction (the homogenizing tendencies of competing fluctuations) and emerges victorious. As the champion continues to grow in size, he spreads his regime of physical order throughout the surrounding ether empire; the critical fluctuation divides the homogeneous ether state and forms an ordered field pattern, the material particle. The newly emerging explicit order is born.

The self-organizing materialization process is an example of what thermodynamicists and systems scientists call *order through fluctuation*. Alternatively, we may call it *parthenogenesis* (virgin birth), a word used in biology to refer to the spontaneous occurrence of pregnancy without prior intercourse. This term more appropriately acknowledges the feminine nurturing aspect of this order-creation process.

The general system theorist and psychiatrist Dr. William Gray has developed his own theory and terms for describing parthenogenesis, which he calls "general system formation."[1] Although he has applied his terminology primarily to the fields of psychology and sociology to describe human thought formation and social interaction processes, he notes that the concepts are also generally applicable to a variety of open systems. Elements of his theory are essentially the same as those presented above. His term "organizing focus," which he defines to be a sponta-neously arising entity that has the potential for forming an ongoing system, is equivalent to the concept of "fluctuation." To "system-form," Gray says that this focus must come in contact with a "relevant nurturing environment," a region that sets up positive feedback relations that cause the organizing focus to intensify. In other words, this fluctuation must arise in a supercritical region. He regards both the organizing focus and the relevant nurturing environment as equally important to the system formation process and refers to these two aspects as "system precur-sors." Moreover, he also recognizes the importance of the protective environment around the system precursors. Such a "system-forming space," he says, "provides the necessary degree of isolation needed for system-forming to take place."

THE REGENERATION OF FORM

The ether metaphysics of ancient mythology solves a fundamental problem plagu-ing contemporary field theory in that it explains how a subatomic particle field pat-tern is able to retain its form despite the eroding action of entropy. Basically, the ether accomplishes this feat by continually regenerating the particle's wave pattern. Just as the biochemical processes in a living organism keep entropy at bay by regen-erating new cells and protein structures to replace aged ones, the functionally ordered ether transmutation processes continuously add and remove etherons from appropriate parts of a particle's concentration pattern, thereby maintaining a subatomic particle's structure. No superordinating principle is needed to "guide" this process because this wave pattern emerges as a result of the natural operation of the ether reaction-diffusion processes.

A wide variety of reaction-diffusion systems, including Model G and the Brusselator, maintain their dissipative structures through such form regeneration. Consider the two boxes shown in figure 5.5, which depict high and low X and Y ether concentrations found in two adjacent shells inside a proton's dissipative structure. Owing to the differences in concentration between these shells, X-ons will continuously diffuse to the right from shell 1 to shell 2 and Y-ons will diffuse to the left from shell 2 to shell 1. Just as fast as Y-ons diffuse into shell 1, however, they are converted into X-ons by the nonlinear ether reaction processes therein, thereby making up for the X-ons lost through diffusion into shell 2. Thus the shells' X and Y concentrations remain unchanged at their steady-state values. A similar dynamic equilibrium is maintained in shell 2. Here, because Y is the predominant species rather than X, the ether reactions instead preferentially convert X-ons into Y-ons. So again, transformation processes counteract diffusion and thereby main-tain the concentrations at nonuniform steady-state values. In a similar fashion, the structural coherence of the particle's entire wave pattern can be attributed to a bal-ancing of reaction and diffusion processes.

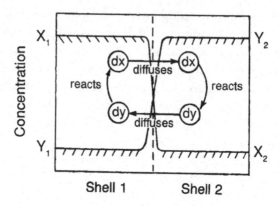

Figure 5.5. An illustration of how X and Y ether concentrations can maintain a nondispersing pattern despite the continuous action of diffusion.

Consequently, a subatomic particle's ability to maintain its form in spite of entropy may be attributed to the operation of the ether transmutation flux, the same all-permeating flux responsible for bringing the material particle into existence in the first place. In effect, the particle's explicit order is continuously regenerated from the implicit order inherent in the underlying ether reaction processes. Unlike explicit order, implicit order is immune to the eroding effects of diffusion. This enduring hidden order, which initially spawned our physical universe and continually sustains it, is the true mother of creation. We find this sustenance notion symbolically portrayed at the point in the Osirian passion where the resurrected Osiris (the regenerating flux) bestows his *ka*, or vital energy, upon the victorious Horus, the symbol of the emerged physical order.

Like Shiva the Cosmic Dancer, who maintains the universe through a balance of ether creation and destruction, the transmuting ether sustains material particles through a balance between reaction processes that continually create and destroy etherons and diffusion processes that continually move etherons to and fro. If some disturbance were to cause the concentrations at a given location in this pattern to temporarily depart from their steady-state values, the etheron levels would automatically readjust so as to once again attain their original values. Such seemingly purposeful behavior is exhibited by a variety of open-system phenomena, one example being the tenacious ability of the B-Z reaction to stick to its limit-cycle orbit despite outside disturbances.

Rather than destroying order, etheron diffusion helps to create order. In fact, the prevailing rate of etheron diffusion directly determines the wavelength of the subatomic particle's dissipative structure. Higher X and Y etheron diffusion rates have the effect of shortening the Compton wavelength of the subatomic particle. So entropy-increasing diffusion is an important foundation upon which the ether builds its edifice of explicit order. The Osirian myth's insightful depiction of the submissive Set bearing aloft the resurrected Osiris, the newly emergent physical order, provides a brilliant metaphorical description of this important interdependence between order and disorder.

Surprisingly, this open-system reaction-kinetic physics of ancient times surpasses quantum mechanics in its ability to model subatomic particle structures. Quantum mechanics, first developed in the mid-1920s by the Austrian physicist Erwin Schroedinger, represents a subatomic particle as a "wave packet" made up of electromagnetic energy potentials (figure 5.6). Schroedinger used the physical principle of linear wave superposition to synthesize these packets by mathematically superimposing long trains of sinusoidal energy waves with slightly differing frequencies and phases. When combined, the energy amplitudes of these waves constructively and destructively interfere with one another and cancel out everywhere except in one small region where they form a localized wave packet. This packet was presumed to define the location of a subatomic particle. But the model had a serious flaw: the wave packet tended to spread out rapidly and dissipate with the passage of time. Thus, unlike the ancient dissipative-structure subatomic particles, this wave packet model failed to resist the dispersing effects of entropy.

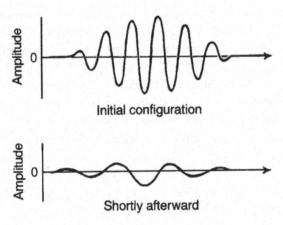

Figure 5.6. The wave packet model proposed in contemporary quantum mechanical field theory. Over time the packet gradually dissipates.

The French physicist Louis de Broglie suggested in 1950 that the problem might be solved if nonlinear field interactions were somehow to "glue" together these traveling waves. One such nonlinear field formula was proposed in the 1970s by a group of physicists and shown to produce a *soliton* waveform that maintained some degree of coherence. Their idea was based on a discovery made in the nineteenth century by a Scottish engineer named John Russell. One day Russell had been watching a boat being tugged along a canal waterway and found when the boat stopped that a water wave produced from the prow of the boat continued to move down the length of the canal for a few miles as a solitary wave impulse. It has since been found that such soliton waves maintain their coherence owing to nonlinearities in the motion of their fluid medium. Theorists have suggested that similar nonlinear field interactions might be at work forming a subatomic particle wave packet. Yet solitons still suffered from the dispersion problem, although to a lesser extent than quantum mechanical wave packets. Solitons also had the disadvantage that they always had to be on the move: the higher the wave, the greater its speed. So a soliton could not model a stationary particle. By drawing from the theory of nonlinear reaction chemistry rather than from hydrology, cosmologists of ancient times seem to have succeeded where modern physicists have failed.

WHY MATTER AND NOT ANTIMATTER?
Self-Stabilization

Any cosmology that attempts to explain the origin of the universe must answer the following fundamental question. Why do we live in a world composed of matter and not antimatter? From observing high-energy particle collisions, physicists know that almost every particle of matter has a complementary antimatter particle and that if the two are brought together they will annihilate one another, leaving only radiant energy. They also know that when a gamma ray of sufficiently high energy collides with an atomic nucleus, it transforms with equal probability into a particle of either matter or antimatter, a phenomenon known as *pair production*. A gamma ray having an energy of slightly over 1 million electron volts will produce a positron-electron pair, whereas a gamma ray with an energy of about 2 billion electron volts will produce a proton-antiproton pair. Modern big bang theorists have assumed that such an energy-to-matter conversion process must have occurred when the big bang fireball formed into matter. But astronomers have always been nagged by the question of how this matter became created, for if gamma rays in the fireball produced equal amounts of matter and antimatter, these particle complements should have annihilated one another soon afterward and reverted back into energy.

The reaction kinetic physics of ancient mythology resolves this long-standing cosmological problem. Material particles do not arise from gamma ray collisions; rather, they autonomously nucleate from fluctuations in the ether, a process that is strongly biased toward the creation of matter, as opposed to antimatter.

A negative gravity potential fluctuation emerging in the ether would create a nurturing supercritical environment that promotes the emergence of explicit order. By its nature, however, this G-well fluctuation is transitory. Eventually it subsides, and as it does, the fertile region it was forming disappears. The ether reverts back to the infertile subcritical state, leaving the particle to fend for itself. Unless it has the ability to generate its own sufficiently deep *G*-well (for example, by consuming *G*-ons from its surroundings), the particle will gradually dematerialize. Order-regenerating processes will not be able to keep up with order-destroying entropy processes, and as a result the particle's wavelike concentration pattern will disperse. As it turns out, a particle having a positive electrostatic charge, such as a proton, would generate a gravity well and so maintain a local nurturing environment conducive to its perpetuation. The Model G reaction system predicts that a positively charged particle would set up the self-stabilizing causal loop shown in figure 5.7.

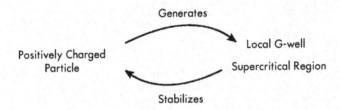

Figure 5.7. A self-stabilizing causal loop produced by a proton.

Because of its greater mass, a proton's *G*-well should be two thousand times deeper than the *G*-well produced by a positron (figure 5.8). Consequently, a proton should have a much greater chance of surviving in a subcritical environment. Negatively charged particles such as an electron or antimatter proton would instead generate subcritical gravity potential hills and hence promote their own destruction. In the case of a neutral hydrogen atom formed of a single proton nucleus and orbital electron, the deep supercritical *G*-well of the proton would counteract the low subcritical *G*-hill of the electron to produce a net supercritical *G*-well. The opposite situation would prevail for an atom of antimatter hydrogen made up of an antiproton nucleus and orbital positron. The extremely high *G*-hill of the antiproton nucleus would prevail, creating a subcritical environment that would cause the entire antimatter atom to dematerialize. This difference in the stability of neutral matter and antimatter could explain why we live in a world made up predominantly of matter.

Bifurcation Diagrams

The tendency for the Model G matter-creation process to favor protons over antiprotons is also evident when we solve mathematical equations representing the Model G reaction system. These solutions may be displayed graphically by a

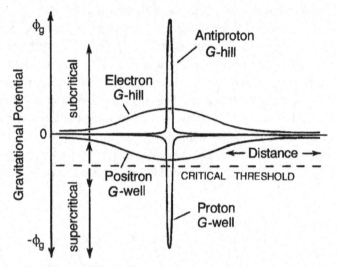

Figure 5.8. A comparison of G-wells and G-hills produced by various subatomic particles. A proton is more likely than a positron to become self-stabilizing in a subcritical environment. The vertical scale for the proton and antiproton has been reduced four hundred-fold to permit a full view of their profiles.

diagram (figure 5.9) that charts the steady-state concentration values allowed for the *X* ether at various levels of reaction system criticality. Because an initially solitary solution splits, or bifurcates, into two possible solution states, systems theorists call this a *bifurcation diagram*. The point at which the solution divides is alternately termed the bifurcation point, point of instability, or critical threshold. A variety of nonlinear reaction systems, including the Brusselator, generate forklike mathematical

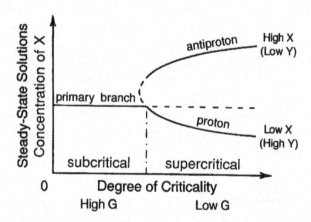

Figure 5.9. Bifurcation diagram for the Model G reaction system displaying hypothetical solutions for X. Past a certain critical threshold, the concentration of X must choose between two possible solutions. The dashed lines denote regions of instability.

solutions similar to this.[2] This bifurcation is particularly important in the forthcoming discussion about the Tarot (chapter 8) because Arcanum 6 of the Tarot metaphorically depicts such a diagram in every detail.

The horizontal axis plots increasing levels of system criticality resulting from increasingly negative values of gravity potential (*G* ether concentration). The solitary solution, called the *primary branch*, corresponds to the condition in which the reaction system is subcritical and the ether reactants are homogeneously distributed. In other words, it represents the ether in its primordial "vacuum" condition where no physical form is present. As the ether's level of criticality increases (as gravity potential becomes more negative), eventually the critical threshold is reached beyond which the reaction system begins to become supercritical. As the reaction system enters this fertile regime, the concentration of *X* may change to either the lower or upper curve depending on the size and polarity of the emerging fluctuation. A change to the lower branch (low *X*/high *Y*) represents the formation of a positive particle such as a proton, whereas a change to the upper branch (high *X*/ low *Y*) represents the formation of the proton's negatively charged complement, the antiproton.

While the change to the lower branch solution requires an *X* fluctuation large enough to overcome the weathering competition of neighboring fluctuations, the transition to the upper branch solution requires an *X* fluctuation sufficiently energetic to broach the large gap in the solution curve (dotted curve in figure 5.9). Thus matter creation would be strongly biased toward the creation of protons as opposed to antiprotons.

A Model G ether can actually generate a number of different types of subatomic particles (and their antiparticles) of various masses. Particles of increasingly greater mass (and increasingly shorter wavelength) would emerge at specific critical thresholds at an increasing distance from equilibrium (toward the right in figure 5.10). Ranking some of the known particles in order of increasing mass, these consecutive branchings would correspond to the electron and positron ($e^{\pm}$), muons ($\mu^{\pm}$), pions ($\pi^{\pm}$, π°), *K*-mesons ($K^{\pm}$, K°), proton and antiproton ($p\pm$), neutron and antineutron (n°), and a series of more massive particles known as hyperons (Λ°, $\Sigma^{\pm}$, Ξ, . . .). Most of these matter states would be unstable, even in our currently supercritical ether environment.

In summary, Model G predicts that in the case where fluctuations emerge out of "empty" space, only those of *positive* charge polarity capable of nucleating a stable, high-mass particle would succeed in producing matter. This suggests that protons would have been the first particles to emerge. The G-well of each such primordial proton would have provided a fertile environment conducive to the production of additional matter. A neutron arising in the proton's G-well would either bond with the proton to form a heavy hydrogen nucleus or would separate from it and quickly decay into a proton and electron pair that would combine to form an electrically

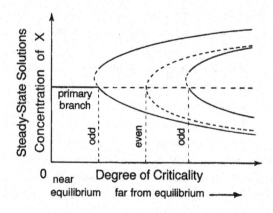

Figure 5.10. Hypothetical bifurcation diagram for Model G showing consecutive bifurcations leading to increasingly massive particle solutions. Only odd bifurcations produce stable solutions.

neutral hydrogen atom. As the matter-creation process continued at each of these sites, these two atomic nuclei would soon become four, which would soon become eight, and so on ad infinitum. As a result of this polarity bias in the matter-creation process, a universe would eventually develop that was made up of matter, as opposed to antimatter. The ancient ether metaphysics, which utilizes an ether reaction system similar to Model G, leads to this same sort of continuous creation scenario.

Ilya Prigogine has noted that a "nonsymmetrical" universe in which matter completely dominates antimatter could be explained if matter creation proceeded under *nonequilibrium* conditions.[3] Working within the framework of conventional cosmology, he envisioned that the universe was in its most nonequilibrium state in the early stages of the Big Bang when subatomic particles were first being formed. Nevertheless, his observation about the connection between nonequilibrium conditions and the asymmetrical creation of matter and antimatter applies just as well to the continuous creation process envisioned in the ancient physics. As we have seen, the transmuting ether, which in the ancient cosmology continuously gives birth to matter, always maintains itself in a state far from thermodynamic equilibrium.

THE CREATION OF THE UNIVERSE

Compared with the explosive chaos of the Big Bang, the ancient parthenogenic cosmology proposes a much gentler beginning for the physical world. The ancient version involves an initiating impulse of potential energy that is far smaller than the rest-mass energy of a single subatomic particle. The initial energy fluctuation could be less than one ten-thousandth of that required to move a grain of dust a hair's breadth (less than one ten-million-billionth of a calorie). By comparison, the Big Bang requires the simultaneous emergence of an amount of energy sufficient to create all of the matter in its universe of some 10 billion light-years' radius. Such a fluctuation would exceed the amount of energy released in one million, billion, billion, billion, billion, billion one-megaton hydrogen bomb explosions, or approximately 10^{66} calories.*

Comparing these two scenarios, it is obvious which of the two kinds of energy fluctuations would have a greater chance of occurring. In an age when the motto "More is better" has almost led the world to the brink of global thermonuclear extinction, it is refreshing to find that the ancient ether metaphysics leads to a "Small is beautiful" cosmology. It likens the miracle of creation more to the germination of a seed or to the blossoming of a flower than to the detonation of a cosmic bomb.

The ancient creation science indicates that the matter making up our physical universe came into being gradually through a process of continuous creation, a process that continues even today. Unlike the big bang theory, which advocates a single creation event at a unique zero point in "space time," the ancient view suggests that matter emerged through innumerable particle materialization events occurring at a multiplicity of sites scattered throughout the vast reaches of space. This process would have a random character. The particular place and time of occurrence of each materialization event would depend upon where and when a set of self-reinforcing fluctuations of sufficiently large size happened to emerge.

The precise value of the G etheron critical threshold in Model G is determined by the concentration values of both the A and B ethers. If the concentrations of these ethers were to vary by slight amounts over distances of billions of light-years, they could bring some regions of space closer to the critical threshold than other regions, thereby inducing matter to form preferentially in those more favorable locations. Such patterning of the ether into alternating fertile and infertile regions could explain why galaxy clusters are grouped into filament and sheetlike structures positioned at regular distances to form periodic patterns.

Since the supercritical domains formed by G ether fluctuations are transitory, there would be a very small probability that any of these would coincide with an X-Y seed fluctuation sufficiently large to form a subatomic particle. Nevertheless, considering the enormous numbers of fluctuations occurring in the infinite vastness of space over the immensity of time, the creation of the universe becomes not only plausible but certain. The time required for the first particles of matter to come into being may have been thousands of times longer than big bang cosmologists' present estimate for the age of the universe.

---

*The inflation cosmology proposed by Alan Guth considerably reduces the amount of energy initially required for the big bang explosion. This approach, however, still has several unsolved problems (see chapter 14). Interestingly, the self-amplifying fluctuation growth process that inflation theory proposes in many ways resembles that of parthenogenesis.

Materialization would be expected to proceed much faster in the vicinity of existing matter, since protons would create localized nurturing supercritical regions where fledging fluctuations could rapidly grow in size. Consequently, matter creation would take place in clumps. This evolutionary development from primordial particle to spiral galaxy is illustrated in figure 5.11.[4] The diagram begins with a primordial proton materializing in space of its own accord. This mother particle then serves as a locus for the materialization of additional protons, electrons, and neutrons that combine to form primordial hydrogen atoms. These daughter hydrogen atoms also spawn proton and electron progeny, all of which serve as nucleation sites for the production of second-generation progeny, and so on ad infinitum. Just as the rate of births in a growing population soars over time, matter would be generated at an exponentially increasing rate.

With continuing materialization, a moderate density of hydrogen gas would form, and over time this cloud would develop differences in density that would cause it to condense to form a planet-size mass. As a result of the prevailing supercritical conditions, *genic energy,* a kind of spontaneously created energy, would arise continuously from the planet's interior. Astronomical evidence supporting this unique ether metaphysics prediction is summarized in chapter 15. Owing to the creation of matter and energy in its interior, this gaseous mass would gradually grow in size and increase in temperature, so that it would eventually become a luminous primordial star. As the star increases in mass through the accumulation of matter and attains sufficiently high temperatures and densities at its core, nuclear fusion reactions begin to ignite and begin fusing the star's hydrogen into helium, lithium, beryllium, carbon, oxygen, nitrogen, and other heavy elements.

Stellar wind gases driven from the star's surface by the pressure of its light radiation condense near the star to form orbiting gaseous planets. As these daughter bodies acquire additional matter from the mother star's stellar wind and generate matter spontaneously in their interiors, they too grow in size. They eventually evolve into stars and, like their mother, begin spewing gases into their space environment. These expelled gases eventually condense into second-generation planetary masses that ultimately grow into second-generation stars. As a result, the number of stars grows exponentially, and eventually a primordial star cluster forms.

The primordial mother star, residing at the center of the star cluster, has by this time grown very dense and massive. Because it has a much deeper gravity well, the mother star's matter-creation rate greatly surpasses that of its progeny stars. As increasing amounts of matter are produced, this primordial star cluster grows in size until it has reached the size of a dwarf elliptical galaxy, at which point it contains billions of stars. Its central mother star by this time may have achieved a mass of perhaps several hundred thousand to several million solar masses and a luminosity of several million solar luminosities. The star in the core of our own galaxy,

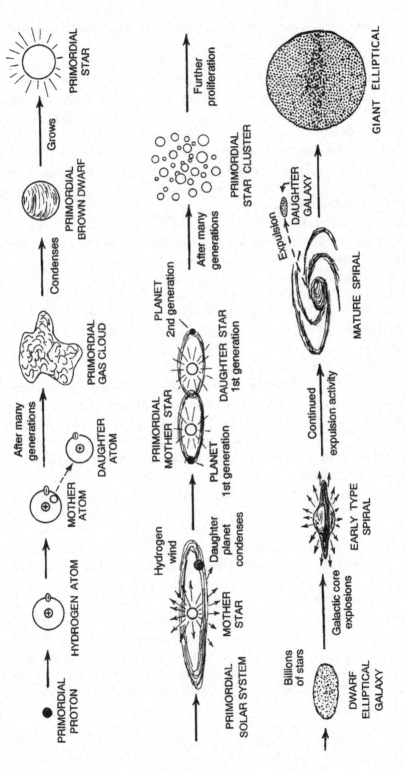

Figure 5.11. Galactic evolution, the development from primordial particle to galaxy.

for example, is believed to contain up to 1 million solar masses and to be about 20 million times as luminous as the Sun.

As an increasing number of progeny stars form around the core, deepening its gravity well, the mother star's rate of genic energy production progressively increases. Its internal temperature and energy production rate become so high that it radiates its energy primarily in the form of cosmic ray particles. The galactic core's energy output eventually becomes so great that instabilities develop and cause it to explode. During this temporary active mode, its luminosity increases millions of times over, and it releases an intense volley of cosmic ray particles and high-energy radiation that travels radially outward in the form of an expanding shell called a *galactic superwave.*

As this superwave propagates outward, it propels large quantities of dust and gas away from the galactic core region, and as a result, the magnitude of the ambient gravitational potential field in the galaxy's center decreases. This, in turn, causes the rate of genic energy production in the galaxy's core to decrease below the critical point and allows the core to once again operate in its quiescent energy generation mode. But this low activity state is only a temporary phase. As the galactic core continues to generate matter, the gravitational potential at the galaxy's center eventually increases past the critical point, causing the core to once again enter its explosive phase. Consequently, the cycle repeats.

With galactic core explosions recurring perhaps as frequently as once every ten thousand years, successive superwaves would propagate away from the core region, propelling gas and dust toward the galaxy's periphery. This would cause the dwarf elliptical galaxy to develop spiral arms projecting outward from its equator and eventually evolve into a mature spiral galaxy similar to our own. Every so often with the occurrence of a particularly violent central outburst, a massive star cluster would be ejected from the core region with sufficient force that it would leave the galactic nucleus and go into orbit around the mother galaxy. Such clusters would become autonomous matter-creation centers, or "embryos," that with growth would develop into satellite galaxies.

The Large and Small Magellanic Clouds in orbit around our own galaxy are examples of such satellite galaxies. Dwarf satellite galaxies are also visible as two bright spots orbiting Andromeda, our nearby sister galaxy (figure 5.12). Studies of the dwarf elliptical at its upper right (NGC 205) reveal that it has a condensed mass at its center like that found at the center of Andromeda and our own galaxy.

With further matter creation and ejection, star clusters would increasingly populate high-latitude regions above and below the spiral arm disk, forming a spherical halo of stars. Consequently, a mature spiral galaxy would gradually evolve into a giant elliptical galaxy.

A mother galaxy and its associated satellite galaxies may be thought of as a kind of "island universe." Each such island of matter would have been nucleated by a

Figure 5.12. The Andromeda galaxy and its two satellite galaxies NGC 205 and M32 (the two nearby bright spots). Andromeda, our closest neighboring galaxy, lies 2 million light-years away. Photo courtesy of Mt. Palomar Observatory.

single primordial particle that might still exist in the mother galaxy's core. Other galaxies scattered throughout space would be numerous examples of such island universes, each creation initiated by its own warrior-god hero. All can be expected to be composed of matter rather than antimatter.

According to one estimate, it would take about 150 trillion years for a galaxy like our own containing some 100 billion solar masses of matter to grow from a single primordial proton. By comparison, ancient Hindu scriptures teach that one Brahma lifetime, the time between a successive materialization and dematerialization of the universe, spans 311.04 trillion human years. Because of the exponential nature of the matter-creation process, over 99 percent of the matter in our galaxy would have been produced within the past 10 to 15 billion years or so.

Space Telescope observations have now confirmed the continuous-creation prediction of subquantum kinetics, and the ancient cosmology, which proposes that dwarf elliptical galaxies evolve into spirals and that spirals in turn gradually evolve into giant ellipticals. In 1985 and 1994 I had pointed out that if this is the case, then very distant (younger) regions of space probed by the Hubble Space Telescope should be found to contain greater numbers of galaxies in an earlier stage of evolutionary development, that is, more isolated dwarf ellipticals than spiral galaxies and more spiral galaxies than giant elliptical galaxies.[5] Space Telescope observations do in fact corroborate the continuous-creation prediction. In 1995 astronomers announced that deep field studies of distant galaxies revealed that small irregular galaxies called "blue dwarfs" were more numerous several billion years ago, whereas spirals and giant ellipticals were much rarer.[6] They reported an excess of irregular blue dwarf galaxies at redshifts greater than 1 and an absence of spirals at moderate to high redshifts. Consequently, they were led to conclude that spirals began forming more recently, becoming more abundant at redshifts of less than 2. Conventional cosmology, on the other hand, does not anticipate such evolutionary trends. Big bang cosmologists assume that elliptical and spiral morphologies arose because primordial gas gravitationally contracted in differing ways. But they have no clear idea of why such differing modes of contraction would have arisen. Moreover, they have no reason to suspect that the relative numbers of these different galaxy types would have changed appreciably since primordial times.

In recent years astronomers have discovered isolated stars drifting alone in the intergalactic void and have found that the region between galaxies is filled with a tenuous hydrogen gas medium. These findings have led them to conclude that there may be as much matter residing in the "voids" between galaxies as resides in the galaxies themselves. This finding poses a problem for the big bang theory because it indicates that there is far more ordinary matter in the universe than big bang models predict. This view of distributed creation, however, poses no problem for the subquantum kinetics cosmology which predicts that matter creation has been taking place throughout all of space.

THE DISSOLUTION OF MAYA

Reaction systems such as the Belousov-Zhabotinskii reaction or the theoretical Brusselator system must maintain a certain level of flux to sustain their explicit order. The same is true of the transmuting ether. If the ether's transmutation rate were to fall below the critical threshold necessary for maintenance of the ordered state, the material world would begin to dissolve. The ether substrates whose warp and woof compose the physical world would return to a state of uniform distribution, their field-free "vacuum" state. Thereafter, if the ether reaction system were to once again adopt concentration values bringing it close to the critical threshold, cosmogenesis would begin afresh. New primordial particles would emerge and eventually new galaxies would form, complete with an array of life forms, including intelligent life.

Eastern mystics describe a very similar kind of termination and regeneration for the cosmos. For example, the Indian Sufi master Meher Baba, in his book *The Everything and the Nothing*, explains:

> In what is called space numberless universes are continuously created, sustained and destroyed. This procession of creation continues so long as God goes on imagining. And when God's imagination is suspended, as it is at moments in Eternity when God withdraws Himself into His Sound Sleep State (just as a man's imagination ceases when he is in deep sleep), the Creation is withdrawn and dissolved (Mahapralaya). . . . The cosmos is a dream. God alone is Real.[7]

The nineteenth-century Hindu saint Lahiri Mahasaya has passed down a story of how in 1861 he had the opportunity to wander through an immense gem-studded golden palace that his guru Babaji had materialized for him in the midst of a Himalayan jungle. Inquiring of his guide how such a beautiful structure had been brought into being, he was answered as follows:

> There is nothing inexplicable about this materialization. The whole cosmos is a projected thought of the Creator. The heavy clod of the earth floating in space is a dream of God's. He made all things out of His mind, even as man in his dream consciousness reproduces and vivifies a creation with its creatures.
>
> The Lord first formed the earth as an idea. He quickened it; atomic energy and then matter came into being. He coordinated earth atoms into a solid sphere. All its molecules are held together by the will of God. When He withdraws His will, all earth atoms will be transformed into energy. Atomic energy will return to its source: consciousness. The earth idea will disappear from objectivity.
>
> In tune with the infinite all-accomplishing Will, Babaji is able to command the elemental atoms to combine and manifest themselves in any form. . . . Babaji created this beautiful mansion out of his mind and is holding its atoms together by the power of his will, even as God's thought created the earth and His will maintains it . . . when this structure has served its purpose, Babaji will dematerialize it.[8]

Hindu scriptures such as the Upanishads describe a similar scenario. During the interval of manifestation, the all-generating, self-transforming divine substance, MAYA, gives rise to *maya*, physical creation, only later to dissolve it once again. Heinrich Zimmer writes the following about this:

> All the universes co-existing in space and succeeding each other in time, the planes

of being and the creatures of those planes whether natural or supernatural, are manifestations from an inexhaustible, original and eternal well of being, and are made manifest by a play of *maya*. In the period of nonmanifestation, the interlude of the cosmic night, *maya* ceases to operate and the display dissolves.[9]

We even find this notion in the creation mythology of ancient Egypt. Chapter 175 of the Coffin Texts, dating from around the twenty-second century B.C.E., records a conversation between Osiris, the Lord of Order, and Atum, the Supreme Creator. From his bleak underworld abode, Osiris inquires of Atum how long he must endure his hapless plight. Atum responds:

> *You will live more than millions of years, an era of millions,*
> *but in the end I will destroy everything that I have created,*
> *the earth will become again part of the Primordial Ocean,*
> *like the Abyss of waters in their original state.*
> *Then I will be what will remain, just I and Osiris,*
> *when I will have changed myself back into the Old Serpent*
> *who knew no man and saw no god.*[10]

The "Abyss of waters in their original state" and the "Old Serpent" refer to the homogeneous state of the ether that prevailed at a time prior to the emergence of physical form. Consequently, the ancient Egyptians, like the Hindus, taught that the physical realm was an outward manifestation of the primordial flux and would one day disappear, leaving only the flux. This "ending" of things was not something to be feared, but rather something to look forward to, like the reunion with an old friend. As with Hindu teachings, which note that the current separation between creator and created, MAYA and *maya*, will disappear with the dissolution of the physical universe, Atum proclaims that his separation from Osiris will cease at the end of the present dispensation. Osiris, who in his underworld role personifies the coming into being of physical form, regains his true identity with the creator as all become one.

All of these accounts, whether they arise from the teachings of contemporary mystics, the scriptures of Eastern religions, or the allegorical symbologies of ancient creation myths, describe the nature of creation in the same fashion. It has often been said that with the advancement of scientific knowledge, West will eventually meet East, those who have been following the "outward path" will eventually come to comprehend with their mathematical equations and diagrams what those on the "inward path" have intuitively, and perhaps more directly, understood for a long time. We are now nearing that time. With the development of the general theory of open systems and subquantum kinetics, we now have a way of understanding the wisdom contained in these ancient truths.

PART 2

Examining the
Ancient Record

6

THE EGYPTIAN CREATION MYTHS

THE CREATION MYTHS ANALYZED HERE carry two levels of meaning, an outwardly apparent literal meaning and a less obvious metaphoric meaning. As mentioned earlier, these myths tell stories that have literal significance in terms of everyday life experience, making them comprehensible to the average person so that they might be passed on from one generation to the next. At the same time, however, they appear to have been carefully crafted so that their characters and events metaphorically convey an advanced science of cosmic creation. The preceding several chapters, which describe how certain kinds of open systems spontaneously produce ordered wave patterns, provide sufficient scientific background that we may now take a closer look at some of these ancient myths.

To recognize the metaphorical meaning in these stories of creation, it is important to avoid the presumption that they were created by primitives and instead to view any evidence of possible scientific advancement with an open mind. Schwaller de Lubicz, a leading proponent of the symbolist school of thought, has emphasized the importance of adopting the proper mind-set:

> Olympiodoros, Zosimos, and Democritos the Philosopher, among others, never ceased to proclaim the existence of a sacred science in ancient Egypt, but because our own science knows nothing of it and does not believe in it, scholarly Egyptologists relegate such affirmations to the realm of fable. Consequently, this great ancient empire is rationalized, translated, and adjudged through the wrong mentality: through ours, and not through that of the sages of yore. As a result allegorical and metaphorical writings concerned with matters beyond our ken are labeled as being conjuring spells. Texts implying the highest science given to man are called "magical," in the pejorative sense of "sorcery."[1]

Traditional Egyptologists have consistently underestimated the scientific knowledge of the ancient Egyptians. For example, in the field of mathematics, it is often claimed that the Egyptians were content with using the fractional value $^{256}/_{81} = 3.16$ as an approximation for pi, a value 0.6 percent larger than the true value of 3.141592+. Evidence of this approximation was found in mummy wrappings dating from around 1700 B.C.E. The dimensions of certain pyramids and temples indicate, however, that the Egyptian masons had a much more accurate knowledge of pi (π) as much as a millennium earlier. For example, in the case of the Great Pyramid of Giza, the ratio of the perimeter of its base to its height is found to express a value of pi several orders of magnitude more accurate than the fractionally derived value. Schwaller de Lubicz has also noted that the height and width of the doorway facade of the temple of Karnak (circa 1400 B.C.E.) had a ratio equal to 3.141641, a value that deviates from the true value of pi by just sixteen parts per million.[2] The discrepancy was apparently intentional, since this pi ratio was calculated from phi, the golden section ratio. The masons used a geometrical method to calculate a precise phi ratio [$\phi = \frac{1}{2}$ $(1 + \sqrt{5}) = 1.6180399+$] and then used this to obtain a value for pi by calculating the relation $\pi \approx 1.2 \, (\phi + 1)$. These pi and phi ratios are also apparent in the proportioning of a mummy fresco displayed in the Tomb of Ramses IX at Thebes.[3]

Historians usually attribute the discovery of the golden section and an accurate derivation of pi to the ancient Greek Pythagorean geometers. It is likely, however, that Pythagoras obtained much of his mathematical knowledge from the Egyptian priests, since it is known that he had been initiated into their mysteries. Like the Pythagoreans, the Egyptian magi made a habit of keeping their mathematical methods secret from the general public, transmitting them orally from teacher to student and keeping no written record of their knowledge. The same tradition shrouded the transmission of their astronomy and medicine and the esoteric scientific significance of their creation myths. Nevertheless, just as the pi and phi ratios in certain of their temple constructions and wall art can be discerned through careful observation, it is possible through careful study also to discern the creation science expressed in their myths.

The earliest written record we have of Egyptian mythology, and in particular of myths describing the creation of the universe, come from the Pyramid Texts. These inscriptions, which date from the end of the Old Kingdom (circa twenty-fifth to twenty-second centuries B.C.E.), are regarded as one of the largest single collections of religious writings yet recovered from such an early period. The myths they describe, however, are believed to originate from a much earlier time, from the prehistoric era preceding the rise of Egyptian civilization.

THE MYTH OF ATUM

The story of how the high god Atum and his descendants came to create the universe relates the following sequence of events: Atum's self-emergence from the primordial etheric ocean; his subsequent generation of twin progeny, the god Shu and goddess Tefnut; the coupling of Shu and Tefnut to generate the initially embracing siblings, Geb (earth) and Nut (heaven); and finally the separation of Geb and Nut by Shu, an act that the Egyptians identified with the birth of the physical universe. As shown below, this sequence symbolically expresses in relatively sophisticated scientific terms how physical order may have emerged from a transmuting ether.

The Primeval Waters (The Ether)

The story of Atum starts by establishing the existence of the primordial ether, a vast, amorphous, waterlike wasteland called *Nun*. The Egyptologist R. T. Rundle Clark describes it as follows:

> Every [Egyptian] creation myth assumes that before the beginning of things the Primordial Abyss of waters was everywhere, stretching endlessly in all directions. It was not like a sea, for that has a surface, whereas the original waters extended above as well as below. There was no region of air or visibility; all was dark and formless. . . . Water is formless, it has no positive features and of itself assumes no shape. The Primeval Waters being infinite, all dimensions, directions or spatial qualities of any kind are irrelevant. Nevertheless the waters are not nothing. They are the basic matter of the universe and, in one way or another, all living things depend upon them. The waters are, then, "the waters of life" and the Primordial Ocean, known to the Egyptians as Nun, is "the father of the gods."[4]

Like the transmuting ether of subquantum kinetics, the primeval waters are invisible and of infinite extent, and their substance forms the basis for the created universe. Although Nun is described as initially existing in an inert state, it becomes vivifying and transmutative following the emergence of Atum, the "Becoming One."

The Emergence of Atum (The Onset of the Transmuting Flux)

Atum's first appearance out of Nun is marked by the coming into being or "emergence" of the primeval mound or hill (figure 6.1). The rising of the primeval mound should not be taken in the spatial sense of something rising in an upward spatial direction, for the primeval waters are conceived to extend endlessly in all directions and hence should have no upper surface. Rather, this emergence was understood to represent the self-creation of light within the absolute darkness of the primordial waters, an event symbolized by the rising sun. This "coming of light" does not refer to the creation of visible light (physical energy), since the material physical universe does not exist at this early stage in the myth. Atum's emergence instead refers to the appearance of the central source of vital energy, the prime mover that powers and activates the ether, irreversibly driving its transmutation processes.

Figure 6.1. Hieroglyphic symbols for the primeval mound. Adapted from Rundle Clark, Myth and Symbol, fig. 2.

Just as the sun serves as the principal activator for life processes on earth, Egyptian lore conceives of Atum as the primary activator of the cosmos. He traditionally symbolized the archetypal solar principle, the principle of radiation and of giving forth, the notion of continual irreversible change. The appearance of the primeval mound, then, would symbolize the coming into being of the primal flux, an event that the ancient Egyptians associated with the beginning of time.

Atum was not regarded as an entity separate from Nun; rather, the High God portrayed the activated aspect of Nun. This dynamic trait is also implied in the word *Nun*, which in ancient Egyptian also meant "young," "fresh," or "new," words that connote self-renewal. Thus, with the appearance of Atum, the waters of Nun became self-renewing much like the ever-living fire of Heraclitus.

The Pyramid Texts describe Atum as bringing himself into being through the autogenic act of masturbation. Although this arousal metaphor may sound crude to the sensitive mind, it brilliantly depicts the archetypal notion of creative flux, using the living human body for the purpose of illustration.

At the ancient site of Heliopolis, where the story of Atum's emergence held a prominent place in religious rites, the primeval mound was represented by a sacred stone known as the *Benben*. The Benben stone, probably conical in shape, was placed on top of an erect pillar. Heliopolitans marked the dawn as the time when the first rays of the rising sun reflected from the top of this obelisk structure.[5] Pyramid tombs such as those found at Giza were also symbols of the primeval mound. Like the Benben obelisk, they signified the emergence of existence.

When the primeval mound first arose, Atum is said to have alighted on it in the form of a phoenix, a mythical, heronlike bird with beautiful plumage of purple-red and gold color. The phoenix is said to live a long life of five hundred years, at the end of which it cremates itself on a funeral pyre only to rise from its ashes in the freshness of youth. By periodically repeating this ritual it passes from one life to the next, thereby achieving immortality. As such, the phoenix symbolizes the concept of continual transmutation, the passage from one state of existence to the next. It is also significant that its transformation occurs through combustion, since fire is an example of a nonequilibrium reaction process. But we are given to understand that this fire is of a metaphysical rather than a physical nature. Instead of transforming higher states of order to lower states, this etheric fire instead resurrects, transforming lower states to higher, like the vivifying flux of the *uas* scepter.

The ancient Egyptians also represented the emerging Atum in the form of a scarab beetle, which like the phoenix was a symbol for spiritual resurrection and immortality. In this form Atum was called *Khopri* (or *Khoprer*), "the Becoming One" or "the Not-Yet-Completed One." He was depicted pushing a ball of dung upward while standing erect on the primeval mound (see figure 6.2). This portrayal beautifully illustrates the concept of anabolic transformation, for the scarab reproduces itself by inserting its egg into a rolled-up ball of dung. On the one hand, dung is a waste product, an example of increased entropy; on the other hand, it is a potential source of food, one that can nourish the egg within it. Hence the egg and dung ball symbolize the emergence of directed life-creating activity.

Figure 6.2. Khopri rising from the primeval mound. Adapted from Rundle Clark,
Myth and Symbol, fig. 3.

These various metaphors—the emerging primeval mound, the resurrecting phoenix, the procreating scarab beetle, and the rising sun—portray Atum as a vivifying transmutative process that, in some mysterious way, came into being and began an unending series of death and rebirth transformations of the ether substance. Thus, Atum's emergence sets forth the primary prerequisite for the creation of the physical universe—the existence of the ongoing ether flux.

The Creation of Shu and Tefnut (Differentiation)

Although texts often refer to Atum as male, he was actually bisexual, "the great He/She." Masculine and feminine aspects emerged only later when he created his son, Shu, and daughter, Tefnut. One passage from the Pyramid Texts (Utterance 527) relates that Atum created his two children through masturbation:

Atum was creative in that he proceeded to masturbate with himself in Heliopolis;
he put his penis in his hand that he might obtain the pleasure of emission thereby
and there were born brother and sister—that is, Shu and Tefnut.[6]

Although Heliopolis is mentioned here in a symbolic context, this act was believed to take place in the primeval waters. Thus like Atum, Shu and Tefnut are processes that emerge in the ether.

Another passage (Utterance 600) relates that Atum created Shu and Tefnut by spitting: "You spat forth as Shu, you expectorated as Tefnut, you put your arms around them in an act of *Ka*-giving, so that your *Ka* might be in them."[7]

Like masturbation, the act of creation through spitting very ingeniously conveys the idea of unidirectional flow using the human physique. Shu and Tefnut, therefore, personify the concept of differentiation of the primal flux (Atum) into two distinct irreversible processes. Atum, who instills in them his *ka*, his vital essence, is the prime mover, the creator and activator of this flow.

Rundle Clark points out that the names Shu and Tefnut in ancient Egyptian are really puns on the word for "spit." *Shu* sounds similar to *ishesh*, meaning "to spit," and *Tefnut* resembles the word *tef*, which has a similar meaning. Thus the two together constitute a verbal assonance that serves to emphasize the fact that they are both similar and different at the same time. Shu and Tefnut are similar in that they are both embodiments of the same process driven by Atum's vital essence. They are different in that they are of opposite gender.

With the generation of Shu and Tefnut, the story of Atum presents the second important criterion for the generation of physical form, the emergence of the two primary ether transmutation pathways necessary for the creation of our physical universe:

$$B + X \longrightarrow Y + Z \text{ (Shu)} \quad \text{and} \quad A \longrightarrow G \longrightarrow X \longrightarrow \Omega \text{ (Tefnut)}$$

The subdivision of the initially unitary Atumian flux is portrayed in the geography of the Nile Delta watercourse. The Delta's geometrical apex would symbolize the "primeval mound" or "headwaters" of this flux, while the downstream effluvial branchings would represent the multiple paths forming the ether transmutation network. It is probably no accident that Heliopolis, the site of the temple of Atum, was built near this Delta apex (see figure 6.3). In fact, the Pyramid Texts state that Heliopolis is the site where Atum made his first appearance as the primeval hill and where he subsequently spawned Shu and Tefnut. It is also fitting that giant pyramid tombs symbolizing the primeval mound were built at Giza in this same locale.

The Birth and Separation of Geb and Nut (Emergence of Explicit Order)

The myth of Atum then relates that Shu and Tefnut subsequently united and gave birth to a son and daughter, Geb (earth) and Nut (sky). Initially, Geb and Nut were born closely embracing each other, but they were later separated by their father,

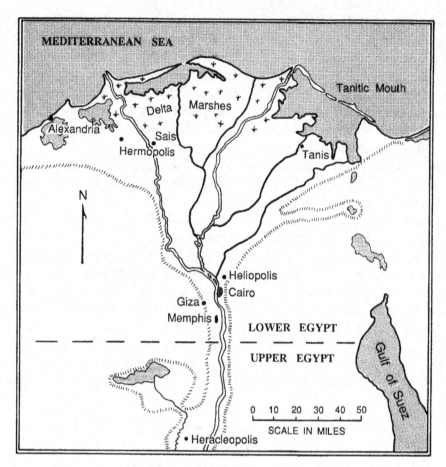

Figure 6.3. The Nile Delta region.

Shu. This separation, depicted in figure 6.4, is an eventful moment in the cosmogony of Atum. It signifies the beginning of physical creation, the moment when (physical) light first appeared in the universe. It is said that upon being held aloft by her father, Nut was able to begin to give birth to the stars and to "take them up," that is, to let them drift across her body, the sky. This pageant is described in Section 777 of the Pyramid Texts, where Geb speaks as follows:

> *O Nut! You became a spirit,*
> *You waxed mighty in the belly of your mother Tefnut before you*
>   *were born.*
> *How mighty is your heart!*
> *You stirred in the belly of your mother in your name of Nut,*
> *you are indeed a daughter more powerful than her mother. . . .*

Figure 6.4. Shu separating the sky goddess, Nut, from the earth god, Geb, to form the firmament. From the papyrus of Nisti-ta-Nebet-Taui in Piankoff, Egyptian Religious Texts, fig. 30.

O Great One who has become the sky!
You have the mastery, you have filled every place with your
beauty,
the whole earth lies beneath you, you have taken possession
thereof,
you have enclosed the whole earth and everything therein within
your arms . . .
As Geb shall I impregnate you in your name of sky,
I shall join the whole earth to you in every place.
O high above the earth! You are supported upon your father Shu,
but you have power over him,
he so loved you that he placed himself—and all things beside—
beneath you
so that you took up into you every god with his heavenly barque,
and as "a thousand souls is she" did you teach them that they
should not
leave you—as the stars.[8]

The sexual union of Shu and Tefnut calls to mind the coupling of the transmutative pathways that form Model G's wave-generating cross-catalytic loop (figure 6.5). Although the myth does not present the same degree of detail found in Model G, it does clearly indicate the coupling of "male" and "female" ether processes. The

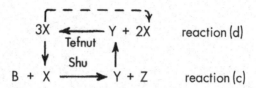

Figure 6.5. The union of Shu and Tefnut as a cross-catalytic loop.

separation of the offspring, Geb and Nut, represents the separation of the X and Y ethers in forming the first particle of matter. Thus if we were to identify Geb and Nut with reaction variables X and Y, the children's initially close embrace would signify the ether's initial homogeneous steady state, in which the X and Y ether concentrations are uniformly distributed throughout space. The subsequent separation of the siblings, with Nut arching over Geb, would graphically depict the local decrease in X concentration and simultaneous increase in Y concentration to constitute the core of a subatomic particle (figure 6.6).

It may be significant that Geb and Nut are separated by Shu, for X and Y ultimately separate into a discrepant state because of the deviation-amplifying tendency of the "Shu-Tefnut" reaction loop. It is also appropriate that Shu, whose name signifies "space," is the one to carry out this separation. For matter creation would take place at a number of sites throughout preexisting space. Moreover, Shu, who was known as the god of the air, is shown in figure 6.4 being assisted in the separation act by birdlike air spirits symbolizing the divine animating principle, *ba*. In other portrayals he is assisted in the separation by ram-headed wind spirits. These air and flight themes are appropriate metaphors for expressing the notion of emergence.

The air spirits and attendants portrayed in figure 6.4 gesture with their arms bent upward at the elbow in a loving embrace. This is the sign for the act of *ka*, the

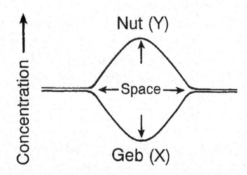

Figure 6.6. The separation of Geb and Nut as depicted in the symmetrical departure of the X and Y ethers from their uniform steady state.

transmission of vital essence or divine power from a god to his progeny or from god to man. Shu gestures in a similar way as he raises Nut. This prevalent symbol in the separation scene indicates that physical form comes into being as a result of the bestowal of divine grace, the same conscious essence that continuously activates the underlying ether processes.

Shu, who is usually depicted as a human-headed god, is shown here with the head of a baboon, reminiscent of the god Thoth, patron of the sacred mysteries. The U-shaped symbol that rests on his head is the "horizon mountains" hieroglyph, which traditionally portrayed the emergence of the first light at the dawn of creation. Often the sun was depicted as rising midway between the two peaks. The symmetrical placement of the peaks and the symmetry of Shu's hand gestures seem to portray a symmetric splitting from an initially unitary state, thereby reiterating the archetypal systems principle of symmetry breaking.

This symmetrical-deviation theme is also found in the back-to-back portrayal of Shu and Tefnut represented as the "horizon lions" (figure 6.7). Together with the horizon mountains, they indicate the emergence of explicit order at the moment when light dawned upon the universe. Like the splitting of Geb and Nut, the opposed stance of Shu and Tefnut conveys the notion that the X and Y ether concentrations depart from their former steady-state values in opposite directions, with the concentration of one species becoming greater as the other becomes less.

The portrayal of the dynamic ether processes in terms of gods capable of both

Figure 6.7. The horizon lions Shu and Tefnut. From the Sarcophagus of Khonsu, Cairo Museum, Piankoff, *Egyptian Religious Texts*, fig. 15.

speech and procreation indicates that the ancient Egyptians regarded the ether and the creation event itself as something that was alive, conscious, and divine. Such metaphors are clearly appropriate to an open-system view of physical creation, since the open-system concept explains the emergence not only of living organisms but also of conscious thought. How refreshing and different the Egyptian cosmology is from the sterile inert conception offered by present-day physics and astronomy.

THE MYTH OF OSIRIS

Although the Osiris creation myth is found in its earliest written form in the Pyramid Texts, artifactual representations of its gods have been traced back to much earlier times. What we know of the story comes not only from oral tradition, such as that preserved by Plutarch, but also from inscriptions and art associated with funerary rites and royal installation ceremonies.

The Osiris myth has five principal actors who sprang forth at the time heaven and earth separated. These include Nut's four children—Osiris, Set, Isis, and Nephthys—and Horus, the son of Isis and Osiris. According to the story, summarized in chapter 1, the jealous and scheming Set slays and dismembers his brother, Osiris, Lord of Creation, in order to usurp rulership of the created world. Isis later recovers the body of her deceased husband, unites with it, and gives birth to a son, Horus. When fully grown, Horus avenges his father's death and wins back control of the empire.

Since the parental gods, Geb and Nut, live in the cosmic ocean Nun, we may infer that Osiris and his siblings also represent universal principles governing the ether's operation. In fact, the story of Osiris appears to repeat the same creation metaphysics but covers details not found in the Atumian myth, such as the battle fought between order and chaos. Thus the Osirian myth closely complements the Atumian creation myth to which it is appended.

Osiris, Set, Isis, and Nephthys are born from Nut simultaneously with her separation from Geb; they come into being at the same time as the first particle of matter. This simultaneity might encourage us to identify the story of Osiris with this very first emergence event. It is also possible, however, to interpret it as portraying subsequent matter creation taking place on an ongoing basis, for the ancient metaphysics teaches that the ether has continued to spawn matter since the time of this first primordial event.

Compared with the story of Atum, the Osiris myth appears to have a more earthly focus; its events are staged in the environs of Egypt and its more local underworld domain, rather than in some far-off place in Nun. Moreover, its gods are associated with natural principles that directly affect people's lives in the real present. For example, the deceased Osiris is responsible for the annual renewal of vegetation, flowing water, the fertilizing Nile floods, and the reproductive power in animals, while Set is associated with a variety of destructive natural forces such as earthquakes, desert winds, and typhoons.

Although the ancient Egyptian populace eventually came to interpret the story of Osiris as a reenactment of the soul's resurrection and entry into the afterlife, the myth was also known to metaphorically convey certain metaphysical ideas. But details of this more esoteric aspect were known only to an inner circle of properly initiated priests. Aided by a knowledge of modern science, we are now able to fathom that part of this secreted metaphysics that deals with the sacred miracle of matter creation.

According to myth, Osiris was Egypt's first ruler. He delivered the Egyptians from their destitute and brutish manner of living by showing them the benefits of cultivation and by giving them laws. Afterward he is said to have traveled throughout the world civilizing humanity. But with his return to Egypt, he was tragically assassinated. Thus began the intriguing and emotionally moving mythical drama that not only captured the hearts of the ancient Egyptians for many generations, but also formed the central organizing theme of their governing rituals. Let us take a moment to read it.*

✳ ✳ ✳

WHILE OSIRIS WAS TEMPORARILY ABSENT FROM EGYPT, Isis, who was in control of the empire, remained vigilant and alert. As a result, their brother Set attempted nothing revolutionary. But when Osiris returned, Set formed a group of seventy-two conspirators and contrived a treacherous plot against him. Having secretly measured Osiris's body and prepared a beautiful chest of corresponding size artistically ornamented, he had it brought into the room where festivities were in progress to celebrate Osiris's return. The company admired it greatly, whereupon Set promised in jest to present it to the man who should find the chest to be exactly his length when he lay down in it. All in turn tried it, but no one fit into it; when Osiris got in and lay down, Set's conspirators ran to it and slammed down the lid, fastening it with nails and soldering it with molten lead. They then threw the chest into the Nile and let it be carried to the sea through the Tanitic Mouth.

When she heard of the disaster, Isis at once cut off a lock of her hair and put on a garment of mourning. Deeply perplexed, she wandered everywhere; she addressed everyone she approached, and when she met some little children, she

---

*The majority of this myth is adapted from "Isis and Osiris" in Plutarch's *Moralia*, trans. F. Babbitt (Cambridge, Mass.: Harvard University Press, 1936), pp. 35–49. The account of the conception of Horus is taken from a version given in R. T. Rundle Clark's *Myth and Symbol in Ancient Egypt* (New York: Thames and Hudson, 1959), p. 106.

even asked them about the chest. As it happened, they had seen it, and they told her of the mouth of the river through which the friends of Set had launched the coffin into the sea.

Isis later learned that the chest had been cast up by the sea near the land of Byblos (in Phoenicia) and that the waves had gently set it down in the midst of a clump of heather. In a short time, the heather formed a beautiful and massive stock, enfolded and embraced the chest with its growth, and concealed it within its trunk. The king of the country admired the great size of the plant and cut off the portion that enfolded the chest, now hidden from view, and used it as a pillar to support the roof of his palace. Isis, informed of these facts by the divine inspiration of rumor, came to Byblos and sat down by a spring, dejected and full of tears; she spoke to no one, but she welcomed the queen's maidservants, whom she treated with great amiability, plaiting their hair for them and imparting to their bodies a most wonderful fragrance. Thus it happened that Isis was sent for, and she became so intimate with the queen that the queen made Isis the nurse of her baby.

After some days, the goddess disclosed herself and begged for the pillar that held up the roof. She removed it and cut away the wood of the heather that surrounded the chest. She then threw herself down upon the coffin with such a dreadful wailing that the younger of the king's sons expired on the spot. Having placed the coffin on a boat, she put out from land.

She brought the coffin to Buto in the Nile Delta marshes. But Set, who was hunting at night by the light of the moon, happened upon it. Recognizing the body, he tore it into fourteen parts and scattered them about, each in a different place. Isis learned of this and went looking for them, sailing through the swamps in a boat of papyrus. The only part of Osiris's body that Isis did not find was the male member, for it had been tossed at once into the Nile, and the porgy and pike had fed upon it.

Gathering together his parts, except for the phallus, Isis reassembled the form of Osiris's body. She then succeeded in reviving Osiris sufficiently to be able to conceive a son by him, who was named Horus. To prevent his discovery by Set, Isis gave birth to Horus in the swamps of the Nile Delta. There she secretly nursed and raised him.

Later Osiris came to Horus from the underworld and trained him for battle. After some time, Osiris asked Horus what he thought to be the noblest thing a man could do, to which Horus replied, "to avenge one's father and mother for evil done to them." Osiris then asked him what animal he considered the most useful for those who would go forth to battle; and when Horus said, "A horse," Osiris was surprised and raised the question why he had not said, "A lion," instead. Horus answered that a lion was a useful thing for a man in need of assistance, but that a horse served best for cutting off the flight of an enemy and annihilating him. When Osiris heard this he was very pleased, since he felt that Horus now had adequate preparation.*

Upon reaching maturity, Horus left his Nile Delta abode to search for Set and challenge him to battle. The struggle lasted many days, and Horus finally prevailed. Set was delivered in chains to Isis; she did not cause him to be put to death, however, but instead let him go, whereupon Set formally accused Horus of being an illegitimate child. The matter was submitted for arbitration to the great council of the gods, and with the help of Thoth to plead his case, it was decided that Horus was in fact legitimate. Set refused to relinquish control and so two more battles were fought. Set managed to claim Horus's left eye, but Horus succeeded in tearing out Set's testicles. Horus thereupon triumphed and regained control of his father's empire.

* * *

This myth is more than just an entertaining story. When examined closely, the various actions or attributes of its gods are found to metaphorically portray a sequence of fundamental natural principles that describe a highly sophisticated scientific theory of how the physical world first came into being out of a primordial vital flux.

The Slaying of Osiris (Entropy Increase)

Plutarch's version of the myth begins with the events surrounding Osiris's assassination,[9] as described in chapter 1. Osiris was traditionally identified with creation, order, light, life, and good, whereas his brother was identified with destruction, violence, disorder, darkness, death, and evil. Hence the murderous suffocation of Osiris (order) and the subsequent rise of Set (disorder) to rulership ingeniously portrays the universal tendency for ordered states to decay when a system is placed in a state of isolation. As mentioned in chapter 2, physicists call this entropy-increase principle the second law of thermodynamics. It is reasonable that the myth should begin by allegorically describing this concept, since the entropy-increase tendency is the key sustainer of irreversible process, and process in turn is a fundamental prerequisite for the evolution of order.

Osiris was commonly identified with the Nile, flowing water, and regenerative life force, all of which convey the idea of nonequilibrium process. Since water is the element of the primordial ether, Osiris, like Atum, esoterically symbolizes the

---

*In funerary text versions of the legend, Osiris's soul does not come to the day world to train Horus but instead remains unconscious in the underworld, regaining awareness only after Horus defeats Set.

transmuting ether. This process concept is further illustrated by the imagery of Osiris's coffin drifting out to sea in the Nile current. His downstream passage reenacts the "journey" of the ether constituents as they evolve along the transformation dimension.

The Column (Emergence)

The ancient Egyptians associated Osiris with the principle of vegetative growth, a metaphor for order creation. Hence the treelike growth that develops around his coffin when it washes ashore at Byblos is actually an expression of his generative aspect. Thus whereas the first part of the myth depicts the "negative," order-destroying aspect of process, this part of the myth depicts the "positive," order-creating aspect.

The tree trunk column used to support the palace roof alludes to the separation of heaven and earth. It calls to mind the cosmic sky support, known as the *djed* pillar, which was traditionally identified with Osiris when he was later revived from death. Hence the Byblos column metaphor is apparently intended to foreshadow the cosmic creation event that becomes actualized later in the myth.

The Dispersal of Osiris (Entropy Increase and Differentiation)

Set's dismemberment of Osiris's body into fourteen pieces in the Nile Delta marshes again illustrates the concept of the tendency for ordered states to decay. At the same time, it portrays the differentiation of the initially unitary ether flux into many branching paths to form the ether transmutation network. This concept is also illustrated by the subdivision of the Nile outflow into numerous crisscrossing channels.

The Union of Isis and Osiris (Reactive Coupling)

The gathering together of Osiris's body portrays the concept of order creation. In effect, it is a reversal of Set's attempt toward dispersal. It is significant that Isis is the one to conduct this reassembly, for esoterically she signifies the archetypal principle of circular form, hence the geometrical ordering of process into an order-creating loop. It is also significant that she is able to recover only thirteen of Osiris's pieces, because this number is the esoteric symbol for death and transformation to a new state of being, a concept depicted in Arcanum 13 of the Tarot. This rebirth concept effectively describes the series of deaths and rebirths that individual etherons would undergo as they transform through their succession of ether states, a principle also portrayed by the Hindu god Shiva in his role as the Cosmic Dancer. Moreover, this rebirth metaphor also anticipates the coming transcendent emergence of explicit order (Horus).

The consumption of Osiris's phallus by fish swimming in the Nile is also significant. Swimming fish convey the notion of autonomous forward movement through a medium, simulating the alchemic displacement of etherons along the transformation dimension. The astrological sign of Pisces and the yin-yang symbol similarly use swimming fish to illustrate this concept allegorically. Moreover, the phallus connotes both fluid flow and regenerative process. Like the fish, it symbolizes the transmutative impulsion of the primordial ether.

The scene in which Isis unites with Osiris's reassembled body portrays the concept of alchemic union (transmutative union). Like the mating of Shu and Tefnut (figure 6.5), their bonding illustrates the coupling of two Model G reaction processes to form a cross-catalytic loop (figure 6.8). Of the two reactions, the upper (d) most appropriately typifies Isis, since its nonlinear positive-feedback character tends to foster the growth of indigenous fluctuations. Isis traditionally portrayed such nurturing traits. She was known for her ability to restore and maintain health and grant immortality. She was also regarded as the bestower of fertility and the nourisher of all creatures.

The absence of Osiris's generative organ emphasizes the noncoital nature of Isis's bonding with Osiris. Thus the seed from which Horus springs does not come from his parents but emerges spontaneously in their midst as a fluctuation arising from the ether itself. Isis and Osiris contribute to his generation by forming a matrix of circular causality that nurtures Horus into being. Isis is the prime actor here in that she consummates her union with Osiris by marginally reviving him from his death state. This emphasizes the feminine, formal aspect of process as being the primary seat of generation.

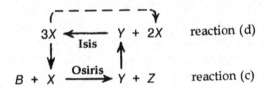

Figure 6.8. The union of Isis and Osiris as a cross-catalytic loop.

The Birth and Raising of Horus (Growth of the Seed Impulse)

The image of Horus being born and raised in the delta swamps is consistent with the idea of the physical universe emerging at a specific location in the ether transmutation network. Moreover, the marshes express two criteria necessary for successful parthenogenesis: they are fertile, and their thick vegetation provides a place to hide from enemies. Similarly, if it is to survive to criticality, an emerging etheric fluctuation must arise in a fertile part of the ether and at a site that is temporarily devoid of competing fluctuations. Such sheltered sites arise in the ether from time to time owing to the statistical nature of the ether fluctuation process.

Horus's name in its complete form is given as Horpi-chrud, which means "emergent child," *Hor* deriving from *hri,* signifying "up, above, on top," and *chrud,* signifying "child." In harmony with this imagery, Horus would represent the positive electric energy pulse (high *Y*/low *X* fluctuation) that spontaneously arises in the ether and is destined to grow into a particle of matter.

The Battle with Set (The Emergence of Explicit Order)

The time when Horus reaches maturity and leaves the delta in search of his foe symbolizes the moment when the emerging ether fluctuation has reached criticality. Figure 6.9 depicts Horus at this stage of his maturity when he is known as Heru-khuti, the dweller in Behutet. As Heru-khuti, one of his greatest and most important of forms, he represents the god of light and typifies the power of the sun prevailing at the time of its greatest midday heat. It is in this warrior form that he engages Set in battle.

It is also significant that Horus was represented in the form of a hawk or falcon. Birds are one of the few categories of animals able to move vertically as well as horizontally. Hawks and falcons, in particular, are known for their ability to soar to great heights, and they also happen to be birds of prey. Similarly, the critical fluctuation not only soars upward, but to manifest explicit order, it must also combat and overcome the order-eroding effects of its erratic environment.

This particular fresco further develops the theme of combative emergence by showing Horus holding an upward-pointing spear in his right hand. The spearhead with its small hawk head seems to suggest that Horus brings light into being through upward growth. By holding this weapon in his right hand, he indicates that his emergence brings about explicit order, because the right side of the body is associated with outward expression. The depiction of Horus seated rather than standing suggests that he has not yet fully emerged at this point. In this rendition he very much resembles the Emperor pictured in Arcanum 4 of the Tarot. As is shown in chapter 8, the Emperor is also a warrior who, like Horus, is about to engage in battle to establish a kingdom of order.

The conflict between Horus and Set depicts the struggle between the emergent critical fluctuation and competing fluctuations threatening to erode its form. Set's emasculation in this battle symbolizes his loss of power and authority. The victorious Horus, with only his *right* eye remaining, symbolizes the emergence of the explicit realm. This triumph of the god of light over the god of darkness signifies the emergence of a particle of matter amid the chaotic fluctuations of the feature-

Figure 6.9. Horus as Heru-khuti, the dweller in Behutet. From Budge,
Gods of the Egyptians, pl. 31.

less ether. It is a moment analogous to the instant when Geb and Nut cleave "heaven" from "earth." In this victorious state Horus is usually depicted in a standing position clutching Set, who is symbolized in the form of the *uas* scepter.

As pointed out in chapter 2, open systems do not annihilate the tendency toward disorder, they merely redirect it so that it spawns explicit order. In a similar fashion, the subdued Set is compelled to support the resurrected Osiris. That is, Set's potentially destructive energetic flux is redirected toward creative ends, thereby sustaining physical creation.

That Set is allowed to survive in subservience to the resurrected Osiris (physical order) is appropriate in another sense. Namely, the newly formed explicit order does not prevent chaotic fluctuations from emerging; instead, it dominates their effect. Random fluctuations continue to erode the emergent material wave pattern but have no major effect, since the underlying etheric flux continuously regenerates its form. It is therefore fitting that Set is made subservient to Osiris—the god of regeneration.

SYMBOLS OF CREATION

Ancient Egyptian lore contains a number of symbols that express concepts related to the creation of the universe. Although these individual symbols do not convey as much detail as the creation myths analyzed above, the metaphorical concepts they express are consistent with the same open-system order-genesis science and are thus worth examining from the standpoint of this modern science.

The Cosmic Serpent

Related to the myth of Atum is a tale about the primeval serpent Iru-to, "Creator of Earth." Like Atum, Iru-to came into being in the midst of the dark primeval waters before any definite thing existed and emerged from the waters coeval with the rising of the primeval mound. Iru-to is therefore an attribute of Atum. In paragraph 1146 of the Pyramid Texts there is a passage in which Iru-to speaks as follows:

> *I am the outflow of the Primeval Flood, he who emerged from the*
> *waters.*
> *I am the "Provider of Attributes" serpent with its many coils,*
> *I am the Scribe of the Divine Book which says what has been and*
> *effects what is yet to be.*[10]

In a hymn from the Coffin Texts (First Intermediate Period, 2250 B.C.E.) the serpent proclaims:

> *I extend everywhere, in accordance with what was to come into*
> *existence,*
> *I knew, as the One, alone, majestic, the Indwelling Soul, the most*
> *potent of the gods.*

> *He [the Indwelling Soul] it was who made the universe in that*
> *he copulated with his fist and took the pleasure of emission.*[11]

Like Atum, the primeval serpent depicts the general systems concept of irreversible, nonequilibrium process. In saying that the serpent extends everywhere, the texts seem to suggest that this emergent state of transmutation pervades the primordial waters. Consequently Iru-to, like Atum, represents the omnipresent transmuting ether. Just as the ether is invisible to human sight, so is the primeval serpent. In this aspect he is sometimes called Amun, the "Hidden" or "Invisible One."

The use of the snake symbol to personify unidirectional process is quite ingenious. For, not only does a snake have an elongated body, suggestive of unidirectional flow, it is also one of the few animals that is incapable of moving backward. Its shinglelike belly scales are able to grip the ground effectively only for forward movement. The primordial serpent, in its form as the "Provider of Attributes," is shown in figure 6.10. The serpent's two heads are reminiscent of Atum's generation of Shu and Tefnut. These two branchings of the primary flux represent the two principal ether transmutation pathways, whose interaction brings about the physical universe.

The Ankh, the Djed Pillar, and the Tyet

The ankh hieroglyph, known as the "Key of Life," is a common motif in ancient Egyptian art. Considered in the context of reaction kinetic science, a striking similarity is noticeable between the ankh's cross-and-loop geometrical form and the functional order of the Brusselator and Model G (figures 3.9 and 3.10). A suggested superposition of the Model G reaction pathways onto the ankh is illustrated in figure 6.11. The horizontal member would signify the "feminine" reaction pathway

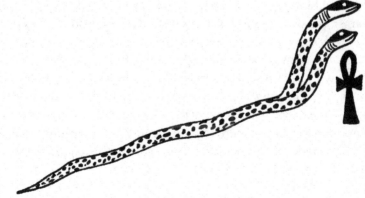

Figure 6.10. The primeval serpent. Adapted from Rundle Clark, *Myth and Symbol*, fig. 7.

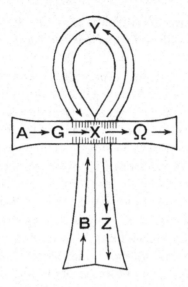

Figure 6.11. An ankh shown with the Model G ether reaction scheme superimposed.

$A \longrightarrow G \longrightarrow X \longrightarrow \Omega$, earlier identified with Tefnut and Isis.* The vertical member would signify the "masculine" reaction process $B + X \longrightarrow Y + Z$, earlier identified with Shu and Osiris. The ligature that binds them together metaphorically conveys the idea of a junction, hence the coupling of these two primary reaction pathways. The ankh's lower vertical portion is usually depicted as being subdivided into two distinct members. One would represent the "upward" input of the B reactant; the other would represent the "downward" output of product Z, with the Y intermediate product being directed along the ascending arc of the upper loop. The reaction loop would be completed by the "feminine" autocatalytic reaction. This converts Y back into X and forms the descending arc of the ankh's loop.

This loop would present an important feature of the ether's implicit order. As indicated earlier, the self-closing X-Y loop allows the transmuting ether to evolve wave patterns with stable spatial periodicities, an essential requirement for producing realistic models of material subatomic particles. The X and Y variables, portrayed at the center of the cross and at the crown of the ankh's loop, respectively, would serve as the principal generators of physical form. Variations in their concentrations would form the material particles, energy waves, and energy fields that make up our universe. Consequently, this cross-and-loop monogram would symbolize that unique node in the ether reaction network where our physical universe

---

*This reaction is "feminine" because its G reactant is instrumental in forming the supercritical pocket that nourishes upcoming fluctuations.

is spawned. Interestingly, the cross symbol is used in esoteric tradition to symbolize the physical plane. It is also the key element in the monogram (⊕) used by both astrologers and astronomers to symbolize planet Earth.

The ankh may be thought of as a glyph that portrays, in stylized form, the implicit order that underlies and permeates our entire universe. It would designate the principal reaction pathways that etherons follow as they transform through our part of the ether transmutation network. The ankh, therefore, symbolizes the etheric womb that gives rise to physical existence. It represents life in the broadest conceivable sense of the word.

One motif that commonly decorates the entrances to temples shows the ankh together with the *uas* and *djed* pillar hieroglyphs positioned above a basket (figure 6.12). The hieroglyphic meaning of these symbols is normally given as "prosperity" *(uas)*, "life" (ankh), and "stability" or "durability" *(djed)*. Schwaller de Lubicz, however, identifies them respectively with spirit, soul, and body. He says that the basket signifies the *all*, the totality, and suggests that the entire pictogram depicts the three principles of beginning, "the Magistery of origin incarnating the divine Word and causing the heterogeneity of a milieu which had previously been perfectly homogeneous *(Nun)*."[12]

So this icon summarizes in a succinct manner the process of physical creation. As noted earlier, the *uas* scepter traditionally symbolized the divine flux, hence the spirit that animates the ether. The ankh, the living "soul" of the universe, may be interpreted as a symbol of the functional ordering of this animating process. Finally, the *djed* pillar may be interpreted as a symbol of the physical form, or "body," produced by this spirit-animated "soul," thus denoting the explicit order that incarnates from the underlying implicit order.

The *djed* was understood to symbolize the cosmic pillar that supports the vault of the sky and thereby maintains physical form in existence.[13] This concept was

uas   *djed*   *ankh*

Figure 6.12. Pictogram showing the uas, djed, and ankh. From Schwaller de Lubicz, *Sacred Science*, p. 145.

sometimes pictured in wall decorations as a series of *djed* pillars supporting a stone window arch. The column was also identified with the backbone of Osiris. Its upright stance was supposed to portray life overcoming the process of death and decay, the cosmic victory of order over disorder. It was associated with the rising up of Osiris's soul following the victory of Horus over Set. On the last day of the rites reenacting the passion of Osiris, the king or chief priest would commemorate Osiris's revival by setting the *djed* pillar upright.[14] This ceremonial erection esoterically signifies the incarnation of divine spirit, the momentous birth of physical form.

The four tiers that cap the top of the *djed* indicate that the pillar symbolizes physical form, the number four being the traditional symbol for solid matter. Similarly, the ancient Egyptians conceived heaven to be supported by four pillarlike gods called the "four Sons of Horus," or the "gods of the four cardinal points." Dwa-mwt-f was dog-headed and represented the north; Kebehsenwf was falcon-headed and represented the south; Hâpi was baboon-headed and represented the east; and Imset was human-headed and represented the west.

The *tyet* is another symbol that has particular relevance in the context of this creation science. It is a strip of cloth or leather knotted in such a way that it looks very much like the ankh; the ankh's stiff cross member is replaced by two drooping ties (figure 6.13). The *tyet* is a symbol for Isis. Usually red in color, it is said to be symbolic of her blood when she gave birth to Horus.[15] It is also the ancient Egyptian sign for "protection" and a symbol associated with the bond of matrimony. In terms of the ancient science of physical creation, the *tyet* represents the fertile ether reaction nexus and protected environment that spawned and nourished the seed energy pulse.

Sometimes the *tyet* and *djed* are portrayed in combination, with the tyet shown tied around the *djed* pillar about halfway up its length (figure 6.13). In this form it presumably indicated the union of Isis and Osiris. Egyptologists in the past have had

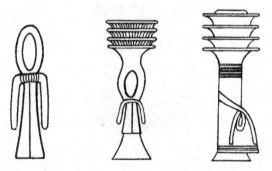

Figure 6.13. Tyet (left), and djed and tyet in combination (middle and right).

Figure 6.14. Ancient clay artifacts portraying a symbol similar to the tyet, but without cross arms. National Archaeological Museum, Athens, Greece.

trouble understanding the significance of this combination in relation to the world pillar symbolism. The meaning of the *djed* and *tyet* becomes quite clear, however, when pondered from the perspective of the open-system creation science. From one standpoint, the combined motif symbolizes the coupling of male and female ether processes into the cross-catalytic loop, the prime generator of physical form. At the same time, the pillar that arises from the midst of the loop's energized embrace symbolizes the explicit order that was earlier born from this union and has now achieved full emergence as matter (Horus victorious, Osiris resurrected). Thus the *djed*-and-*tyet* motif symbolizes both the cosmic sex that generates the physical universe and the product born of that interaction, namely, the physical universe itself.

The ankh and *tyet* are quite similar in that they both portray the cross-and-loop implicit order that underlies the physical universe. But clearly there is a difference, as indicated by the different arrangement of the cross arms of these symbols (see figure 6.14). The *tyet*, which displays a limp loop with drooping protective arms, would portray the feminine principle in the early stage of etheric parthenogenesis where Isis first gives birth to Horus (the seed fluctuation). On the other hand, the ankh, which displays an erect loop structure with arms extended, would portray parthenogenesis at a later stage where the seed has survived chaos and established itself as a fully developed material particle. As such, the ankh would portray feminine and masculine principles combined: Isis and the victorious Horus or alternatively Isis and Osiris resurrected. Consequently, the ankh would be a close equivalent of the *djed-tyet* hieroglyph.*

As the symbol of life, the ankh would signify the principle of resurrection, calling to mind the symbolism of the cross in Christianity. The ankh signifies the mystery of the divine incarnating into physical form. This would account for its frequent appearance in murals depicting the mystery of order creation (figures 6.4, 6.7, 6.9, and 6.10). More generally, it would represent the creative generation of transcendent form, the fundamental principle that repeats in infinite regress generating order at all levels of nature's hierarchy.

The Cosmic Lotus

A myth dating from the time of the Old Kingdom compares the first creation to the opening of a lotus flower. It states that prior to physical creation, there existed an ether resembling a dark sea of limitless expanse. One day from the sea's "surface" emerged an immense luminous lotus bud. With the bud's opening, light and life came into being. This first dawn was personified by the divine child, sitting with its finger in its mouth upon the petals of the open lotus. The Pyramid Text relates that when the lotus appeared, "Order was put in the place of Chaos."[16]

The blossoming lotus serves as an excellent metaphor for the emergence of physical form within the transmuting ether. Floating on the surface of a pond, it appears to take substance from the water itself (figure 6.15). In subquantum kinetics the critical energy impulse similarly springs as a bud from the ether substance, gradually grows and expands, and finally blossoms into a mature subatomic particle.

The papyrus drawing shown in figure 6.16 connects the lotus metaphor with the story of Osiris by cleverly depicting how the transmuting ether gives birth to physical creation and continually sustains its form. It shows the resurrected Osiris seated on a throne, which in turn floats on a foundation of water, symbolic of Nun. Isis and Nephthys stand behind Osiris, giving him their support. A blooming cosmic lotus bearing the four Sons of Horus, the four supports of heaven, arises from the water.

This metaphor is reproduced in almost every detail in the ancient Hindu myth

---

*It is probably no coincidence that Venus, the planet named after the goddess of love, was represented by a glyph (♀) that is believed to have evolved from the ankh. In fact, astrologers' interpretation of the Venus glyph is very similar to the above interpretation of the ankh. That is, the upper circle in the symbol is said to signify the marriage of the male and female forces of the universe into one supreme parent, while the lower cross is said to signify the result of this union—the cross of matter (A. Oken, *As Above, So Below* [New York: Bantam Books, 1973], p. 273).

Figure 6.15. The lotus, or water lily, gives the impression that it arises from the water substance itself.

about Vishnu dreaming the physical universe into existence. Figure 6.17 shows the god Vishnu, "the Preserver," resting on the coils of the seven-headed abyssal serpent Ananta, whose name means "unending." Vishnu, Ananta, and the chaotically fluctuating Cosmic Milky Ocean on which they float together personify the limitless, formless expanse of primordial generative waters. As Vishnu sleeps, a dream lotus grows from his navel and hovers above his body. This flower in turn gives birth to Brahma, the Lord of Light and creator of the universe. He sits on the blossoming lotus, his four faces illuminating the four corners of the universe. The fertility goddess Shri Lakshmi, Vishnu's beautiful consort, stimulates him tactilely as he dreams and maintains the creation, nourishing it through his umbilical stem.*

At the end of one hundred Brahma years, a period equivalent to about 311 trillion human years, Vishnu withdraws his dream of Brahma, and the lotus retracts back into his body. "Like a spider that has climbed up the thread that once issued from its own organism, drawing it back into itself, the god has consumed again the

---

*In some versions of the myth, the chaotic ocean spawns the cosmic lotus when Vishnu directs gods and demons to agitate its milky substance. This turmoil induces the ocean to yield up its valuable elixir of life essence, much as churned milk yields up its butter. The rich substance that emerges is the lotus, personified by Shri Lakshmi. Like Brahma, she is often pictured seated on a lotus (D. Kinsley, *The Goddesses' Mirror: Visions of the Divine from East and West* [Albany: State University of New York Press, 1989], pp. 53–70).

Figure 6.16. A scene from the Hunefer Papyrus (1317–1301 B.C.E.) showing Osiris sustaining the cosmic lotus. From Schwaller de Lubicz, Sacred Science, p. 215.

web of the universe."[17] When Vishnu enters this period of dreamless sleep, all of his dream lotuses simultaneously become reabsorbed. With the dissolution of *maya*, the cosmic dreamer remains absorbed in a dreamless sleep for a period of one hundred Brahma years. Eventually, though, something within him stirs, and Vishnu once again unfurls his lotus dream.

The lotus, a Hindu symbol of fertility and life, portrays this created universe in terms of an organic metaphor. It represents the universe as an entity imbued with vigorous fertile power, continually growing and proliferating and requiring continual sustenance. Through this "vegetative" growth, the life-giving powers of the primordial waters (the ether) become transformed into embodied organic life (physical form). Heinrich Zimmer comments on the lotus metaphor as follows: "Rising from the depths of water and expanding its petals on the surface, the lotus *(kamala padma)* is the most beautiful evidence offered to the eye of the self-engendering fertility of the bottom. Through its appearance, it gives proof of the life-supporting power of the all-nourishing abyss."[18]

As seen in figure 6.17, many lotuses bob on the surface of the Cosmic Sea, all symbolically connected to Vishnu's navel. Thus at the same time that Vishnu dreams "our" lotus into existence, he also dreams into being innumerable other lotuses, each spawning a Brahmic universe of its own. Assuming that a single

Figure 6.17. A nineteenth-century print showing Vishnu and Shri Lakshmi floating on the Cosmic Sea. From Moor, The Hindu Pantheon, pl. 7.

Brahmic universe signifies an entire physical creation consisting of innumerable galaxies, these other lotuses would represent parallel universes residing at other junctures in the ether transmutation network. The Cosmic Sea then would be portraying the ether from a perspective of more than three dimensions.

Evidence that the ancient Hindus viewed existence from such a multidimensional perspective is found in a passage from the *Brahmavaivarta Purana.* This text relates a conversation between Transcendent Wisdom, incarnated as a boy, and Indra, the powerful sky god whose primordial victory over chaos has allowed the creative, nourishing waters of the deep to rush forth and create a fertile, habitable cosmos. To quell Indra's excessive pride, the boy informs him of the other "Indras" that have gone before him:

> Many indeed are the Indras I have seen, one succeeding another, eon after eon, as the great wheel of existence grinds away inexorably. Worlds come into being and dissolve, much as drops of water are thrust up by the ocean's waves and sparkle momentarily in the sunlight, only to return to that formless mass from which they arose.... Who can number the Indras who have ruled in a beginningless series, and shall rule afterwards in the endless future?
>
> ... Hold! I have spoken only of those worlds within this universe. But consider the myriads of universes that coexist side by side, each with its Indra and Brahma,

and each with its evolving and dissolving worlds. Stretching beyond the limits even of your mind, O great Indra, are those universes. Can you presume to know them, count them, or fathom the reaches of all those universes with their multitude of worlds, each with its legions of transmigrating inhabitants?[19]

Indra's once insufferable pride quickly crumbles as he comprehends the magnitude of the vision presented to him.

Although this chapter has focused specifically on analyzing myths and lore dealing with the creation of the universe, it would be incorrect to assume that this is the only kind of advanced knowledge symbolically encoded in myths of ancient Egypt and other cultures. For example, one myth about the eye of Atum appears to display a sophisticated understanding of human brain function. This story relates that mankind was created from the tears shed by the wandering eye of the high god, Atum, which had become enraged upon seeing that it had been replaced on the Great One's face by another. To appease it, Atum changed it into a rearing cobra that he then "elevated" and bound around his forehead. Accordingly, we often see Egyptian gods or kings bearing an aroused cobra protruding from the third-eye position of the forehead. Modern neurophysiologists have only recently delineated the brain's arousal network known as the *ascending reticular activating system,* a fiber bundle that extends upward from the "reptilian brain" (the medulla and spinal cord) to the "mammalian brain" (limbic region), and then forward to the prefrontal cerebral cortex. Moreover, only recently have we learned that this arousal system plays a central role in the process of creative thought formation by modulating and amplifying feeling tones cycling between the limbic and the prefrontal cortex.[20] Viewed in terms of modern scientific understanding, this ingenious allegorical tale seems to explain the origin of human intelligence, the unique characteristic that distinguishes humans from most animal species.

7
THE EGYPTIAN MYSTERIES

THE INITIATION

The sacred mysteries of the ancient Egyptians were kept in the trust of the magi, an initiated inner circle of priests who had dedicated themselves to the study of the universe.[1] The mysteries, as well as the myths, religious rituals, astronomy, geometry, medicine, and laws of governance of the Egyptians were attributed to the god Thoth, and this knowledge was known as the "science of Thoth." In the Egyptian pantheon, Thoth was the scribe of the gods, the messenger of the gods, the lord of divine books, the lord of law, the lord of divine speech, the patron of learning and intelligence, and the announcer of words of wisdom.[2] The magi also considered him to be the founder of their order. He was traditionally represented in the form of an ibis (figure 7.1) or sometimes in the form of a baboon. The Greeks named him Hermes Trismegistus (Thrice-Greatest Hermes), and the Egyptian mysteries attributed to him came to be called the Hermetic science.

The science of Thoth was conveyed from generation to generation under the cover of enigmas, allegories, and symbols whose meanings were known only to the magi. The Russian mystic P. D. Ouspensky offers the following explanation for the use of symbols in transmitting this knowledge:

Unlike despotic orthodoxies, a symbol favours independence. Only a symbol can deliver a man from the slavery of words and formulae and allow him to attain to the possibility of thinking freely. It is impossible to avoid the use of symbols if one desires to penetrate into the secrets (mysteries), that is to say, into those truths which can so easily be transformed into monstrous delusions as soon as people attempt to express them in direct language without the help of symbolical allegories. The silence which was imposed on initiates finds its justification in this. Occult secrets require for their understanding an effort of the mind, they can illuminate the mind inwardly, but they cannot serve as a theme for rhetorical arguments. Occult knowledge cannot be transmitted either orally or in writing. It can only be acquired by deep meditation. It is necessary to penetrate deep into oneself

Figure 7.1. The god Thoth (Hermes), founder of Hermetic philosophy and overseer of the Osirian mysteries. He offers a bowl containing the uas scepter and ankh, a motif depicting the mystery of the resurrection of order from the sacred ether flux. From Budge, Gods of the Egyptians, pl. 11.

in order to discover it. And those who seek it outside themselves are on the wrong path.³

The ceremonial transference of royal authority, which transmitted the Osirian mysteries from one dynasty to the next, encountered its first setback around 2250 B.C.E. The Heliopolitan regime, which had been in gradual decline, finally collapsed at the end of the Sixth Dynasty, bringing the Old Kingdom to an end. A period of feudal anarchy ensued for several generations, accompanied by class struggles and invasions of Asiatic tribes into the Nile Delta region. Order was reestablished around 2050 B.C.E. with the rise of the Middle Kingdom centered in Thebes.

Upon the collapse of the Old Kingdom, the Osirian ritual for the transmission of royal authority entered an uncertain future, although the passion of Osiris continued to be enacted by popular cults to celebrate the birth of the soul into the afterlife. The priests understood, however, that the legends and rituals they were preserving were also symbols for sacred metaphysical concepts, the details of which were kept secret from the general public.⁴ During this period of social unrest, the keepers of the sacred mysteries must have taken extra care to restrict access to their esoteric science in order to protect it from the profane who might otherwise misunderstand and distort its meaning. Those wishing to learn this ancient wisdom had to first prove that they were worthy. The aspirant first had to undergo a series of ordeals designed to test intelligence and strength of character. Each test he passed allowed him to advance one step further forward. But if he failed anywhere along the line, he was not given a second chance. In this way, only those possessing unusually high aptitude and moral strength were able to gain access to this sacred wisdom.

Very few of the priest class had the privilege of attaining the ranks of the magi. On occasion, foreigners were permitted to attempt the initiation rites, but only after first being carefully screened by an assembly of the sacred College of the Magi. Proclus relates that Thales, Pythagoras, Plato, and Eudoxus were among the Greek foreigners who became indoctrinated.⁵ All initiates were compelled by oath and penalty of death to keep silent about what they learned and were permitted to teach their learned principles in fable and parable only. Herodotus, a Greek historian living in the fifth century B.C.E. who had some knowledge of the Osirian mysteries performed in the Temple of Neith at Sais, disclosed his vowed reticence on the subject in the following passage:

> Here, too, . . . at Sais, is the tomb of one who[se] . . . name I prefer not to mention in such a connection; it stands behind the shrine and occupies the whole length of the wall. Great stone obelisks stand in the enclosure and there is a stone-bordered lake near by. . . . It is on this lake that the Egyptians act by night in what they call their mysteries, the Passion of that being whose name I will not speak. All the details of the performance are known to me, but—I will say no more.⁶

Nevertheless, a document relating the principal scenes of the initiation apparently did survive through the Middle Ages, its author thought to be Iamblichus, the fourth-century c.e. philosopher who founded the Syrian school of Neoplatonism. This treatise, entitled *An Egyptian Initiation,* was brought to public attention by the nineteenth-century French esoteric scholar Jean-Baptiste Pitois, who wrote under the pen name Paul Christian. His *History of Magic* (1870) contains a French version of Iamblichus's treatise, whose English translation was published in 1965.[7]

An Egyptian Initiation describes an initiation ordeal conducted in the Temple of the Mysteries, a subterranean labyrinthine sanctuary supposedly located beneath the Sphinx of Giza, which, by means of an underground passage, communicated with the nearby Great Pyramid. The treatise relates that the aspirant began at Memphis, a city located eleven miles south of the Giza temple complex. The two eldest magi, holding the title "Guardians of the Rites," would blindfold him so that "he would not know the distance traveled nor the secret place to which they were conducting him." They reportedly led the novice to the foot of the Great Sphinx, through a door situated between the Sphinx's forelegs, and into a subterranean chamber where the initiation would commence. After surviving a series of formidable trials and successfully traversing a long, tortuous passage with the aid of an oil lamp, the aspirant would arrive at a wooden door that opened into a long, narrow gallery. The walls of this gallery were adorned with a series of twenty-two frescoes, eleven on either side, which bore a striking resemblance to the twenty-two major arcana of the Gypsy Tarot. An initiate holding the rank of "Guardian of the Sacred Symbols" would greet the aspirant at the door and then proceed to instruct him on the metaphysical meanings of their symbolism. As is discussed more fully in the next chapter, these frescoes depicted metaphysical laws governing the universe and explaining its creation.

Iamblichus relates that after the postulant had fully absorbed the meanings of the twenty-two frescoes and committed them to memory, he was allowed to pass on his own through a door at the opposite end of the gallery. This led into a corridor where he was obliged to tolerate ordeals by fire, water, and darkness. After he had passed through this series of symbolic trials, he was greeted by a group of twelve sanctuary guardians bearing lamps. He was then blindfolded, escorted through an underground passage leading to the Great Pyramid, and brought to a crypt hollowed out in its center, where he came before an assembly of the College of the Magi. After first affirming vows of silence about what he had learned, the initiate was bestowed with a title that made him a member of the ranks of the initiated. With this his blindfold was removed.

This ancient initiation ceremony continues to be practiced even today, being a part of the secret rites that some freemason lodges use to initiate their new members. Ranking members dress up in costume and stand in line to represent the frescoes in the ancient gallery, the postulant being brought before each in sequence and instructed on the symbol's associated lore. Although little is known publicly about the particulars of this Masonic cermony as currently practiced, it is likely to correspond closely to the description given by Iamblichus. At least such may be concluded from studying a variation of this initiation treatise published under the title *Egyptian Mysteries* which is believed to have emerged at the end of the eighteenth century from among the circles of the Bavarian Order of the Illuminati.[8] It is something of a mystery where Christian obtained the Iamblichan treatise. The historian Christopher McIntosh writes that in 1839 Christian held a post in the library of the Ministry of Public Instruction, where he was given the task of cataloging books and manuscripts that had fallen into the possession of the French government after the confiscation of monastic libraries.[9] As he delved into this collection, he became fascinated by a number of unusual writings on forbidden esoteric subjects. In these suppressed writings, perhaps, lies the source of Christian's information.

The subterranean initiation ritual described by Christian/Iamblichus finds corroboration in an initiation rite described in an Egyptian Coffin Text that dates from around the twenty-second century B.C.E. This ancient text relates the story of the journey of a messenger, the "Divine Falcon," who enters the underworld by passing through a gate situated in the breast of a sphinxlike beast and there participates in the mystery of Osiris's resurrection. Additional corroboration for this underground initiation is found in the writings of Ammianus Marcellinus, a Greek-Syrian historian who lived around the time of Iamblichus. In book 22 of his history, after describing the pyramids of Giza, Marcellinus writes the following:

> There are also subterranean fissures and winding passages called syringes, which, it is said, those acquainted with the ancient rites, since they had foreknowledge that a deluge was coming, and feared that the memory of the ceremonies might be destroyed, dug in the earth in many places with great labour; and on the walls of these caverns they carved many kinds of birds and beasts, and those countless forms of animals which they called hierographic writing.[10]

THE SPHINX: PORTAL TO ANOTHER DIMENSION

No evidence has been found of the sacred entrance in the front of the Great Sphinx described by Iamblichus. Nor has an underground passage been discovered leading from the Sphinx to the Great Pyramid, which lies some six hundred meters to the northwest. Although seismograph measurements made in 1991 indicate "anomalies" or "cavities" situated deep in the bedrock between the paws of the Sphinx and along its sides,[11] further investigations are needed to determine whether these are an indication of the presence of underground chambers. Nevertheless, it should be kept in mind that the lower part of the Sphinx was usually inaccessible; it was covered by sand during most of the dynastic Egyptian era. Unlike the pyramids, which were assembled from stone blocks, the Sphinx's 65-foot-high, 240-foot-long form was carved from a natural limestone knoll, sculpted in such a way that the lower half of its body lay entirely below ground level in an excavated cavity. If not kept cleared, this cavity fills up with windblown sand in a matter of decades, leaving only the Sphinx's head exposed. Since it is a rather difficult task to clear away this sand, it was not done frequently. For example, John Anthony West estimates that the Sphinx remained unexcavated for at least three-fourths of the 2,400-year period between the time of Chephren and the Ptolemies (2700 B.C.E. to 300 B.C.E.).[12] Consequently, in order to conduct an uninterrupted sequence of initiations, the magi would have had to carry out their ceremonies in chambers situated elsewhere in the pyramid complex and to lead the blindfolded novice to believe that he had actually entered through the Great Sphinx.

Entry through the Great Sphinx, whether real or imagined, would have been very important for reasons of its symbolism. Both in ancient Egypt and in Mesopotamia, sphinx statues were placed at the entrances to temples and in front of city gates, where they stood as guards to ward off unwanted intruders. In a similar fashion, the Giza Sphinx guarding the entrance to the initiation chambers of the Temple of the Mysteries would have served as the "gatekeeper," protecting the inner sanctum where the ancient knowledge was kept. In fact, from its ancient Greek name, *sphinctre*, we have inherited the word "sphincter," which signifies a constriction that controls entrance (or exit) through a passage.

Just as the sphinx in Greek mythology permitted passage only to travelers able to solve its riddle and devoured those who failed, only those pilgrims who demonstrated sufficient acuity by passing a series of tests were permitted to learn the mysteries. The Great Sphinx, however, did not verbalize its riddle: the puzzle it posed to humanity was communicated instead through the symbolism of its stone form. P. D. Ouspensky made the following comment about the Sphinx:

> There is a tradition or theory that the Sphinx is a great complex hieroglyph or a book in stone, which contains the whole totality of ancient knowledge, and reveals itself to the man who can read this strange cipher which is embodied in the forms, correlations and measurements of the different parts of the Sphinx. This is the famous riddle of the Sphinx, which from the most ancient times so many wise men have attempted to solve.[13]

A clue to the meaning of the sphinx cipher may be found in astrology. In its complete representation, the sphinx is composed of the hindquarters of a bull, the forequarters of a lion, the wings of an eagle, and the head of a man. These correspond with the four fixed signs of the zodiac: Taurus (the Bull), Leo (the Lion), Scorpio (the Scorpion), and Aquarius (the Water Bearer), the Eagle being traditionally considered an alternate symbol for Scorpio. As discussed in chapter 9, the twelve signs of the zodiac are found to encode a metaphysics describing how the first particle of matter came into being, the same creation science portrayed in the myths of Atum and Osiris. These four fixed signs depict key stages in this matter-creation process: Taurus, the source of vital energy; Leo, the critical fluctuation; Scorpio, hierarchical restructuring; and Aquarius, the new state of order.

The Great Sphinx at Giza, which is composed of the body of a lion and the head of a man, presents just two of these concepts. Leo symbolically expresses the concept of the critical pulse of electric potential that has emerged from the ether and is about to disrupt the ether's uniform symmetry, while Aquarius represents the condition of the ether after its symmetry has been broken, when this critical fluctuation has grown into a macroscopic wave pattern—the primordial subatomic particle. Like Horus in the myth of Osiris, Leo is the warrior prince who ultimately resurrects the ordered regime—Aquarius. It is significant that the Sphinx's lower, lion half should correspond with the concept of the emerging fluctuation and that its upper, human half should correspond with the notion of the fully materialized primordial particle. The lower half of the Sphinx represents the spirit world, the realm of implicit order, the subquantum domain, while its upper half represents the world of the living, the realm of explicit order, the physical world.

In the story of Osiris, this emergent state of physical order is portrayed by Horus, who is the ruler of the royal kingdom. So it is not surprising to find the head of the Sphinx adorned with royal symbols, such as the trapezoid-shaped Nemis Crown, a white magical wig worn only by royalty (figure 7.2). The uraeus snake, the royal symbol of wisdom and transcendent action, also appears at its brow, although much of the uraeus was defaced in the eighteenth century when the Mamluks used the head as an artillery target. Incidentally, the word *uraeus* is derived from the ancient Greek root word that means "flux." This meaning is quite appropriate in this context, since the Osirian mysteries, said to have been revealed in the bowels of the Sphinx, essentially present a science founded on process.

The Giza Sphinx, then, is a hieroglyph that embodies in its form the advanced concept of how ordered patterns form in nonequilibrium systems, of how explicit order emerges from implicit order. It is appropriate that the Sphinx should stand guard over the sanctuary where initiates were introduced to the mysteries of cosmic creation. To draw a comparison, it is as if modern-day physicists had constructed in front of the entrance to their physics laboratory an immense stone monument expressing a fundamental physical concept they wished to preserve for posterity.

Ancient Egyptian writings reveal an additional reason the Sphinx was specified as the keeper of the subterranean entrance to the Temple of Osiris. After Osiris was murdered by Set, his soul is said to have gone into the Duat, the underworld. There within a cavern he lay helpless on his back, encircled in the coils of the serpent

Figure 7.2. The head of the Great Sphinx adorned by the Nemis Crown and uraeus snake.
Photo courtesy of Paul W. Wallace.

Nehaher, whose purpose it was to protect him from disintegration.* This cavern, or "cocoon," was said to reside within the body of the Aker (figure 7.3), a monstrous Siamese twin–like beast, each half of which was made up of the head of a man and the body of a lion. The Aker's heads and foreparts guarded the two entrances to the underworld at the edges of the earth, one to the east and one to the west. It was believed that at dusk the sun entered the underworld through the Aker's west gate, and after journeying therein would reemerge at dawn from its east gate.

The Sphinx at Giza may have originally been intended to represent one of the two Aker gates. Since it faces to the east, its creators perhaps intended it as a symbol of the eastern gate, the portal from which the sun emerges at sunrise. Hence like the horizon lions Shu and Tefnut in the myth of Atum (figure 6.7), the Great Sphinx signified the principle of creative emergence.

The ancient Egyptian priests must have viewed the Sphinx as a kind of dimensional portal, somewhat like the monolith stargate in Arthur Clarke's *2001*. By symbolically passing through it into the underworld, one crossed from the visible world

---

*The bandages that the Egyptians wrapped about their dead to embalm them may have symbolized this entropy-inhibiting snake (R. T. Rundle Clark, *Myth and Symbol in Ancient Egypt* [New York: Thames and Hudson, 1959], pp. 154–55, 161).

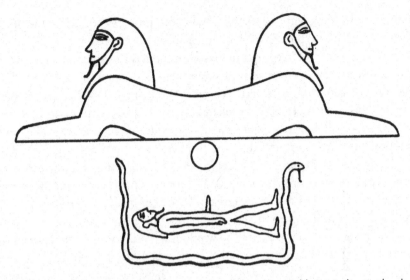

Figure 7.3. The Aker, guardian of the two gates to the underworld. Osiris lies in the dark depths below, enfolded by a serpent. From the Book of Caverns. Adapted from Rundle Clark, Myth and Symbol, fig. 23.

of three dimensions into the higher dimensional realm of the etheric abode. Although the Duat was symbolically below ground level, this was only a metaphor intended to convey the idea that this realm, while closely bordering our physical world, lay "beneath" or beyond it in the sense of not being directly observable from it. The underworld did not carry the same negative connotation as the Christian concept of hell. It was conceived to be an intermediary region of the afterlife in which the soul could work toward its own resurrection.

Besides serving as a gateway to the underworld, the Aker was also directly identified with it; its hieroglyph signifies both "ground" and "underworld." In effect, the Aker was an aboveground extension of the underworld. This symbolism is seen in the architecture of the Giza Sphinx. Carved from the limestone bedrock, its body is essentially an upward extension of the ground below. The lower, leonine half of its body was appropriately designed to lie entirely below ground level.

These insights into the Aker symbolism help to clarify why the ceremonies for initiating priests into the mysteries of Osiris were staged in a subterranean chamber located "below" the Sphinx. The Sphinx, like the Aker, served as the entrance to Osiris's underworld kingdom. In fact, Middle Kingdom funerary texts indicate that Rostau, as the Giza locale was called in ancient times, was considered the gateway to the Duat.[14] It may be significant that the initiation was begun after dusk. Perhaps at such an hour the Sphinx symbolized the western Aker gate, the one through which the setting sun was supposed to pass and through which the underworld could be accessed. In relating Iamblichus's account of the Egyptian mysteries, Christian does not connect the Sphinx with the mythical Aker gate, probably because at the time he was writing in 1870 this hieroglyph had not yet been deciphered.

The magi's practice of keeping the mysteries hidden from public view may not have been just for the purpose of protecting them from wanton distortion. There may also have been a more esoteric reason. The ether physics that the mysteries encode pertains to phenomena that take place at the subquantum level, a domain of existence normally inaccessible to direct observation. The underworld signified the unobservable subquantum ether, the implicit realm, while the daylight world where Horus and Set engage in their battles signified the domain of the observable physical world, the realm of the explicit. In moving downward from the physical world (Aquarius) to the subterranean realm (Leo), the aspirant to this secret order was symbolically passing from the world of observable phenomena into the underlying occult realm of the ether. There he would learn the ether's mysteries. His subsequent journey from the subterranean chambers beneath the Sphinx to the Great Pyramid (the symbol of emergence), and his confirmation there into the Order of the Magi, parallels the story of the resurrection of Osiris and elevation of Horus to the throne. It also celebrates the miracle of physical creation, the sacred emergence of a particle of matter in the midst of the fertile ether.

The Aker-Sphinx symbolism, which was so central to the Osirian creation science mysteries and so intertwined with ancient Egyptian lore, appears to have originated from the same predynastic cultural source that created the science-laden myths of Atum and Osiris. Some support for this suspicion comes from evidence that the Giza Sphinx is of prehistoric age. In 1961 Schwaller de Lubicz called attention to marks of deep-water erosion on the Sphinx's lower surface as evidence that this monument may be much older than modern archaeologists had supposed.[15] The presence of such erosional marks would be difficult to explain if the Sphinx were of a relatively young age, because Egypt has had a very arid climate for the past ten thousand years. Schwaller de Lubicz pointed out that wind could not have been responsible, since the erosion is confined primarily to the lower body, which was normally protected from the wind by the sand cover. He speculated that the cause may have been a prolonged regional inundation, citing Herodotus, who noted that the ancient Egyptian priests claimed that all lands north of Lake Moeris were but marshes at the beginning of their history.

In 1991 John Anthony West led a geological expedition to study the Sphinx to check out Schwaller de Lubicz's water erosion theory.[16] He and geologist Robert Schoch concluded that the two-foot-deep erosional channels cut into the body of the Sphinx and into the walls of the surrounding excavated hollow are due to precipitation-induced weathering. Their findings indicate that, at the very minimum, the Sphinx must be seven thousand to nine thousand years old. West favors a much earlier age predating the period at the end of the Ice Age when the climate in Egypt was much wetter, hence dating it prior to fifteen thousand to twelve thousand years ago. In fact, Lake Magadi, located in the Nile headwater region in Kenya, had an unusually high water level between 14,400 and 11,500 calendar years before the present (B.P.), an indication that this part of the African continent was receiving substantial precipitation during this interval.* Hence the Sphinx may be over twelve thousand years old, with a similar antiquity indicated for the symbols it expresses in its form. A comparably old date of 12,500 to 15,800 years B.P. is indicated for the formulation of the zodiac constellations.

THE JOURNEY OF THE DIVINE FALCON

The story of the journey of the Divine Falcon, found in Coffin Text 312, may be among the oldest evidence we have of the existence of initiation rites into the mysteries of Osiris. This dramatic ritual celebrating the resurrection of Osiris dates from the Intermediate Period following the collapse of the Old Kingdom. The story is particularly worthy of attention because it alludes to an initiatory journey through a sphinx gateway into the underworld.

This Coffin Text version of the Osirian passion differs from the Old Kingdom version in that, following his victory over Set, Horus does not himself travel down into the underworld to tell his father the news but instead dispatches a messenger to go in his place—the Divine Falcon.† The interjection of this proxy is understandable, since, in ancient Egyptian tradition, only the king had the prerogative to represent Horus. The individual who enacted the metaphorical descent into the underworld in this drama obviously was not the king but a priest, possibly one aspiring to become initiated into the Order of the Magi. Out of respect for tradition, he could not himself play the part of Horus.

According to the Coffin Text, after Horus had ascended to the throne, he wished to inform Osiris of his victory over Set, hoping that the good news might encourage Osiris to "set his soul in motion" and "make it come forth in control of its motion." So, taking the advice of his grandfather Geb (earth), Horus sends a messenger to the underworld to inform Osiris. This emissary is the Divine Falcon, "one of those glorious beings who dwell in beams of light" whom Horus has changed into a form like his own. This light being describes his transformation into the

---

*The lake's water depth, which was formerly near zero, rapidly rose to a depth of about thirty meters around 14,400 calendar years (B.P.), rose further to about sixty meters' depth around 13,800 years B.P., fell off to about ten meters' depth 1,100 years later, and rose again to about forty meters' depth by 11,500 years B.P. At present, the lake is only one meter deep (N. Roberts et al., "Timing of the younger Dryas event in East Africa from lake-level changes," *Nature* 366 [1993]: 146–48).

†This messenger theme is also encountered in the Old Kingdom Pyramid Texts in Pyramid Spell 606, where Horus enters the underworld in the capacity of a messenger of the God Above (Atum) (Rundle Clark, *Myth and Symbol*, p. 152).

Divine Falcon as follows:

> *I grew strong and waxed mighty*
> *so that I became different from the other dwellers in the light*
>   *beams*
> *who had come into being along with me,*
> *and appeared as a Divine Falcon.*[17]

The Divine Falcon would symbolize the critical fluctuation that has "waxed mighty" and spawned the ordered state. Like Horus, he would signify the emergence of light (explicit order) into the physical world, a realm formerly ruled by darkness and chaos (Set). In the context of the initiation ceremony, the Divine Falcon would also represent the aspirant who is about to set out on his underground journey to be initiated in the Temple of Osiris.

Just as the aspirant must first pass a series of examinations and become properly qualified before attempting this rite, the falcon must accumulate the proper set of credentials so that he may freely enter the gate of the underworld guarded by the Aker. In the beginning of his journey, he must first pass by the castle of the Leonine One (the Double Lion), where he is required to answer certain questions and give a convincing reason for his trip to the underworld. Once satisfied of the importance of the mission, the Leonine One gives the Divine Falcon the Nemis Crown, the same headdress that is worn by the Giza Sphinx. The Leonine One also gives the falcon wings so that he may fly on his journey to the house of Isis in the delta marshes, where he witnesses "her hidden mysteries" and watches the birth of Horus. Just as the aspirant is sworn to secrecy about the mysteries related to him in the Temple of Osiris, the Divine Falcon states that if he were to disclose what he saw, "columns of air would chase him away as a punishment for his audacity."[18]

The Divine Falcon then soars upward into the sky, where he encounters Nut, goddess of the heavens. Before he can continue on his journey he must obtain her permission for passage. The High God, Atum Re, intervenes by requesting Nut not to oppose the falcon, since he possesses the highest credentials. Nut gives her support and the falcon proceeds to the final checkpoint, the gate to the underworld guarded by the Aker. Atum then addresses the Aker on the falcon's behalf, saying, "Let the Sacred Way be open for him. . . ." The falcon then speaks to the beings who guard the gate for Osiris, saying:

> *I have come with the message of Horus for Osiris,*
> *I have taken the Hoary Heads and assembled the "Powers."*
> *Be off with you, O Gatekeepers!*
> *Here I am, make way for me!*

Let me pass through, O Cavern Dwellers who guard the House of Osiris![19]

The cavern dwellers and keepers of the house of Osiris then show themselves, whereupon the Divine Falcon lists his credentials and initiatory accomplishments and asks them to "open the secret caverns" and "reveal the mysteries" to him. The gods therein inform Osiris that the underworld has been opened to a newcomer, that the "*ways on earth and in the sky* have been revealed to him," and that "none has been able to oppose him." Finally the Divine Falcon meets the recumbent Osiris, urges him to rise up, and tells him the good news that his son, Horus, has been placed upon his throne. This triggers the sacred resurrection of Osiris.

The first phase of the falcon's journey to obtain entry to Osiris's tomb summarizes the key points in the drama of the emergence of the first particle of matter. The Double Lion beast who gives him the Nemis Crown is reminiscent of the horizon lions, which in the myth of Atum signify the dawning of light on the world, the first creation of matter. The visit to the mysteries of Isis indicates that the falcon has witnessed the birth and growth of the critical ether fluctuation. His subsequent flight up to the heavens to gain passage from Nut, the goddess who was separated from her brother, Geb, at the moment of creation, signifies the bivalent splitting of the uniform ether as X and Y diverge to form the primordial subatomic particle-wave.

So, by the time the falcon reaches the gate of the underworld, he symbolizes the "spirit of Horus," one who has challenged the powers of chaos and emerged victorious to establish rulership over a state of order. He represents the critical fluctuation that has emerged in the ether to compose the first particle of matter. Consequently, he has earned the right to wear the royal Nemis Crown, without which the Aker will not allow him to enter, for only those beings who symbolize the nucleation of explicit order are allowed to journey downward to visit Osiris and encourage him to rise up.

Just as the Divine Falcon must obtain the proper credentials to enter the secret caverns in the underworld to have the mysteries revealed to him, priests aspiring to be initiated into the mysteries of Osiris had to pass certain preliminary tests before they were allowed to gain entrance to the underground sanctuary for their formal initiation. The spirits that await the falcon within the underworld, the cavern dwellers and the guardians of the house of Osiris, find their counterpart in the magi, who, according to Iamblichus, play various roles in the subterranean initiation rites and who bear titles such as "Guardians of the Rites," "Guardians of the Sacred Symbols," and "Preservers of the Sanctuary." The "Sacred Way" that the falcon must tread on his underworld journey is the conceptual path that leads to Osiris. Esoterically, it comprises the sequence of metaphysical concepts that as a whole explain the mystery of the creation of ordered form.

The underworld journey of Horus described in the Pyramid Texts and the journey of the Divine Falcon described in the Coffin Texts both convey the important idea that Horus and Osiris are intertwined in a state of synergistic interaction. That is, on the one hand, Horus encourages Osiris through his divine messenger to rise up. On the other hand, the resurrected Osiris bestows upon Horus his nourishing divine essence, *ka*, thereby assuring that his son will have a fruitful reign. Since the enthroned Horus symbolizes the newly formed subatomic particle and Osiris symbolizes the underlying ether processes, this drama portrays the interdependent relation that exists between a physical particle and the ether processes that sustain its form.

When a proton comes into being, its core consumes G etherons at a high rate, thereby forming a surrounding G concentration well (gravity potential well) and a fertile supercritical pocket. So, by increasing its consumption of G, a primordial proton is able to create a local nurturing condition that stabilizes its form. The victorious Horus symbolizes the emerged particle, and the recumbent Osiris symbolizes the subcritical state of the reaction system. The news of his victory, which Horus wishes to convey to Osiris through a messenger, symbolizes the G-well that proceeds downward into the underworld as a result of the particle coming into being. Just as the lowering of the G concentration fires up the nonlinear properties of the ether reaction system and fosters the emergence of order, the messenger's good news encourages Osiris to rise up, become revitalized, and bestow his nourishing essence upon Horus.

This *ka*-bestowing gravity well also facilitates the emergence of other particles, thereby initiating the matter evolution scenario pictured in figure 5.11. Moreover, at later stages in galactic evolution, this same gravity potential effect spontaneously creates energy within celestial bodies, thereby accounting for the "excess" energy observed to radiate from stars, planets, and massive galactic cores.

Thus the mystery of the resurrection of Osiris encodes a highly sophisticated concept that has broad implications for a reaction-kinetic science of physical creation. The phenomenon whereby physical form (Horus) augments the reaction processes (Osiris) that originally gave birth to it is one of the most sophisticated concepts encountered in understanding the behavior of the Model G reaction system. This property was discovered only recently by pondering the implications of complex differential calculus calculations performed on the Brusselator reaction system.[20] How did the ancient authors of these symbologies acquire such technical sophistication?

There are other inscriptions from the Coffin Texts that comment on the soul's journey through the underworld and the mystery of the resurrection of Osiris. These too are of interest both for understanding the esoteric scientific significance of the Osirian mysteries and for comprehending the symbolic meaning of the initiation rites described by Iamblichus. These texts describe the underworld as subdivided into sections, each of which is guarded by frightening monsters. On its transformational journey through this netherworld, a soul must pass through a series of obstacles or gates to travel from one section to another, traversable only by means of the proper magic formula.[21] In a similar fashion, Iamblichus relates that the aspirant must pass through a series of doors in his trek through the labyrinth, some of which are opened by officers of the temple "only after having received a password and a recognition sign."[22] The ordeals that the novice must go through during his initiation in the Temple of Mysteries beneath the Sphinx, then, are symbolic of the frightening adventures that a soul must endure in journeying through the underworld to the cavern, or "Temple," where Osiris lies.

The Egyptian magi may have understood that the writings and rituals that described the soul's journey through the underworld also represent the transmutative etheric processes that sustain the universe. Just as the underworld is divided into sections separated by gates, the transmuting ether segments into an infinite series of states: A'', A', A, ... Ω, Ω', Ω'', and so on. The soul's transformational passage from one section to another effectively portrays an etheron's fourth-dimensional journey as it repeatedly disposes of its former etheric form to don a new one that characterizes its next state of etheric existence.

Subsequent texts, such as the Book of What Is in the Underworld (the Duat), substituted the journey of the soul through the underworld with the passage of the sun's disk (Atum).[23] Parchments show the solar disk being carried on a barque that navigates a river through the underworld. The text repeatedly describes this journey as "mysterious" and "hidden." The forward-progressing solar principle and river metaphor together evoke the notion of irreversible transmutation, a theme further portrayed by the description of the metamorphosis of the scarab beetle Khopri, "the Becoming One," who also travels along the river on a boat. The text traces his development from birth, through his larval stage as a water snake, and finally to his emergence out of primordial darkness. This development has clear links to the Egyptian creation cosmology, since Khopri's emergence from Nun was identified with the self-creation of the first god Atum, the ultimate creator of the universe. The identity of the etheric realm with the "underworld" domain of spiritual evolution indicates that the ancient Egyptians regarded the creation of the universe as a sacred process.

The Book of Caverns from the New Kingdom era, preserved on the walls of the Osiris temple at Abydos as well as in the tombs of Ramses VI and IX, gives us a glimpse of the drama of the revival of Osiris from his recumbent state.[24] The book describes the journey of the sun's disk as it passes through a succession of caves in the underworld. As the sun enters the dark cavern containing the body of Osiris and bestows upon him its energizing rays, Osiris's sexual organ becomes erect, a sign that the god has begun to revive. To arise, however, Osiris must first uncoil Nehaher, the serpent that binds him. The ancient Egyptians identified the coiled serpent with the concept of preservation and resistance to change, and the uncoiled

Figure 7.4. The straightening out of Apopis and the resurrection of Osiris. From the wall texts of Sarcophagus Hall 1. Adapted from Piankoff, Tomb of Ramses, fig. 127.

serpent with the unrestricted operation of universal process and life force. The primordial serpent in Egyptian mythology signified the concept of irreversible process in the ether; hence the serpent coiled around Osiris symbolizes the storage of etheric energy—the accumulation of a source of transmutative force. Its uncoiling symbolizes the release of that stored vital energy, the change from a condition of restricted etheric flux to one of free transmutation.

Since the victory of Horus and rising up of Osiris signify the emergence of ordered form, this uncoiling episode brilliantly encodes a concept well known to modern systems theorists. Namely, ordered form emerges only when the reaction rate of the open reaction-diffusion system is sufficiently high, that is, only when the system is operating sufficiently far from equilibrium to maintain itself in a supercritical state.

Osiris at first has difficulty freeing himself, for Nehaher transforms himself into Apopis, the dragon serpent of darkness, who is a manifestation of Set. While serving as the conservation principle (Nehaher), the serpent earlier aided Osiris by preventing his deterioration. Now it opposes him by resisting his efforts to rise from his current state of lifelessness.[25] Fortunately, Osiris is assisted by two ram-headed attendants who, just before the moment of dawn, straighten out Apopis (see figure 7.4). Osiris then is able to stand up, a sign of his resurrection from his deathlike near-equilibrium state. It is also significant that this event takes place at dawn, just as the sun is leaving the underworld to enter the world of the living. This suggests that the increased transmutative flux and consequent emergence of order brings light (explicit order) into the world.

8

THE TAROT
A Key to the Ancient Metaphysics

THE EGYPTIAN CONNECTION

The Tarot fortune-telling card deck differs from the modern playing card deck in that it has seventy-eight cards, rather than the familiar fifty-two. It consists of twenty-two picture cards, called the "major arcana," which are numbered 0 through 21, and fifty-six suit cards, called the "minor arcana," made up of four suits, each having fourteen cards (numbers 1 through 10, horseman, youth, queen, and king). The modern deck is believed to have descended from this sequence of minor arcana, with the Fool (or Joker) trump card being the sole remnant of the major arcana.

The Tarot card deck has been traced back to Renaissance Europe. The earliest known reference to playing cards is found in an essay by a Swiss monk of Brefeld who wrote:

A certain game called the game of cards has come to us in this year of Our Lord 1377. In which game the state of the world as it is now is most excellently described and figured. But at what time it was invented, where, and by whom, I am entirely ignorant.[1]

Some have speculated that the Tarot may have been carried to Europe from the East by merchant ships, which could explain why the earliest decks were disseminated from Marseilles and Venice. The theory most popular with taroists, however, is that it originally came to Europe from ancient Egypt: that Gypsies migrating from India to Egypt and then on to Europe had acquired knowledge of it during their passage through the Nile Valley. This idea was first suggested in 1781 by Antoine Court de Gébelin. According to Court de Gébelin, the word *tarot* was derived from the Egyptian *tar*, meaning "path," and *ro*, meaning "royal," hence

143

translating as "the royal path."[2] This calls to mind the Egyptian royal "path" that marked the symbolic sequence of events relating to Horus's ascent to the throne.*

Alliette, a French Hermeticist who also claimed to know of its Egyptian origin, subsequently declared that the Tarot must be a remnant of the legendary *Book of Thoth* that conveyed the ancient science of Thoth. Following this line of thought, the nineteenth-century French physician Gérard Encausse, writing under the pen name of Papus, conjectured that the Egyptian magi may have purposefully created Tarot cards as an alternate way of ensuring the survival of their sacred knowledge:

A time followed when Egypt, no longer able to struggle against her invaders, prepared to die honorably. Then the Egyptian savants held a great assembly to arrange how the knowledge, which until that date had been confined to men judged worthy to receive it, should be saved from destruction.

At first they thought of confiding these secrets to virtuous men secretly recruited by the Initiates themselves, who would transmit them from generation to generation. But one priest, observing that virtue is a most fragile thing, and most difficult to find, at all events in a continuous lineage, proposed to confide the scientific traditions to vice.

The latter, he said, would never fail completely, and through it we are sure of a long and durable preservation of our principles.

This opinion was evidently adopted, and game was chosen as the preferred vice. Small plates were then engraved with the mysterious figures which formerly taught the most important scientific secrets, and since then the players have transmitted this Tarot from generation to generation, far better than the most virtuous men upon earth would have done.[3]

Court de Gébelin's intuitions of an Egyptian origin for the Tarot were placed on a more secure footing by the French esoteric scholar Jean-Baptiste Pitois (Paul Christian), whose rendition of a treatise attributed to Iamblichus was mentioned in chapter 7. According to Christian/Iamblichus, initiation into the Temple of the Mysteries took place in a long, narrow gallery whose ceiling was sustained by twenty-four caryatid columns shaped in the form of sphinxes, twelve on each side. The walls of the gallery were said to be adorned by twenty-two frescoes positioned between the columns, eleven on each side, and illuminated by eleven sphinx-

---

*William Postel and J. Vaillant have suggested that the Tarot has also gone by the names Rota and Thora (Papus, *The Tarot of the Bohemians: The Most Ancient Book in the World*, trans. P Morton [Hollywood, Calif.: Wilshire, 1889, 1973], p. 9). The first is traced to the Latin word meaning "wheel," while the second word has been associated with the Torah, the body of wisdom and law contained in the Jewish scripture.

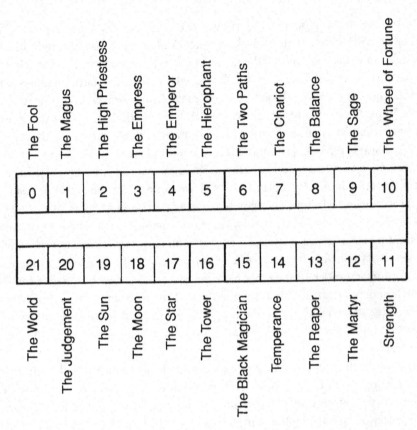

Figure 8.1. The probable arrangement of the Tarot frescoes in the gallery of the sphinxes found in the legendary Egyptian Temple of the Mysteries. The corresponding names are those given in Christian's account of Iamblichus's treatise.

shaped crystal oil lamps placed along the center of the corridor (figure 8.1).*

The Iamblichan treatise's fresco descriptions closely match the appearances of the correspondingly numbered major arcana in the Marseilles Tarot, the card deck considered to be the most exact reproduction of the oldest Gypsy Tarot decks. Many taroists thus argue that the European Tarot preserves a pictorial record of these ancient temple frescoes, collectively termed the "Egyptian Tarot."

---

*Wirth has suggested a similar layout, except that he placed the Fool at the end of the sequence, following Arcanum 21 (O. Wirth, "Essay upon the astronomical Tarot," in *The Tarot of the Bohemians*, pp. 242–43). The layout proposed here would instead place this card at the beginning of the sequence. With this arrangement, the pictogram series describing physical creation would run along the north wall and would ascend from west to east, while the series describing human evolution and spiritual development would run along the south wall and would ascend from east to west.

According to Iamblichus, the twenty-two Egyptian frescoes depicted "the Science of Will."[4] Since "will" signifies the act of attaining a goal through directed effort, we may translate this as "the science of directed effort" or, alternatively, "the science of irreversible process." In fact, as we will discover, the Tarot metaphorically encodes the same process-based creation metaphysics conveyed in the myth of Osiris. In particular, many taroists teach that Arcana 1 through 10 portray principles governing the creation of the physical universe, while Arcana 11 through 21 describe matters relating to human evolution and spiritual development.

Aided by the recent development of theories that explain the emergence of ordered patterns in nonequilibrium systems, we can now for the first time resurrect the Tarot's ancient wisdom. The symbolic images portrayed in Arcana 1 through 10 present a precisely ordered array of systems principles describing how a primordial transmuting ether spawns physical form, subatomic matter. These naturally divide into two parts. The first five arcana (1 through 5) appear to describe the functional order, or implicit order, of the ether reaction system, and show how a growing seed fluctuation develops from this ether at a time when the ether is still uniformly distributed. The second five arcana (6 through 10) appear to describe how this growing fluctuation bifurcates the uniform state to form an ordered wave pattern, the ether's explicit order.

The arcana illustrated throughout this chapter are taken from the Marseilles Tarot as it appears in Papus's *Tarot of the Bohemians*. Displayed to the right of these images, for the purpose of comparison, are drawings that attempt to reconstruct the appearances of the Egyptian temple frescoes according to the descriptions given in the Iamblichan treatise.* The names of the frescoes and the accompanying descriptive quotes are from that document. The meanings assigned to the major arcana and their symbols are taken from several sources (for example, Papus, Zain, and Doane and Keyes). Finally, the arcana are matched with archetypal systems principles that express similar ideas. These concepts join smoothly together to describe a science of order genesis.

In recording the meanings that are traditionally assigned to each Tarot arcanum, we must rely on relatively recent sources, whether they are written or oral. That is, there are no ancient written records like the ancient hieroglyphic texts describing Egyptian myths. This is a lore that has been transmitted from earlier times primarily

---

*Along with the twenty-two Marseilles Tarot arcana, Papus also portrays a slightly modified version of the Tarot designed by Oswald Wirth that incorporates some symbolic modifications suggested by the research of Eliphas Levi. Wirth's changes mimic many of the features disclosed in Iamblichus's description of the Egyptian frescoes (*An Egyptian Initiation,* trans. P. Christian and G. S. Astley [Denver: Edward Bloom, 1965]). Rather than reproduce Wirth's plates, however, I chose to design a new set of illustrations based on descriptions given in the Iamblichan treatise. These fresco portrayals are similar to those found in Zain's *Sacred Tarot,* with some simplification and a few other modifications.

by word of mouth and, as a result, could have suffered some degree of distortion during the course of its conveyance. Any such distortions must have been minimal, however, for if errors had crept in and substantially altered the meaning of the arcana, we would have been left with a nonsensical conglomeration. We find instead that the arcana convey in a clear, orderly manner universal principles that together portray the process of how wave order self-organizes in a nonequilibrium reaction system, essentially the same cosmic science that is portrayed in the creation myth of Atum and the story of Osiris.

Contrary to the more popularly held view that the Tarot is of ancient origin, the esoteric scholar Richard Cavendish has speculated that the Tarot may be a product of the intellectual climate of Renaissance humanism, and that it was invented in the north of Italy as a means of conveying teachings about the world that could not be easily expressed using a fifty-two-card deck.[5] He proposes that the twenty-two major arcana were developed separately from the minor arcana, as a later addition, and that the card deck probably got its name from its use in the valley of the Taro River, a tributary of the Po between Genoa and Bologna. This Italian-origin theory seems rather implausible in light of the present discovery that the Tarot portrays an advanced creation science. For then one would have to explain how, more than one hundred years before Galileo, Renaissance Italians could have concocted a science that anticipated twentieth-century nonequilibrium thermodynamics concepts and at the same time kept its knowledge secret from the rest of the world. Since the Tarot's creation metaphysics so closely parallels that found in ancient creation myths, an earlier origin seems more likely. Moreover, Iamblichus's treatise on the Egyptian mysteries suggests that the Tarot is of considerable antiquity.

THE SYMBOLOGY OF THE TAROT, PART I

Arcanum 0

Before deciphering the Tarot's creation science, let us study Arcanum 0, variously named the Fool, the Materialist, or the Crocodile. Although this arcanum does not appear to be part of this story of order emergence, it is useful to analyze it because it portrays the antithesis to order creation, order dissolution.

There is some ambiguity as to the placement of Arcanum 0 in the Tarot sequence; some decks number it 0, others 22. Its uncertain numerical position partially stems from the tradition of regarding this card as an out-of-sequence maverick standing in isolation from the other twenty-one major arcana. As in the case of the Joker or "wild card" in modern playing card decks, Arcanum 0 enjoys the status of a wanderer.

This arcanum (figure 8.2) depicts a blind wayfarer who is cut off from perceptual contact with his environment. Rather than using his walking stick as a blind

The Fool, The Materialist
Attributes described by taroists

Overemphasis on material life with neglect of the spiritual

One who is slave to desire and blinded by self-interest

Failure, folly

The Crocodile
Egyptian fresco description from Iamblichus:

You see here a blind man, carrying a full wallet and who is going to stumble against a broken obelisk, upon which is resting a crocodile with gaping jaws (*An Egyptian Initiation*, p. 36)

Figure 8.2. ARCANUM 0. Systems concepts: the isolated system; dissipation.

person would do, he instead drags it at his side. In so doing, he isolates himself tactilely as well as visually. In proceeding, he appears to be indifferent to dangers in his environment. Although a dog attacks him, clawing his leg and ripping his pants, the man pays little attention. The Egyptian fresco version shows him about to stumble over a broken obelisk and encounter a crocodile who waits with open jaws ready to devour him. All of these metaphors indicate that he is both mentally and sentiently isolated from his environment. Also, the wayfarer is seen to carry all of his belongings with him in a bag slung over his shoulder, suggesting self-sufficiency and independence from his social milieu. Hence the Fool appears to be socially cut off from his environment as well.

From a systems science standpoint, this arcanum expresses the concept of a *closed or isolated system,* a system totally cut off from any exchange with its environment. In a cosmological context, Arcanum 0 could represent the etheric realm at a time when it was in a totally self-contained unexpressed state, comparable to the ancient Egyptian concept of the primordial wasteland Nun prior to the emergence of Atum, the Prime Mover.

Arcanum 0 also depicts the important concept expressed in the second law of thermodynamics, namely, that ordered configurations in isolated systems ultimately tend toward disorder. There are numerous signs in the traveler's environment that connote the destruction of order: the degenerate condition of the man's clothes; the fallen obelisk; the crocodile with open jaws; and in the Marseilles version, the dog scratching or biting the Fool's leg. The erect obelisk was equated with the Benben stone, the Egyptian symbol for the emergence of the primordial order-creating flux. A broken and downfallen obelisk would instead portray the destruction of order, the state of ruin that prevails when this regenerative flux is absent. Moreover, the crocodile was the Egyptian symbol for death. It was also the form taken by the god Sebek, who signified the abyss. Set sometimes took a crocodile form, although more commonly he was depicted as doglike, possibly in connection with the dog seen in the Marseilles version. By means of these simple images, Arcanum 0 brilliantly depicts the principle of dissipation or entropy increase, a concept of central importance to any theory of order creation.

Arcanum 1

Like the principle of isolation that it expresses, Arcanum 0 is isolated from the arcanum sequence that follows it. The story of creation makes its true start with Arcanum 1, which describes the concept of the open system undergoing continuous exchange with its environment—the antithesis to the isolated system concept. Here the principle of dissipation is applied to constructive ends in the form of directed process.

The Marseilles arcanum depicts a magician wearing a hat in the form of an infinity sign (figure 8.3). He gestures upward with a magic wand that he holds in his left hand. Before him on the table lie artifacts that he uses in his tricks of transformation, an assortment of cups, coins, and knives.

The Egyptian fresco depicts a similar scene. A magus (priest) stands before a cubical stone altar. With one hand he holds a golden scepter upraised toward the heavens, while with the other he points downward toward the ground.* On the altar rest a

---

*Iamblichus describes the magus as holding the scepter upraised in his right hand and pointing down with his left, just the reverse of the handedness shown in the Tarot of Marseilles. Of these two renditions, the Marseilles version is more nearly correct because it is traditional to regard the left half of the body as the receiving side and the right half of the body as the giving side. Hence the magus should be directing the transmuting flux from his left arm down through his right arm, as is shown in the version of the fresco reconstructed on the following page.

The Magician

Attributes described
by taroists:

The Prime Mover, the creative force
The unity principle, singleness
Aspiration, will, the action principle, initiative
Beginning, potential, impulsiveness
The masculine quality

The Magus

Egyptian fresco description
from Iamblichus:

Arcanum 1 is expressed by a Magus, archetype of the perfect man, that is to say, in full possession of his moral and physical qualities. He is represented standing; it is the attitude of the Will that precedes action. His robe is white, image of purity, original or reconquered. A serpent biting his own tail serves him for a girdle.... His forehead is girt with a circle of gold. The right hand of the Magus holds a scepter of gold. . . . He raises it toward heaven. . . . The left hand extends the index finger to the earth. . . . Before the Magus upon a cubic stone are placed a cup, a sword and a shekel, a piece of gold money in the center of which is engraved a cross. (*An Egyptian Initiation*, pp. 14–15).

Figure 8.3. ARCANUM 1. Systems concepts: irreversible process;
the open system.

sword, a goblet, and a large gold coin engraved with a cross. The Magus wears a white robe that is secured about his waist by a serpent biting its tail, the ancient Egyptian symbol of eternity and also the symbol for the cosmic ocean, Nun.

Attributes normally assigned to Arcanum 1, such as "creative force, divine will, the principle of action, potential, unity," together convey the general principle of directed process. Translated into systems terminology, the Magus (or Magician), depicts the concept of *irreversible process*. Various symbols portrayed in this arcanum, which are analyzed below, further imply that this directed action is fundamentally alchemical in character and indicate that the Magus signifies the primordial transmuting ether.

Papus equates the Magician of Arcanum 1 with Osiris, the Lord of Order.[6] This is a reasonable identification, considering that the Egyptian Tarot depicted the esoteric significance of the mysteries of Osiris and Arcanum 1 expresses the masculine principle of action and creative force, characteristics portrayed by Osiris. As noted in the previous two chapters, Osiris esoterically symbolized the transmuting, regenerative primordial ether. A similar meaning would apply to Arcanum 1.

The creative ether concept is further indicated by the scepter that the Magus raises heavenward. The *uas* scepter was the Egyptian symbol that signified the vivifying primordial ether and the concept of universal creative activity (chapter 2). Egyptian frescoes that depict the erect "resurrected" Osiris usually show him gripping such a scepter, which signified the god Set in his tamed state (see figure 2.4). At the time of his revival from his state of unconsciousness, Osiris is advised to "keep hold" of Set. In Osiris's grip, the tamed Set becomes transformed from the principle of decay (Arcanum 0) into regenerative flux, and Set's destructive nature is redirected toward constructive ends. The Magus (Osiris) holding his scepter portrays a similar concept.

It is also significant that the Egyptian Magus holds a scepter made of gold. The ancient *uas* scepter, although made of wood, was gilded to symbolize this sacred element. In ancient Egyptian Hermetic lore, gold signified spirit and immortality. In the tradition of alchemy, it symbolized the elemental state of divine perfection that could be attained through the value-enhancing transmutation of base metals into gold. Similarly, the golden scepter would symbolize the notion of ascendant transmutation, an idea emphasized by the Magus's upward-extended arm.

The Magus encircled by the snake biting its tail is reminiscent of frescoes that depict Osiris enclosed and protected by the serpent Nehaher. The enclosing snake also calls to mind the cosmic serpent Iru-to (chapter 6), who arose from the primeval waters at the dawn of time to create the primeval outflow with his self-closing form delimiting the creation. Hence the snake that surrounds the Magus would signify the eternal irreversibly transmuting ether, as would the Magus himself. Both Magus and snake are one and the same, an identity implied in Spell 321 from the Coffin Texts, where the cosmic serpent states: "I extended everywhere, in accordance with what was to come into existence. I bent right around myself. I was encircled in my coils; one who made a place for himself in the midst of his coils."[7]

The cubical altar, or table, before the Magus is said to represent the physical plane, the cube being the symbol for three-dimensional space and physical form. The knives, cups, and coins on its surface correspond with three of the four Tarot card suits: swords (spades in modern card decks); cups (hearts); and coins (diamonds). Taroists associate swords with the element earth, symbolic of physical things; cups with the element water, symbolic of life; and coins with the element air, symbolic of ideas and values. So the objects on the table signify the three primary vectors of physical evolution: the inanimate material vector, the life vector, and the mental vector (see figure 3.2). Their presence on the altar surface indicates that these evolutionary pathways all unfold within the confines of the physical plane.

The Magus's scepter depicts the remaining Tarot card suit, scepters or wands (clubs in modern card decks). Taroists traditionally identify the suit of scepters with the element fire, a metaphor for the transmuting ether. This evokes the story of the self-cremating, self-regenerating phoenix, the form Atum assumed when he first emerged from Nun. Appropriately, the scepter is displayed in the Magus's upraised hand rather than on the altar or table, for the transmuting ether extends beyond the dimensions of the physical plane. By holding the scepter aloft and gesturing downward with his other hand, he indicates that the ether's transmutative flux sustains the physical universe through which it passes.

Just as a magician, through his tricks of transformation, makes objects appear, disappear, and change in form, the Magus of Arcanum 1, who portrays the universal etheric flux, continuously recreates the changing physical world. The objects on the table, which represent various categories of physical manifestation in the illusory realm of *maya*, continuously evolve and transform through his touch.

It is fitting that the Magus is a male figure and that taroists traditionally associate Arcanum 1 with the masculine principle operating in nature. The masculine stereotype and male procreative act traditionally connote outwardly projected, goal-oriented activity, a suitable metaphor for depicting unidirectional creative process. The Magician holds his wand in his *left* hand, however, which denotes a feminine quality. That is, the left side of the body traditionally symbolizes the feminine, unconscious, or hidden aspect of a person's personality (table 8.1), a convention that accords with findings on left-right brain specialization. In this way, the Magician seems to be indicating that the ether transmutation is unobservable from our physical perspective.

It follows from this that the Magus and his altar (or table) with its displayed items portray the thermodynamic concept of an open system. Whereas Arcanum 0 depicts dissipative process in its destructive mode as it erodes order in isolated systems, Arcanum 1 depicts process in its constructive mode, where it is harnessed in an open system to create order.

Although the open-systems concept is generally applicable to a variety of natural systems, the Arcanum 1 symbology refers more specifically to reaction or transformation-type processes taking place in a universal ether. In the context of Model G, the Magus would represent the transmuting ether that extends "above" and "below" physical creation along the transformation dimension (figure 4.4). The

TABLE 8.1
Masculine and Feminine Symbolic Stereotypes

FEMININE	MASCULINE
left-hand side	right-hand side
the moon	the sun
passivity	activity
hidden activity	expressed activity
reception	projection
attraction	volition
involuntary	voluntary
invisible	visible
the implicit	the explicit
the unconscious	the conscious
intuition	logic

altar would signify the *G-X-Y* juncture in the ether transmutation network where the two primary transmutation pathways ($A \longrightarrow \Omega$ and $B \longrightarrow Z$) cross and interact with one another to spawn our physical universe.

Arcanum 2

Arcanum 2 (figure 8.4) depicts a high priestess seated with an open book in her lap. According to Iamblichus, her face is covered by a semitransparent veil, and the book in her lap is half-hidden by her cloak. Her tiara is surmounted by a waxing crescent moon, denoting the feminine principle. She sits at the threshold of her sanctuary, the Temple of Isis, and is flanked by two columns. Iamblichus describes these columns as being red and black, with red signifying "pure spirit and its ascension over matter" and black signifying "the captivity of spirit in matter." Hence the columns symbolize the polar opposite principles of light and dark.

The High Priestess represents Mother Nature. Just as the majority of an iceberg lies submerged and out of sight, nature's multifarious aspects for the most part lie hidden from direct view. The following passage from W. Y. Evans-Wentz's book *Tibetan Yoga and Secret Doctrines* expresses quite well the esoteric meaning of the veil of Isis and how it might be penetrated:

> *Maya* is the Magic Veil, ever worn by Nature, the Great Mother Isis, which Veils Reality. It is by *yoga* alone that the Veil can be rent asunder and man led to self-knowledge and self-conquest, whereby Illusion is transcended. Ultimate Truth is illusorily ever associated with Error; but like an alchemist of things spiritual, the

The High Priestess, Veiled Isis

Attributes described by taroists:

Duality, polarity, form
Receptivity, passivity
Perception, discrimination
Knowledge, insight, intuition
Occult science
The feminine quality

The Door of the Occult Sanctuary

Egyptian fresco description from Iamblichus:

Arcanum II is expressed by a woman seated at the threshold of Isis, between two columns. The column on her right is red. . . . The column on her left is black. . . . The woman is crowned by a tiara, surmounted by a lunar crescent and enveloped in a veil, the folds of which fall over her face. She wears upon her bosom a solar cross and carries upon her knees an open book which she half covers with her mantle. (*An Egyptian Initiation*, pp. 16–17)

Figure 8.4. ARCANUM 2. Systems concepts: duality; differentiation.

master of *yoga* separates the dross, which is Ignorance, from the gold, which is Right Knowledge. Thus, by dominating Nature, he is liberated from enslavement to Appearances.[8]

Iamblichus expresses similar ideas in his discussion of Arcanum 2:

The veil enveloping the tiara, and falling over the face, announces that truth hides itself from the gaze of profane curiosity. The book, half-hidden under the mantle, signifies that the mysteries are revealed only in solitude to the sage who meditates in silence in the full and calm possession of himself.[9]

Consequently, Arcanum 2 appears to depict the Hindu doctrine of *maya*, also known in Tibetan Buddhism as the doctrine of the illusory body.

Papus and other taroists equate the High Priestess with Isis; this is consistent with their identification of the Magus with Osiris. Arcana 1 and 2, then, signify the masculine and feminine universal principles, activity and form, yang and yin. Ancient cosmogonies often depict these polar opposite principles as two aspects of a single cosmic quality. For example, the Hindu concept of *maya* involves such a duality. Evans-Wentz states:

> Primitively, *maya* denoted a form of intelligence, energy, power *(shakti)*, and deception. It chiefly implied a mysterious will-power, whereby Brahman wills and as *maya*, form is realized. As denoting deception, *maya* refers to that magical glamour of appearances which causes the unenlightened percipient of them to conceive multiplicity and duality as being real. As the supreme magician, *maya* produces the great cosmic illusion, the universe of phenomena.[10]

Consequently, the Magician represents the causative, divine-will aspect of *maya* as *shakti* (wondrous power), whereas the High Priestess represents the effective aspect of *maya*, the phenomenal creation, the world of illusion and appearance. The standing Magus signifies the active, creative aspect, whereas the seated High Priestess denotes its passive, formative aspect.

The two columns flanking the High Priestess signify the duality of appearance and reality. Seated at the threshold to the Temple of Isis, she separates the illusory realm before her from the realm of truth within. Arcanum 2, then, portrays the process of distinguishing among alternatives. Taroists acknowledge this theme by associating the High Priestess with perception, discrimination, and knowledge, traits symbolized by the book lying open in her lap. This perception theme is very appropriate, for the mind perceives forms and acquires knowledge about the physical world through the process of distinction, that is, by discriminating differences of kind, such as light versus dark or red versus blue. Consequently, Arcanum 2 metaphorically expresses the archetypal concept of *differentiation*.

In terms of the Tarot's ether physics, Arcanum 2 represents the differentiation of the initially unitary primal flux (Arcanum 1) into numerous distinct transformation pathways that together form the ether transmutation network. Such differentiation is important to the ether's ability to develop an implicit order. Arcanum 2 also expresses the principle of duality, a concept closely related to differentiation and very important to a reaction-kinetic ether physics. To generate ordered subatomic particle concentration patterns, an ether reaction system must contain two interacting ether transformation pathways (for example, $A \longrightarrow G \longrightarrow X \longrightarrow \Omega$ and $B + X \longrightarrow Y + Z$) and must include two mutually interacting reactants (X and Y).

By having her face veiled and the book partly hidden, Isis indicates that her implicit reaction-kinetic order is not outwardly evident. That is, we can become aware of the presence of this implicit order by perceiving the explicit order that it creates. Moreover, just as conscious experience arises from numerous subtle distinctions and associations residing in the unconscious, physical form similarly arises from differentiations in the ether that are beyond direct perception. At this early stage in the ether metaphysics, explicit order is not macroscopically apparent, since matter has not yet formed. Nevertheless, it is present at a microscopic level in the form of subquantum ether fluctuations, the seeds that give rise to physical form.

Arcanum 3

Having established the principles of transmutative process and transmutative duality, we are now ready to consider the next logical step, namely, the union of separate transmutation pathways expressed by Arcanum 3.

Arcanum 3 (figure 8.5) depicts an unveiled crowned woman seated on a throne. This is the Empress. Her left hand holds a scepter, symbol of the ether element, while her right arm protects a shield emblazoned with an eagle. The Egyptian version described by Iamblichus pictures a woman seated at the center of a blazing sun, her feet resting upon a crescent moon, the symbol for birth. One arm holds a scepter, and the other bears an eagle. The woman is identified as the celestial goddess Isis and is said to personify universal fecundity. Pictured together with the sun (symbol of Osiris) and moon (symbol of Isis), the Empress portrays the union of masculine and feminine polarities. Arcanum 3 is identified with the archetypal concept of process coupling, a principle necessary to the construction of any order-generating reaction system.

The orb-and-cross symbol that surmounts the Empress's staff portrays a similar coupling idea. This motif is an inversion of the Egyptian ankh, which traditionally signified the union of the male and female principles and denoted universal fecundity and abundance.[11] Since the cross is the esoteric symbol for matter, its placement at the crown of the scepter suggests that this union of opposites is about to give birth to physical form.

As in Arcanum 1, the scepter signifies the transmuting ether. In combination with the orb and cross, it depicts the union of reaction processes taking place in a transmuting ether. By holding the scepter in her left hand, the Empress indicates that the reactive coupling she represents occurs at a subtle level and is invisible to our physical senses. Interpreted in an ether kinetics context, she portrays the union or coupling

The Empress, Isis Unveiled

Attributes described by taroists:

The active union of male and female principles
The union of polar opposites
Marriage, partnership
Fecundity, gestation, generation

Isis

Egyptian fresco description from Iamblichus:

Arcanum III is figured by a woman, seated in the center of a radiant sun; she is crowned with twelve stars and her feet rest upon the moon. It is the personification of universal fecundity. The Sun is the emblem of creative power. This woman, celestial Isis or Nature, carries a scepter surmounted by a globe. . . . In the other hand she carries an eagle. (*An Egyptian Initiation*, p. 18)

Figure 8.5. ARCANUM 3. Systems concepts: coupling; supercriticality.

of ether transmutation processes involved in the creation of the physical universe. For example, in Model G (figure 4.2), X couples the two primary reaction pathways:

$$A \longrightarrow G \longrightarrow \boxed{X} \longrightarrow \Omega$$
$$B + \boxed{X} \longrightarrow Y + Z$$

Both X and Y also couple two reactions to produce a self-closing loop:

$$3X \longleftarrow Y + 2X \quad \text{(“feminine” reaction)}$$
$$B + X \longrightarrow Y + Z \quad \text{(“masculine” reaction)}$$

Such a two-variable loop is a necessity if a reaction system is to produce a dissipative space structure with regular periodicity. A similar bivalent alchemical union is indicated in the nonsexual merging of Isis and Osiris described in the Osiris creation myth.

The eagle that the Empress cradles in Arcanum 3 would signify Horus, the warrior son that Isis bore as a result of this union.* Thus, as in the story of Osiris, the Empress and her eagle child depict the process of virgin birth. Like Horus, the eagle represents the spontaneously arising ether fluctuation, the electric potential impulse that eventually grows in size to form the first particle of matter. The eagle appears on her right, the side traditionally associated with explicit order.

In nonlinear reaction systems, when a positive feedback process is present in the form of a self-closing reaction loop, incipient concentration fluctuations can be nurtured and caused to grow in size, it is probably no coincidence that this arcanum combines the notion of the alchemic union of opposites (the cross-catalytic loop concept) with the theme of the mother nurturing her child. It is also significant that the Empress shelters the eagle with her arm. This may mean that growth is facilitated when the incipient fluctuation emerges in a serene environment, protected from the eroding action of adversary fluctuations.

Arcana 1 through 3 consecutively present the concepts of transmutation, differentiation, and coupling. These three constitute the basic elements needed in forming a structured ether transmutation network. By means of these three fundamental systems principles, the implicit order of the universe comes into being. Having explained the essence of the ether's implicit order and shown how this "etheric womb" nourishes and protects the emergent fluctuation child, the Tarot is now ready to trace the development of this fluctuation as it grows in size and reaches maturity.

Arcanum 4

Arcanum 4 of the Marseilles Tarot portrays the Emperor as a bearded man wearing the helmet of a warrior surmounted by a crown (figure 8.6). He holds a scepter in his right hand and partially sits or leans upon a chairlike throne with his right leg bent over his left. His throne, in turn, is elevated from below by an eagle with spread wings. The corresponding Egyptian fresco depicts a similar scene, except that the crowned man sits instead upon a cubical stone.

The eagle supporting the Emperor's throne is the same one emblazoned on the shield the Empress embraces. Consequently, the Emperor is of the same lineage as the Empress; he is, in fact, her son, Horus, the son of Isis. The eagle child, which was

---

*Although Horus is usually represented in Egyptian mythology as a hawk, on occasion he also appears in the form of an eagle.

The Emperor, The Sovereign

Attributes described by taroists:

Realization, result, solution
Triumph over obstacles
Confidence, authority, abundant creative energy
Knowledge gained through experience
The setting of things in order

The Cubic Stone

Egyptian fresco description from Iamblichus:

Arcanum IV is figured by a man, on his head is a crowned helmet. He is seated upon a cubic stone; his right hand raises a scepter and his right leg bent, rests upon the other in the form of a cross. This ruler is in possession of the scepter of Isis. (*An Egyptian Initiation*, pp. 19–20)

Figure 8.6. ARCANUM 4. Systems concepts: the critical fluctuation; critical mass.

being nurtured and protected by the Empress in Arcanum 3, is displayed here full grown as the mature sovereign. Hence the Emperor and the eagle are one and the same. Noted for its ability to soar to great heights, the eagle, like the hawk, connotes the notion of emergence. As a symbol of prowess, the eagle also connotes the independent, undefeatable warrior. By supporting and elevating the throne, the eagle confers upon the Emperor these traits of emergence and invincibility.

There is a close resemblance between Arcanum 4 and Egyptian temple paintings of Horus as the mature warrior, Heru-khuti. Comparing the Egyptian fresco with figure 6.9, in each case the hero is seated on a cubical throne, holding a weapon upraised in his right hand. Although Heru-khuti is depicted here with a spear, some ancient frescoes show him grasping a club or mace like that held by the Emperor. Another similarity is that both figures are crowned, the Emperor with a helmet-crown headpiece and Horus with a solar disk encircled by a serpent, or in some cases with a tall pointed hat. It is in the form of Heru-khuti that Horus, the champion of order, wages war against Set, the god of disorder. Like Horus, the Emperor represents the seed of physical form ready to emerge. Such parallels between the Tarot major arcana and ancient frescoes of Osirian creation gods suggest that the two are closely connected and possibly of similar origin, with both allegorically encoding an identical creation science.

The Emperor portrays the concept of *critical mass,* the accumulation of a departure from the steady state that is sufficient to bring about change in a system's state or order. He also signifies that the related notion of the *critical fluctuation,* the unique impulse that has spontaneously emerged from the ether's incessant noise, has triumphed over that chaos, and stands ready to manifest explicit order. Applied to Model G, the Emperor signifies the concentration fluctuation (energy pulse) that has grown sufficiently large that it is now ready to bifurcate the ether's uniform steady state.

The scepter that the Emperor wields is again symbolic of the ether. Whereas the Magician and Empress held this ether symbol in their left hand, the Emperor instead holds it in his right. This right-handedness suggests that Arcanum 4 is concerned with the explicit and observable, as opposed to the implicit and hidden. He is the "result" or "realization" of the unseen gestative process portrayed by Arcanum 3. The explicit mode is further suggested by the use of a masculine, as opposed to feminine, figure. Thus the Emperor would represent the nucleation of explicit order in the ether. The empire over which he rules is the region in which the emerging fluctuation has achieved dominance. It is the volume of space in which the X and Y ether concentrations have begun to deviate noticeably from their formerly uniform steady-state values.

As mentioned earlier, the cross is the esoteric symbol for the material realm. The cross crowning the scepter and the Emperor's crossed-leg posture seem to suggest that the purpose of the Emperor's conquest is to bring the physical world into existence. The Marseilles version pictures him partially seated rather than fully standing, as if to indicate that this explicit order has not yet achieved full emergence and dominance. His erect posture, warrior's helmet, and scepter held upraised in his right hand suggest that he is prepared to engage in battle to overthrow the existing homogeneous regime and that emergence is imminent. In the Egyptian fresco version of this arcanum, the cubical stone throne signifies the imminent emergence of physical form, the cube being the symbol for the physical plane.

Arcanum 5

Arcana 1 through 4 define a logical progression of concepts, beginning with Arcanum 1 (irreversible process); proceeding to Arcanum 2 (differentiation); then to Arcanum

3 (coupling and the supercritical state); and finally to Arcanum 4 (the emergence of the critical fluctuation). Together this tetrad relates processes taking place in an ether that has thus far maintained generally uniform concentrations and is devoid of physical form. This concept of uniformity is summarized by Arcanum 5.

Arcanum 5 (figure 8.7) depicts the Pope, or Hierophant (chief priest), seated with a staff bearing three cross bars in his left hand. This triple cross symbolizes the penetration of the creative power through the divine, intellectual, and physical realms.[12] Mystics sometimes refer to these "three worlds" as the causal, astral, and physical planes. The Hierophant looks down at two men who obediently kneel before him and invites them to meditate by describing the sign of peace and silence with his right hand. The man on his left wears a single wig, and the man on his right bears two headpieces, a light-colored wig and a dark-colored hat that rests on his left shoulder. Iamblichus refers to the two figures as the "genii of light and of darkness." The Hierophant sits between two columns that flank the entrance to the sacred inner sanctuary, the right symbolizing divine law, the left symbolizing the freedom to obey or disobey. While the Hierophant entreats obedience to the divine law, he at the same time allows his subjects the freedom of disobedience.

The staff with the triple cross represents the transmuting ether. Like the Magician, the Hierophant holds this ether symbol in his left hand, suggesting that the creative flux is unseen. The attributes normally ascribed to him, such as "keeper of peace," "preserver of the status quo," and "regulator of being," call to mind a transmutation process that proceeds smoothly and uniformly throughout all of space. Hence the Hierophant portrays the ether in its initial homogeneous steady state.

The "genii of light and of darkness" signify X or Y concentration fluctuations emerging in the ether in either of two possible polarities, higher than average value (light) or lower than average value (dark). The two bowing figures thus depict a duality of quantity, higher concentration versus lower concentration. Arcanum 5 of the Marseilles Tarot provides one additional useful detail. Its light-toned man wears a light wig with a dark hat hanging below it. In the context of Model G, this light and dark combination could portray the reciprocal interdependence of X and Y; that is, when Y adopts a high concentration (light wig), X adopts a low concentration (dark hat below). The opposite polarity is not shown on the dark man. Nevertheless, this symbolism illustrates that each of the two polarities portrayed here involves the reciprocal interplay of two reactants, X and Y, as with the Taoist yin and yang. Since the wig-hat is on the light man to the Hierophant's right, this bipolarity is connected with the emergence of explicit order.

Interpreted in systems theory terms, Arcanum 5 signifies the established state of *uniformity*, the *homogeneous steady state*, the state of *dynamic equilibrium*. It represents the condition of uniformity that precedes the emergence of explicit order. In the context of Model G, Arcanum 5 represents the operation of the transmuting

The Pope, The Hierophant

Attributes described
by taroists:

Religion, the universal law

The regulator of the infinite manifestations of being

The preserver of the status quo, the keeper of peace

Fortunate influences

The Master of the Arcana

Egyptian fresco description
from Iamblichus:

Arcanum V is figured by the image of the hierophant, master of the sacred mysteries. This prince of the occult doctrine is seated between two columns of the sanctuary; he leans upon a cross of three bars, and with the index finger of his right hand he traces upon his chest the sign of silence; at his feet two men are prostrated, one dressed in red, the other in black. . . . His gestures invite you to meditation. The right column symbolizes divine Law, that on the left symbolizes the liberty to obey or disobey. The cross with three bars is the emblem of GOD, penetrating three worlds in order that birth may be given to all the manifestations of universal life. (*An Egyptian Initiation,* p. 21)

Figure 8.7. ARCANUM 5. Systems concept: the homogeneous steady state.

ether in the era preceding the appearance of physical form, when the concentrations of the ether substrates were everywhere uniform.

The presence of two kneeling figures in Arcanum 5 suggests that, for the present, the two men are subservient to the law. Although fluctuations arise in the ether, the prevailing state of uniformity persists. The two columns, however, suggest the potential for disobedience, wherein a major departure from the norm could emerge and initiate the creation of a macroscopic wave pattern.

Although silence reigns on the face of the deep, the Emperor continues to grow and gain strength. Under the Hierophant's rule, which allows the expression of nonconformance, the law of uniformity will ultimately be broken. Just as established political regimes eventually fall and are replaced by new governments advocating a different kind of order, the regime of uniformity ultimately falls, and through the act of creation, explicit order emerges.

Arcanum 16

Standing before Arcanum 5 in the gallery of the frescoes, we catch a glimpse of the fate of the steady state by turning to Arcanum 16, displayed on the opposite wall (see figure 8.1). Arcanum 16 belongs to the second sequence (11 through 21), which relates matters regarding human evolution and spiritual development. Because Arcana 5 and 16 both fall at the geometrical midpoint of the gallery and serve as dividers or "bifurcation points" for their respective Tarot sequences, it is fitting to consult this complementary fresco here.

Arcanum 16 displays a stone tower whose battlements are struck by a bolt of lightning (figure 8.8). Two men are thrown from the heights along with fragments of broken masonry. The tower would symbolize the established regime, the regime of the Hierophant, in which obedience to the law of uniformity prevails. By picturing the tower as a defense fortification, Arcanum 16 suggests that the established regime is under attack. The destruction of the tower and the casting down of its defenders portray the overthrow and destruction of this regime by the champion of order. That this ruin is brought about by a lightning bolt suggests that the overthrow comes unexpectedly and without warning. What is a catastrophe for the old regime, however, is an opportunity for the new. This wisdom was appreciated by the ancient Chinese, who represented these seemingly antithetical concepts with the same pictogram.

The archetypal systems concept portrayed here is that of *system restructuring* or *revolution*. Arcanum 16 represents the quantum jump that brings about the dissolution of the old state of order and the birth of the new. The sudden destruction of the tower therefore signifies the spontaneous disruption of the ether's homogeneous steady state due to the growth of the insurgent critical fluctuation. It marks the moment when the emptiness of space becomes disrupted, as the precursor of explicit order begins to emerge. The nature of this sudden event can be understood by studying Arcanum 6, the next in line in this sequence. In so doing, we continue our investigation into the mystery of creation to study the final stages in the emergence of explicit order.

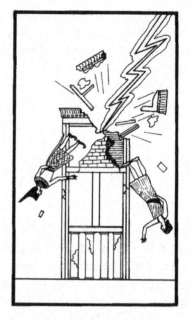

The Tower of Destruction

Attributes described
by taroists:

Accident, catastrophe
Sudden occurrence
The ruin of fortune, frustrated plans, abortive
   enterprises

The Lightning-Struck Tower

Egyptian fresco description
from Iamblichus:

Arcanum XVI is figured by a tower, which the lightning has unroofed. A crowned and an uncrowned man are thrown from its heights with the rest of the debris. (*An Egyptian Initiation*, p. 31)

Figure 8.8. ARCANUM 16. Systems concepts: hierarchical restructuring; revolution.

THE SYMBOLOGY OF THE TAROT, PART 2

Arcanum 6

Arcanum 6 (figure 8.9) portrays a young man, the Lover, who must decide between two maidens seeking his companionship. The crowned one on his right, who beckons him with open arms, represents Virtue, while the one on his left with disheveled hair, who greets him with crossed arms, represents Vice. Cupid hovers behind the Lover with bow and arrow drawn and takes aim at Vice. The Egyptian version depicts this encounter at a fork in the road, with two maidens attempting to bring a young man onto one or the other of two paths and the youth undecided as to which way to go. According to Iamblichus, the maiden on the man's right wearing a headband of gold personifies modesty and virtue, and the maiden on his left wearing vine leaves in her hair represents ostentation, vanity, and vice. Behind them, a winged spirit borne on a nimbus of blazing light aims an arrow for punishment toward Vice.*

After choosing between the two alternatives before him, the bachelor will leave his state of singleness and follow one of the two paths to commit to one of two possible marriages. This male-female symbology suggests that Arcanum 6 is portraying a choice between alternative dualities, each consisting of a pair of opposed states. Reaction-diffusion systems like Model G or the Brusselator must choose between dualities of this sort when their two interdependent reactants produce ordered spatial patterns. Identifying the Y ether with the groom and the X ether with the bride, two alternative marital pairings would be possible: low X/high Y (virtue) or high X/low Y (vice). Arcanum 6 illustrates the important truth that in reaction systems like Model G, only the low X/high Y polarity can spawn a stable primordial particle.

A similar low versus high quantitative duality was indicated by the two subjects (fluctuations) kneeling before the Hierophant. Whereas in Arcanum 5 this duality maintained a state of obedience to the law of uniformity by not deviating appreciably from the uniform steady state, in Arcanum 6 the personages depicting duality are instead standing. They have turned their backs to the path of uniformity and are now tempting disobedience of its law.

It seems appropriate to identify Arcanum 6 with the archetypal systems concept of *bifurcation*. The two paths stretching out before the youth signify the two polarities that an emerging fluctuation can adopt. The single path behind him that he is about to leave signifies the ether's initial vacuum state, the homogeneous steady state represented by Arcanum 5. Since the Lover stands at a key decision point between two possible dualistic alternatives, Arcanum 6 also depicts the archetypal systems concept of *critical point*.

Arcanum 6 is essentially an allegorical bifurcation diagram. The branched path it portrays bears a striking resemblance to the forking mathematical solutions of a nonlinear reaction system (see figure 5.9). The path of modesty and virtue to the youth's right corresponds to the stable lower branch of an odd-type bifurcation, which spawns a positive subatomic particle, a dissipative structure with a low X/high Y core concentration. The path of ostentation and vice to the youth's left corresponds to the unstable upper branch that represents the complementary antiparticle.

---

*Some early Tarot decks, such as the Lombardy Tarocchi deck from Renaissance Italy, picture instead just two human figures, a man and a woman, with an angel hovering between them, "The Lovers" (S. R. Kaplan, *The Encyclopedia of Tarot*, vol. 2 [New York: U.S. Games Systems, Inc., 1986], p. 15). Some taroists, such as Paul Foster Case (*The Tarot: A Key to the Wisdom of the Ages* [Richmond, Va.: Macoy, 1947], pp. 84–87), who adhere to this version claim that it signifies the "union of complementary opposites," a concept that actually should be attributed solely to Arcanum 3. This Italian version, however, appears to be a muddling of the original Tarot. Altering the Marseilles Tarot symbology in this fashion breaks the continuity of the carefully wrought chain of concepts and destroys what otherwise would be an ingenious portrayal of a highly sophisticated reaction-kinetic physics.

The Lovers

Attributes described by taroists:

A time to decide between opposing alternatives

Temptation

The Two Ways

Egyptian fresco description from Iamblichus:

Arcanum VI is figured by a man standing motionless at the angle formed by the junction of two roads. His looks are fixed on the ground, his arms are crossed upon his chest. Two women, one at his right, the other at his left, each place a hand on his shoulder, showing him one of the two roads. The woman placed at his right has a circle of gold about her brow; she personifies *virtue*. The one on the left is crowned with a vine branch full of leaves; she represents *vice*, the temptress. Above and back of this group, the genius of justice, hovering in a flashing aureole, draws his bow directed toward vice, the arrow of punishment. (*An Egyptian Initiation*, p. 22)

Figure 8.9. ARCANUM 6. Systems concepts: bifurcation; the critical point.

The hovering spirit with drawn arrow indicates that of the two alternative polarities, the negative, ostentatious polarity ultimately meets its death. Just as a morally loose, superficial partnership ultimately breaks up, the negatively charged (high *X*/low *Y*) polarity is unstable. The particle eventually dematerializes, leaving the

Figure 8.10. The Star of David or yantra.

ether concentrations to return to their uniform vacuum state. Only the virtuous, positive polarity option leads to a stable subatomic particle. This biasing of the parthenogenic creation process inevitably leads to the creation of a universe made up of matter, as opposed to antimatter. Thus Arcanum 6 explains in a simple fashion the existence of matter-antimatter asymmetry in the universe, a problem that has tormented the minds of modern physicists for decades.

When considering Arcanum 6, it is also appropriate to consider the six-pointed star symbol (figure 8.10), which is numerologically associated with the number 6. In Old Testament scriptures it is known as the Star of David or Solomon's Seal, and in Hindu mysticism it is considered a yantra. The upward-pointing, light triangle signifies the positive masculine polarity, while the downward-pointing, dark triangle signifies the negative feminine polarity. The star's design, formed of two opposed interlocking triangles, effectively portrays the union of the masculine and feminine principles, as well as the principle of symmetry breaking, the oppositional deviation of the X and Y ethers from their steady-state values.

A similar symbol is found in Hindu art, usually displaying at its center an upraised phallus *(lingam)* framed by a downward-pointing triangle, symbolic of the female genitalia *(yoni)*.[13] The Tibetan Buddhist yantra in figure 8.11 also presents a geometrical form similar to the Star of David. The crossed *dorjes* (swords) at its center are said to symbolize the cosmic ether (blue air), whose warp and woof form the universe. To the modern systems theorist, this emblem represents an elegant geometrical "equation" summarizing the essential features of a sophisticated reaction-physics cosmology. As in the ankh of ancient Egypt, these two crossed members may signify the two primary ether transmutation pathways at whose intersection the physical universe emerges. The surrounding upward- and downward-pointing triangles would portray the emergence of explicit order at this intersection point, reminiscent of the separation of Geb and Nut.

Figure 8.11. A Tibetan Buddhist six-pointed star containing crossed dorjes, symbolizing the ether. From Evans-Wentz, Tibetan Book of the Dead, p. 63.

The symbolism expressed in Arcanum 6 calls to mind the ancient Phoenician myth of the beautiful youth Adonis and the two goddesses, Aphrodite and Persephone, who both desired his love (see chapter 11). After Adonis's untimely death, the gods permitted his resurrection and decreed that he alternately spend half of the year in the skies with his lover Aphrodite and half of the year in the underworld with his other lover Persephone. A similar fate of sinusoidal alternation is described in the next four major arcana. Arcanum 7 portrays one aspect of this oscillation, the growth of the emerging primordial wave and its dissemination into the surrounding territory of space.

Arcanum 7

Arcanum 7 (figure 8.12) portrays a crowned warrior, the Conqueror, who holds a scepter in his right hand as he rides a square-shaped war chariot. Over his head stretches a baldachin (silk canopy) supported by four columns rising from the chariot's corners. According to Iamblichus, the baldachin was studded with stars. The Marseilles Tarot shows the chariot being drawn by two horses pulling in opposite directions, while the Osirian temple fresco portrays the chariot being drawn by two sphinxes, one white and one black.

The cubical chariot would symbolize matter, and its forward motion would portray the emergence of the material realm. The star-studded tapestry suspended above the charioteer's head symbolizes the heavens, calling to mind the arched starladen body of the Egyptian goddess Nut at the moment when she separated from Geb (earth). The four columns call to mind the four sons of Horus who were said to serve as the supports for heaven. Consequently, the canopy (heaven) raised above

The Chariot, The Conqueror

Attributes described by taroists:

Victory, conquest, attainment
Success earned by hard work or active effort

The Chariot of Osiris

Egyptian fresco description from Iamblichus:

Arcanum VII is figured by a war chariot of square form, surmounted by a starry canopy sustained by four columns. Upon this chariot advances a conqueror, armed in a cuirass and carrying both sword and scepter. He is crowned with a circle of gold from which rise three pentagrams, or golden stars with five points. . . . Upon the square face which the front of the chariot presents, is traced a sphere, sustained by two outspread wings. . . . Three square rules are traced upon the cuirass. . . . Two sphinx, one white, the other black, are harnessed to the car. (*An Egyptian Initiation*, pp. 23–24)

Figure 8.12. ARCANUM 7. Systems concepts: exponential growth; spontaneous symmetry breaking.

the cubical chariot (earth) portrays the separation of the ethers at the time of primordial creation, the soaring of *Y* and the plunging of *X*.

The Conqueror who rides in the chariot is actually the Emperor (Arcanum 4) seen at a later stage in the story of creation as he rises in power to conquer regions formerly under the control of the Hierophant (Arcanum 5). In Arcanum 4 the warrior was seated, and his conquest of the homogeneous steady state had not yet begun. In Arcanum 7 he is standing, and his platform is moving forward; these symbols both connote a state of activity. The battle has not only been launched, it is in the process of being won.

Like the Emperor, the Conqueror holds the scepter, the symbol of the ether, in his right hand. Hence he indicates that explicit order is emerging in the ether. His invasion represents the emergence and outward expansion of nonuniformity, a duality of more versus less, into surrounding regions where the ether concentrations had formerly maintained a peaceful state of spatial uniformity. In a cosmological context, Arcanum 7 would signify the emergence of the energy potential arch that forms the core of the primordial subatomic particle and the expansion of this wavelike energy field into surrounding space.

The light and dark horses (or sphinxes) pulling the chariot symbolize the positive and negative polarity of the emerging order. In Arcanum 5 the polarity symbols were pictured as subjects bowing in submission to the law of uniformity. Here in Arcanum 7, they are leading the invasion to induce the surrounding territory to rebel against this condition. The dark and light animals represent the alternately suppressed and enhanced X and Y ether concentrations that together form the rising wave pattern.

In the Marseilles Tarot version, the front of the chariot bears a monogram that makes it the only arcanum to display lettering. Lettering signifies conscious thought, which in turn signifies explicit order. Just as words give outward expression to inward thought, the emerging wave pattern here begins to outwardly express the ether's underlying implicit order.

Arcanum 7 depicts the archetypal systems concept of spontaneous symmetry breaking, the disruption of "symmetry" that occurs when the reaction medium passes from a state where its reactant concentrations are spatially uniform to one in which they deviate from uniformity in one or the other of two possible polarities (more versus less). Such symmetry breaking is spontaneous because its occurrence depends on where and when a fluctuation (the Eagle) is able to emerge and grow to a sufficient size (the Emperor) so as to be capable of driving the system to a new state of order. Moreover, because fluctuations grow in an exponential fashion, when emergence occurs, it occurs suddenly, like the lightning bolt striking the tower in Arcanum 16.

Arcanum 7 also portrays the archetypal systems concept of *exponential growth,* the "snowball effect." Through exponential growth, a critical fluctuation is able to increase in size and expand its explicit order outward to bifurcate the surrounding territory. But, as is the case with most things in nature, exponential growth cannot go on indefinitely. Balancing forces come into play that tend to limit growth. This concept of counterbalance is expressed by the next arcanum.

Justice, The Balance

Attributes described
by taroists:

Justice, equilibrium, balance
Every action brings forth a reaction

Divine Justice

Egyptian fresco description
from Iamblichus:

Arcanum VIII is figured by a woman, seated upon a throne, her forehead encircled with a crown of lance-heads; she holds in her right hand a sword, the point raised; in her left hand is a balance. . . . The eyes of Justice are covered with a bandage to show that she weighs and strikes without taking into account the conventional differences that men have established for themselves. (*An Egyptian Initiation,* p. 25)

Figure 8.13. ARCANUM 8. Systems concepts: the limits to growth; counterbalance.

Arcanum 8

Arcanum 8 is represented by a crowned woman (Justice) seated on a throne (figure 8.13). In her right hand she holds an upward-pointing sword and, in her left hand, a set of scales. The double-pan balance conveys the concept of equilibrium, while the sword adds the notion of atonement. According to Iamblichus, the Egyptian fresco depicted Justice blindfolded.

Considered in relation to Arcanum 7, Arcanum 8 represents the general principles of *counterbalance* and *the limits to growth (the law of diminishing returns)*. It signifies balanced opposition to the exponential growth process portrayed by

Arcanum 7. In both Model G and the Brusselator, the concentration of a given reaction intermediate *(X* or *Y)* cannot grow in size indefinitely. Owing to the self-closing reaction loop formed by $B + X \longrightarrow Y + Z$ and $2X + Y \longrightarrow 3X$, the concentration of one species tends to limit the growth of its partner (see figure 4.2). Because of this bidirectional conversion of *Y* into *X* and *X* into *Y,* the growth of *Y* and depletion of *X* represented in Arcanum 7 will eventually be brought to a halt.

Just as the arm of a balance tends to rock back and forth in seeking its median equilibrium point, the *X* and *Y* partners reciprocally vary up and down in relation to one another, thereby achieving a kind of dynamic long-run equilibrium—a result of the simultaneous operation of both growth and balance principles. The alternation of *X* and *Y* could manifest either spatially as a periodic steady-state stationary wave pattern, temporally as a limit-cycle oscillation, or spatiotemporally as a propagating reaction-diffusion wave. In each case the manifested ordering would be periodic in nature. Hence periodic behavior owes its existence to the countervailing action of growth-promoting and growth-limiting principles.

Arcanum 8 dictates that the *X* and *Y* concentrations must alternate from one polarity extreme to the other; what began as a male-dominant relation in Arcana 6 and 7 (low *X*/high *Y*) must evolve into a female-dominant relation (high *X*/low *Y*), then change back into a male-dominant relation, and so on in reciprocating fashion. In so doing, the system generates a periodic wave pattern of the sort characterizing the steady-state dissipative structures displayed in chapter 5.

The positive-negative duality theme, which may be traced through Arcana 2, 5, 6, and 7, here achieves its fullest expression, with both *X* and *Y* taking their turn at dominance. Now that the basic principles involved in the generation of periodicity have been presented, the way is made clear to comment generally about the periodic process itself. This appears to be the purpose of Arcanum 9.

Arcanum 9

Arcanum 9 (figure 8.14) is represented by a bearded man, the Hermit or Sage, who walks forward using a stick for support. With his right arm he raises a lantern half-sheltered by his cloak. The lantern's light sheds sufficient illumination for him to see his way through the surrounding darkness.

The lantern in the Sage's right hand represents the dawning of explicit order, the emergence of light (physical order) in the midst of the darkness of space. Whereas Arcana 7 and 8 describe the *X*-on and *Y*-on concentrations deviating from their homogeneous steady-state values to form the expanding periodic field pattern, the concentrations in Arcanum 9 have stabilized at specific inhomogeneous steady-state values. Thus in Arcanum 9, the wave pattern attains a perfection of form characterized by a unique wavelength and amplitude.

It is also significant that the Sage partially shelters his lantern within his cloak; the

The Hermit, The Sage

Attributes described
by taroists:

Wisdom, prudence
Discretion in all things
Perfection of form
Circumspection, protection
Experience gained in the journey of life

The Veiled Lamp

Egyptian fresco description from
Iamblichus:

Arcanum IX is figured by an old wanderer lean-
ing on a staff and carrying before him a lighted
lamp which he half conceals under his cloak. (*An
Egyptian Initiation*, p. 26)

Figure 8.14. ARCANUM 9. Systems concepts: adaptive self-stabilization; autonomy.

lit flame, in turn, is protected by the lantern's transparent enclosure. By means of these caring measures he ensures that the flame is not extinguished by random environmental disturbances, such as gusts of wind. Similar protective, self-stabilizing behavior characterizes dissipative structures produced by Model G and the Brusselator; for example, a positively charged material particle creates conditions favorable to its continued existence by generating a G-well supercritical region around itself. Just as the prudent Sage avoids wide extremes to seek an epicurean mean, in much the same way the X and Y concentration oscillations are neither too large nor too small; if some disturbance were to cause these concentrations to deviate from their limit cycle attractor, the system would respond in a compensatory fashion to restore them to their proper values. In fact, the subquantum kinetics ether physics suggests that the phenomena of force field attraction and repulsion arise because of such adaptive behavior. Furthermore, by maintaining its own nurturing supercritical region, a subatomic particle becomes an autonomous entity. Like the Sage, the primordial particle is able to survive on its own as an independent entity.

The archetypal systems terms that best express such open-system wisdom are *dynamic self-regulation* and *adaptive self-stabilization*. Arcanum 9 also appears to depict the concepts of *autogenesis* (self-creation) and *autonomy*. Note the difference between the isolated autonomy depicted by the Fool of Arcanum 0 and the wary autonomy depicted by the Hermit of Arcanum 9. Whereas the Fool is blind, ventures forward carelessly over rough terrain, and drags his walking stick behind him, the Sage has vision (foresight), ventures forward prudently over smooth terrain, and wisely places his walking stick before him. The Sage carries no excess baggage or material belongings on his journey. Instead, he uses his free hand to hold high the lantern to light his way. Like a subatomic particle, he interacts with his environment while at the same time preserving his autonomy.

Arcanum 10

Having analyzed the second tetrad of the Tarot, Arcana 6 through 9, we now examine Arcanum 10, which summarizes the main aspects of the ether's emerged periodicity. Arcanum 10 portrays a six-spoke rotating wheel whose axle is suspended between two columns (figure 8.15). The figures in Arcana 5, 6, and 7 representing the polar opposites of light and dark (good and evil) reappear here in the form of two spirits clutching the wheel's rim. Iamblichus identifies the ascending figure as Hermanubis, the spirit of good, and the descending figure as Typhon, the spirit of evil. Balanced at the top of the wheel is a crowned sphinx holding a sword in its paw. A symbol for the zodiac, the sphinx personifies the passage of time and destiny.

Arcanum 10 further develops the bipolar interaction theme presented in preceding arcana, with the two interacting species appearing on opposite sides of a rotating wheel. The wheel's vertical orientation and its designation as the "Wheel of Fortune" both suggest a cyclic alternation between a state of greater and subsequently lesser quantity. As the bound spirits alternately ascend and descend, they illustrate the simultaneous operation of the principles of growth and atonement presented consecutively in Arcana 7 and 8. Because the growth of one species counters the growth of the other, the system as a whole achieves a kind of overall seesawlike equilibrium, like that achieved by the sphinx as it balances itself on a board perched atop the turning wheel. Like the Charioteer and justice, the sphinx wields a sword, perhaps implying that this balance is achieved through the two parties mutually limiting each other's growth.

The two beings clutching the wheel represent the mutually coupled X and Y reaction intermediates whose concentrations, or "fortunes," vary cyclically in reciprocal fashion. As the wheel turns, the two beings continuously interchange fates, alternating between boom and bust in sinusoidal fashion. In so doing, they

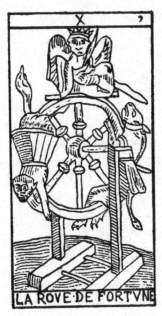

The Wheel of Fortune

Attributes described by taroists:

Change of Fortune
The inevitable cycles of nature
The completion of one cycle and the beginning
of the next

The Sphinx

Egyptian fresco description from Iamblichus:

Arcanum X is figured by a wheel suspended upon its axis between two columns. At the right Hermanubis, genius of good, strives to mount to the summit of the circumference. At the left, Typhon, genius of evil, is precipitated. The Sphinx, in equilibrium on this wheel, holds a sword in the claws of a lion. (*An Egyptian Initiation*, p. 27)

Figure 8.15. ARCANUM 10. Systems concept: the periodic state.

trace out the X and Y concentration pattern that forms the first particle of matter. At the particle's center, X would be at a minimum and Y at a maximum; further out, X would become maximum and Y minimum, and so on, in a regular wavelike fashion. Consequently, the symbology of Arcanum 10 expresses with stunning accuracy the periodic behavior of a two-variable reaction-diffusion system similar to Model G.

In summary, Arcanum 10 is best identified with the archetypal systems concept of the *ordered periodic state,* and more generally, with the state of *emerged explicit order.* Considered in the context of a cosmology of physical creation, Arcanum 10 depicts physical form in its fully emerged state as the fully formed primordial subatomic particle. This notion of describing physical form in terms of wave patterns is quite surprising. In Western science, the wave nature of matter became known only as recently as the late 1920s.

AN OVERVIEW

When interpreting the meanings of the Tarot's symbols there is always the danger of unconsciously projecting one's own conceptual biases. These hieroglyphic portrayals, however, are far from being amorphous ink blots. Each arcanum is made up of an array of pictorial symbols and verbal interpretations that combine in a coherent manner to specify a particular concept or set of concepts. When a detailed study is made of the Tarot, on the one hand, and the theory of open systems, on the other, it becomes apparent that the Tarot depicts the process of matter creation in a nonequilibrium ether reaction system of a type similar to Model G.

Moreover the Tarot presents these systems concepts in a logical sequential order. In fact, in explaining the functioning of such two-variable nonlinear reaction systems and describing how they spawn explicit order, the best way to sequence these concepts would be in the way they are found in the Tarot. For example, the concept of a unitary transforming flux must be established before proceeding to the concept of a two-component flux; the existence of two flux pathways must be established prior to process interaction; the existence of a steady state must be established before bifurcation of the steady state.

Arcanum 0 describes an isolated system subject to the ravages of decay, perhaps depicting the chaos that might have existed prior to the commencement of etheric transmutation. The first portion of the story of creation, Arcana 1 through 5, describes the subquantum etheric realm, its transmutative character, its implicit order, and the growth of the critical fluctuation. Arcanum 1 portrays the irreversibly transmuting ether, the first prerequisite for the creation of cosmic order. Arcanum 2 portrays the subdivision of the ether flux and the development of two reaction pathways. Arcanum 3 represents the intersection and mutual interaction of these two transmutative pathways and their tendency to nurture or amplify fluctuations arising spontaneously in the ether. Arcanum 4 depicts the fluctuation at the stage of criticality, when it is invulnerable to attack and ready to launch an insurrection to bring order to surrounding regions. Arcanum 5 portrays the homogeneous steady state that prevailed prior to the emergence of physical form in which ether substrate concentrations were initially uniform throughout space, the vacuum state.

The second portion of the sequence, Arcana 6 through 10, describes the emergence of a primordial subatomic particle. Arcanum 6 represents the critical point the growing fluctuation must reach when it is about to bifurcate the unitary homogeneous steady state. Although fluctuations can emerge in either of two possible polarities and follow either of two possible bifurcation paths, only one path is permitted, the one leading to the low X/high Y alternative. Arcanum 7 depicts the moment of symmetry breaking when the exponentially growing critical fluctuation creates an expanding region of nonuniformity (explicit order), causing the X and Y ether concentrations to depart from their former values. Arcanum 8 indicates that because of the mutual linking of X and Y, the increasing production of one species is curtailed by the increasing production of the complement species, and the two tend to counterbalance one another. Arcanum 9 represents this oscillation when it has stabilized into an orderly pattern, with neither X nor Y growing to excess. Finally, Arcanum 10 portrays the fully emerged wave pattern in which the X and Y concentrations vary through space in reciprocal fashion, one species achieving its maximum where the other achieves its minimum. Correspondences between the first eleven major arcana of the Tarot and the main stages in the creation myths of Atum and Osiris are presented in table 8.2.

Several key aspects of this physics of cosmogenic process are summarized in the Rose Cross, an emblem consisting of a single circular rose at the center of an equal-armed cross (see figure 8.16). During a lecture in 1785, Alessandro Cagliostro, a renowned Sicilian scholar of the esoteric, reportedly stated that the ancient magi wore the Rose Cross around their necks on a golden chain. Its origin was attributed

---

TABLE 8.2

Correspondences between the Tarot Arcana and Ancient Egyptian Creation Myths

THE COSMOGONY OF ATUM

Arcanum 0:	Primordial waters prior to emergence of Atum
Arcanum 1:	Emergence of Atum, the primeval mound
Arcanum 2:	Atum's creation of Shu and Tefnut
Arcanum 3:	Union of Shu and Tefnut
Arcana 4 and 5:	Creation of Geb and Nut in the unseparated state
Arcana 6–10:	Geb and Nut being separated by Shu

THE MYTH OF OSIRIS

Arcanum 1:	Osiris
Arcanum 2:	Isis
Arcanum 3:	The joining of Isis and Osiris and the birth and nurturing of Horus
Arcanum 4:	Horus as Heru-khuti
Arcanum 5:	Set's reign of chaos
Arcana 6 and 16:	The battle between Horus and Set
Arcana 7–10:	The victory of Horus and the emergence of light

Figure 8.16. The symbol of the Rose Cross. From Christian,
History and Practice of Magic, p. 145.

to the god Hermes-Thoth, founder of the Order of the Magi. It is said that four
letters were written between the branches of the cross, each letter expressing a mystery of nature. According to Cagliostro, these letters, taken from the twenty-two-character "Alphabet of the Magi," have the following meanings:

[I or *Ioïthi* in the sacred language] symbolizes the active creative principle and the manifestation of divine power that fertilizes matter.

[N or *Naïn*] symbolizes passive matter, the mold of all forms.

[R or *Rasïth*] symbolizes the union of these two principles and the perpetual transformation of created things.

[I or *Ioïthi*] symbolizes again the divine creative principle and signifies that the creative strength which emanates from it ceaselessly returns to it and springs from it everlastingly. [14]

So as not to reveal this sacred word to the profane, the magi replaced these letters on the Rose Cross with the four zodiac figures that compose the sphinx: the bull (Taurus), lion (Leo), eagle (Scorpio), and water bearer (Aquarius). Christian's diagram (figure 8.16) apparently mistakenly interchanges the last two sacred symbols, shown flanking Aquarius. That is, *Rasïth* should instead be matched with Scorpio and *Ioïthi* with Aquarius.

The meanings ascribed to these four sacred letters of the magi alphabet correspond to the four sacred Hebrew letters *yod he vav he* (יהוה). According to the ancient *Sepher Bereshith*, the Genesis of Moses, these four letters designated the divinity (יהוה) and were said to contain the key to all the sciences, divine as well as human. *Yod* signified the masculine unity principle; *he*, the feminine duality principle; *vav*, the dynamic linking or uniting of these active and passive principles; and the second *he* signified the transition from one world to the next, from the noumenal to the phenomenal.[15]

In both the Hermetic and Hebrew quartets, the first three process concepts, I-N-R and *yod-he-vav*, clearly express the principles behind Arcana 1, 2, and 3, and the fourth letters, I and *he*, express the notion of emergence described by Arcanum 4 and further developed in Arcana 5 through 10. Consequently, like the zodiacal symbols of the sphinx, this alphabetical cryptogram summarizes the mystery of the creation of the universe, of the emergence of the explicit from the implicit.

The symbols making up the Rose Cross also relate to this creation theme. The cross signifies the implicit order, *yod-he-vav*. Like the Egyptian ankh and Tibetan *dorjes*, it symbolizes the crossing and coupling of the two primary ether transmutation streams, that unique juncture in the ether transmutation network that gives birth to our physical universe. The rose, which is superimposed on the cross, apparently signifies the higher-level manifestation represented by the second *he*. It depicts the explicit order emerging at this juncture. Like the blooming cosmic lotus of Egyptian and Hindu mythology, it represents physical creation. Applied to spiritual evolution, the Rose Cross also signifies the rebirth of the spirit into the afterlife. What a noble symbol these ancient sages earned the honor to wear.

The Rose Cross very much resembles the Crown of the Magi (figure 8.17), the sign with which the magus decorated himself when he reached the highest degree of initiation. It also appeared as Arcanum 21, the very last fresco in the subterranean gallery series. According to Iamblichus, it depicted a star surrounded by a circular garland of roses around whose circumference appeared the heads of a bull, lion, eagle, and man. In Tarot this arcanum, called the World, is quite similar, with the exception that the wreath is oval shaped and appears to be giving birth to a nude woman, who is portrayed within it in place of the star. The arcanum is said to signify the highest possible life attainment, spiritual awakening.

Given that the Tarot symbolism does in fact describe the emergence of order in certain types of open reaction systems, how is it possible that such an advanced level of knowledge was in the possession of the ancients? It was only as recently as the 1930s that an understanding of open systems and their ordering characteristics began to take form. Moreover, the concept of order through fluctuation appeared in scientific literature for the first time around 1971. Are scientists now rediscovering principles that were known to our forefathers thousands of years ago? If so, this necessitates a radical revision of our conception of the level of sophistication of

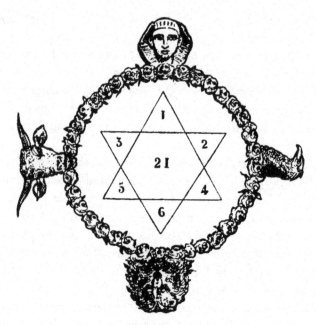

Figure 8.17. The Crown of the Magi.
From Christian, History and Practice of Magic, p. 111.

ancient science. Without a doubt, the Tarot is one of the most valuable ancient scientific documents to survive until modern times. The concepts encoded in the Tarot and the ancient Egyptian creation myths compose only the tip of the iceberg. This same creation metaphysics is embedded in a variety of creation myths from other cultures (see chapter 11), as well as in a closely related esoteric system: astrology.

9
THE THERMODYNAMICS
OF ASTROLOGY

SECRETS OF THE CIPHER

Hermetic tradition holds that astrology, like the Tarot, conveys an ancient natural science and that the symbolism of astrology and the Tarot are closely related. In fact, they are regarded as the golden and silver keys that, when properly understood, can unlock the secrets of nature's mysteries.[1] Astrology was a part of the ancient Hermetic science that, together with the Tarot, was learned by the Egyptian priests during their initiation into the Osirian mysteries. In his explanation of the Egyptian initiation, Iamblichus states that after being instructed on the meanings of the twenty-two arcana in the sacred subterranean gallery, the aspirant was led via an underground passage to a crypt inside the Great Pyramid.[2] There he was brought before a great circular silver table engraved with the twelve signs of the zodiac (figure 9.1). As the supreme test of his initiation, he was asked to explain the zodiac's significance. Indeed, if the aspirant properly understood the physics of the Tarot, he could decipher the esoteric meaning of the zodiac, which metaphorically expresses the same basic principles.

Tradition maintains that astrology was handed down to Egypt, China, India, and to former South American civilizations from an antediluvian civilization that flourished before the sinking of Atlantis, prior to 9000 B.C.E. Knowledge was probably transmitted orally from master to pupil with attempts made to keep its particulars secret from the general public.[3] Astrology's more ancient roots are not as evident in Egyptian lore as in Chaldean lore, but can nonetheless be found. For example, Schwaller de Lubicz has called attention to the sequential change of dominant zodiacal themes in Egyptian religion and architecture with the successive passage from the Age of Gemini to the Age of Taurus to the Age of Aries.[4] This indicates that the Egyptians not only knew the placement of the zodiac constellations from very ancient times but were also aware of the progressive advancement of the equinox.

The myth historian Oral Scott was among those who disagreed with the common

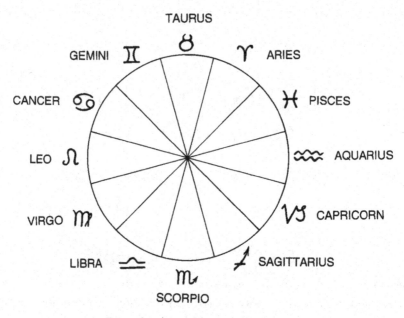

Figure 9.1. The astrological wheel.

belief that the constellations were traced out by idle shepherds biding their nights outdoors. Such an explanation scarcely accounts for the planned arrangement of the star groups, the similarity of their myths in widely separated countries, and the astronomical knowledge necessary for the zodiac's construction. In his book *The Stars in Myth and Fact,* Scott suggests that the zodiac constellations were purposely created to illustrate basic concepts of nature:

> Perhaps, after all, if we would recognize the probability that some of our remote ancestors knew more about nature than we have ever given them credit for knowing; and accept certain traditions that the star groups were arranged by them to illustrate spiritual conceptions of great natural laws, and that they sought in this way to convey certain information that they had acquired; we would in all probability, be much nearer the truth.[5]

Scott is essentially correct; the same open-system creation physics evident in the first half of the Tarot is also metaphorically encoded in the personality traits associated with the twelve signs of the zodiac When properly arranged, these signs allegorically convey an ether physics that describes how matter and energy first came into being and how they continue to come into being today.

One clue about the nature of the zodiac's metaphysics comes from the creation myth of the Babylonians dating from before the second millennium B.C.E. that describes how the hero, Marduk, defeated the forces of chaos and thereby brought the universe into existence (see chapter 11). After creating the sun, moon, and stars, Marduk finally fashioned a zodiac with twelve constellations depicting "the likenesses of the great gods." Since these gods are actors in a pageant whose express purpose is to describe the origin of the universe, the zodiac constellations, which are "likenesses" of these divine beings, similarly depict the story of cosmogenesis.

As far back as several millennia B.C.E. Babylonian astronomers tracked the progressions of the planets through the constellations to divine the influence of cosmic forces on the fate of their kingdom. The personal use of astrology may have emerged as a more recent development. The earliest known natal horoscope dates from 410 B.C.E., during the early Hellenistic period.[6] The ancient Greeks are known for their interest in the workings of the individual personality, so it is not surprising to find this more personal use of astrology emerging at that time.

Just as the Gypsies were for the most part unaware of the open-system physics encoded in their Tarot decks, since ancient times horoscope astrologers have been largely unaware of the specifics of the open-system physics hidden in the lore of their trade. With our present understanding of self-organizing phenomena made possible by advancements in modern science, however, we are now in a position to reconstruct this lost physics.

Unlike the Tarot, the zodiac has no self-evident starting point, and its signs do not present its systems principles in the proper sequence. On closer examination, however, the zodiac takes the form of a cipher that has a way of informing us when its signs have been placed in the proper order. That is, each sign is paired up with a specific set of attributes (see table 9.1). When the signs are given their normal sequence around the ecliptic, the polarities and qualities collectively display an ordered repeating pattern. Most alternate arrangements will scramble this regular order. But when the signs are rearranged in the order that properly expresses the science of order genesis, the attributes adopt a new ordered pattern. Since there is only one chance in 4.79×10^8 that a new arrangement of the signs would yield such a polarity and quality ordering, we may conclude that we have properly interpreted the scientific meanings of the signs.

A survey of the astrological literature indicates that astrologers have been largely unaware that the system of astrology is structured as a cipher. This lack of awareness is not surprising, for the cryptographic aspect of astrology becomes apparent only after one has successfully rearranged the zodiac's signs so that they express a creation metaphysics. This decoding of astrology was first accomplished in the mid-1970s and was possible only because recent advances in the fields of general system theory and nonequilibrium thermodynamics finally allowed astrology's creation metaphysics to be recognized.[7]

TABLE 9.1

SIGN	POLARITY	QUALITY	ELEMENT	PLANET
Aries (the Ram)	+	Cardinal	Fire	Mars
Taurus (the Bull)	-	Fixed	Earth	Venus
Gemini (the Twins)	+	Mutable	Air	Mercury
Cancer (the Crab)	-	Cardinal	Water	Moon
Leo (the Lion)	+	Fixed	Fire	Sun
Virgo (the Virgin)	-	Mutable	Earth	Mercury
Libra (the Balance)	+	Cardinal	Air	Venus
Scorpio (the Scorpion)	-	Fixed	Water	Mars
Sagittarius (the Archer)	+	Mutable	Fire	Jupiter
Capricorn (the Goat)	-	Cardinal	Earth	Saturn
Aquarius (the Water Bearer)	+	Fixed	Air	Saturn
Pisces (the Fishes)	-	Mutable	Water	Jupiter

With the sequentially numbered Tarot, it is a relatively easy matter to find the beginning point. This is not so with astrology. The circular arrangement of the zodiac constellations around the ecliptic presents no beginning or end. So which zodiac sign is best placed first to properly present the zodiac's physics? Although astrologers have traditionally designated Aries to be the first sign of the zodiac, this is not the best sign to start with.

The sphinx, long considered a key to esoteric knowledge, provides a clue to the proper starting point by narrowing our choice down to four of the twelve zodiac signs: Taurus, Leo, Scorpio, and Aquarius. An allegorical passage in the second biblical Midrash also favors Taurus as a starting point for this sequence: "When Solomon wished to sit on his throne, the Ox took him gently on his horns and handed him over to the Lion. Finally the Eagle raised him and placed him on his seat."[8] Hence the proper order appears to be Taurus (the Ox), Leo (the Lion), Scorpio (the Eagle), and Aquarius (Solomon on his throne).

Taurus emerges as a likely choice as sign number one for several other reasons. First, the word for bull in both Hebrew and Hindu scriptures is *aleph,* which is also the first letter in the Hebrew alphabet. In fact, this letter (א) bears a strong resemblance to the astrological symbol for Taurus (♉). Moreover, when the Taurus sign is turned on its side, it becomes the first letter in the Greek and Phoenician alphabets (α). It seems that in ancient times the Bull must have been associated with beginnings.*

The calendar system used by the ancient Egyptians also places its starting point in Taurus. The Egyptian priests kept track of two calendars. Their primary calendar was timed according to the heliacal risings of the star Sirius (Sothis), an event that occurred every 365.25 days, when Sirius rose at the same time as the Sun. This interval, variously termed the *fixed year, Sothic year,* or *Sirian year,* chronicled the true duration of the year. In addition, the Egyptians kept a civil calendar based on the *vague year,* or *moving year,* made up of 365 days. Owing to its slightly shorter length, this vague year lost one day every four years relative to the Sothic year. Rather than add an extra leap year day every four years as we do, however, the Egyptians allowed the discrepancy to accumulate so that the first day of the Sothic year would increasingly lag behind the first day of the vague year. But after accumulating a disparity equal to a full 365 days, the vague year would once again come into coincidence with the Sothic year. This rare event, which occurred once every 1,460 years (365 × 4 years), was celebrated by the Egyptians as their New Year. Double dates recorded in ancient manuscripts have convinced Egyptologists that New Year dates occurred on 4240 B.C.E., 2780 B.C.E., 1320 B.C.E., and C.E. 140. This calendric system must therefore have begun on one of these dates of coincidence; a number of Egyptologists have concluded that 4240 B.C.E. is the most likely date for the calendar's establishment.[9] Yet this date also marks the beginning of the Age of Taurus, when the precessing vernal equinox was about to enter the Taurus constellation, hence this calendric beginning is associated with Taurus. It may also be significant that the exceptionally accurate star charts displayed on the ceilings of Egyptian tombs depict the positions of constellations as they would have appeared on this same Taurean zero date.[10]

This 4240 B.C.E. date also coincides with a fundamental galactic benchmark that may explain why this particular position was chosen to mark the beginning of both

---

*A friend recently brought to my attention the work of astrological historian Cyril Fagan, who had come to the similar conclusion that Taurus was intended to be the first sign of the zodiac. In his book *Astrological Origins* ([Chicago: Llewellyn, 1971], p. 27), Fagan cites historian Richard Hinckley Allen *(Star Names and their Meanings,* [New York: G. E. Stechert, 1899], p. 381), who similarly notes: "As first in the early Hebrew it [Taurus] was designated by "A" or ALEPH, the first letter of that alphabet, coincidently a crude figure of the Bull's face and horns . . . "

Also Fagan writes (p. 14): "So, what was the original, and hence, authentic zodiac of remote antiquity? . . . modern archaeological research has proved decisively that it was a sidereal zodiac, or fixed zodiac of the constellations, the initial point of which was Taurus 0 degrees and which was precisely 15 degrees, 00 minutes, 00 seconds from the fixed star Aldebaran, otherwise known as the 'Bull's Eye' in the mathematical center of the constellation Taurus."

Further he writes (p. 25): "Even among the Romans of the Aries age (1955 B.C. to 221 A.D.) when the vernal point was gliding backwards through that constellation, Taurus was still hailed as *Princeps armenti* or leader of the zodiacal herd!"

the Egyptian calendric system and the zodiac. On this date the vernal equinox was within 0.25° of arc from its closest approach to the galactic anticenter, and the autumnal equinox was within a similar proximity to its closest approach to the Galactic center. Thus the Taurus constellation was carefully positioned so that its starting point, as well as the zodiac's starting point, designates this very special galactic zero-point direction. No similar fundamental alignment can be claimed for the placement of Aries.

Schwaller de Lubicz pointed out that evidence for a 4240 B.C.E. origin for the Egyptian calendar indicates that its originators were scientifically quite advanced. For the development of such a sophisticated calendar, one that rivals our modern calendar in accuracy, presupposes many centuries of astronomical observation. But this dramatically underestimates the Egyptians' level of intellectual achievement. The choice of this particular date as a chronometric starting point indicates that the designers of the zodiac constellations and the originators of the ancient Egyptian calendar system had an intimate knowledge of the galaxy and knew that our solar system resides in a much vaster galactic system of stars that has a center. More important, it indicates that they knew where this center was located, something that modern astronomers first learned in the past several decades, following the development of large radio telescopes.

Decoding the zodiac to arrange its signs properly is somewhat like turning the dial of a combination lock. If we start with Taurus and proceed around the zodiac wheel in a consistent clockwise (retrograde) or counterclockwise (prograde) direction, the signs will not present their systems concepts in a meaningful order. The correct sequence is obtained by proceeding first clockwise from Taurus, then counterclockwise from Gemini, then counterclockwise from Scorpio, then clockwise from Libra, and ending with a jump over to Aquarius. When the signs have been properly rearranged, the zodiac's anticryptographic check mechanism then gives a reassuring indication that the decoding was correctly executed.

When placed in their proper order, the twelve signs subdivide naturally into two sets, each comprising six signs and starting and ending with one of the four signs that compose the sphinx:*

I: [Taurus]-Aries-Pisces-Gemini-Cancer-[Leo]
II: [Scorpio]-Sagittarius-Capricorn-Libra-Virgo-[Aquarius]

---

*A clue to this decoded sequence is found on an ancient drawing of the Esna zodiac sketched from the ceiling of the Temple of Khnum in Esna, Egypt. Cyril Fagan (*Astrological Origins,* p. 25) writes that the scribe had inserted a winged ram symbol in a vertical position between Taurus and Gemini and another between Scorpio and Sagittarius. The Egyptians usually used winged rams to represent the four cardinal points of the zodiac. Here they denoted that the first half of the zodiac commences with the constellation Taurus and the second half with Scorpio.

Bounded by Taurus and Leo, the lower half of the sphinx, the first six signs describe the lower half of the creation process, the hidden gestative processes that must take place in the ether if physical form is to emerge. Bounded by Scorpio and Aquarius, the upper half of the sphinx, the second set of six signs describes the upper half of the creation process, the stage in which physical form becomes outwardly manifest.

The segregation of the zodiac signs into two sets of six describing the implicit and explicit aspects of creation parallels the division of the first half of the Tarot into two arcana sequences: 1 through 5 and 6 through 10. It is appropriate that each set is composed of six astrological signs. Numerologically, six denotes the concept of bifurcation (Arcanum 6), or the moment when the initially uniform state of the ether begins to polarize as explicit order emerges.

Each sign is associated with one of two polarities (+ or −); one of three qualities (fixed, cardinal, or mutable); one of four elements (fire, air, water, or earth); one of seven traditional ruling planets; and one of twelve horoscope houses. Regardless of whether astrology is effective as a predictor of personality or events, these attributes and their associated personality traits traditionally confer specific symbolic meanings to each sign. These meanings, in turn, express specific systems concepts that together convey a coherent science of creation.

THE FIRST SEXTET: THE IMPLICIT ORDER

Taurus (the Bull) ♉

People born under the sign of Taurus (figure 9.2) are said to have powers of concentration, captivity, retention, and conservation. They are associated with solidity, strength, concentrated physical energy, determination, preservation, and self-sufficiency, all of which suggest a large reserve of strength. They are also said to have an instinct for gathering material wealth. Together, these concepts convey the notion of the accumulated value or a reserve of energy. The bull is an appropriate symbol for this concept of accumulated value, for in agrarian cultures it represents capital or wealth.

In a systems theory context, Taurus signifies the accumulation and storage of reactive driving potential, which, upon release, is able to produce some kind of physical, chemical, or electrical change. Thermodynamicists refer to such driving potential as thermodynamic force and equate it with chemical affinity, the non-equilibrium potential difference that drives a reaction forward, transforming a reserve of input reactants into reaction products. In terms of an ether physics, Taurus signifies the accumulation and storage of the potential that drives the universal ether flux. Taurus can be equated with the primeval mound, the self-created etheric source potential.

April 21 to May 21

Polarity: Feminine, negative

Quality: Fixed (stability, stored-up potential energy)

Element: Earth (solidity, stability, nurturing)

Planetary Ruler: Venus—attraction (the lower aspects: acquisitiveness and selfishness)

Second House: Financial rewards, financial security, property, material values

Figure 9.2. Taurus (the Bull). Systems concepts: energy reserve; source; driving potential; thermodynamic force.

Taurus is not easily matched up with any single Tarot arcana. Although the Bull's penchant for accumulating wealth is reminiscent of the Fool in Arcanum 0, unlike the Fool, who hoards his material possessions, the Taurean acquisitiveness is not a permanent state. For in Aries, this accumulated energy pours out with great force. Thus in some respects Taurus expresses the concept of beginning encountered in Arcanum 1.

March 21 to April 20

Polarity: Masculine, positive

Quality: Cardinal (directed activity)

Element: Fire (force, activity, creative energy)

Planetary Ruler: Mars—energy (the lower aspects: violence, dynamism, impulsiveness)

First House: Self-expression, the projected image

Figure 9.3. Aries (the Ram). Systems concepts: irreversible process; energy release; reaction flux; entropy production.

Aries (the Ram) ♈

Aries (figure 9.3) is said to signify chaos, disruptiveness, explosiveness, extravagance, force, strength, energy, vigor, impulsiveness, assertiveness, impatience, and spontaneity. All of these concepts together convey the notion of an energetic flux violently or profusely discharging from a central source. The ram is a very appropriate symbol of forward process. For at the least provocation, the male sheep has a penchant for charging forward head first. The glyph for Aries (♈) also conveys the notion of flux. Besides schematizing the ram's horns, it suggests a spouting fountain, which is symbolic of the tremendous outpouring of life force.

Aries depicts the systems principle of irreversible (nonequilibrium) process, the notion of *potential energy release.* It effectively portrays the universal tendency for entropy to increase over time, a phenomenon described by the second law of thermodynamics, and exemplifies the reaction-kinetic concept of reaction flux. The Taurean thermodynamic force is causally related to the Arian flux in that the former drives the latter, with the magnitude of the flux depending directly on the magnitude of the force. In a cosmological context, Aries represents the transmuting ether.

Aries matches best with Arcanum 1 of the Tarot. Like Aries, the Magus expresses the concept of universal flux. Interestingly, Osiris and Atum, gods who both symbolized process, were sometimes represented in combined form as a ram god known as the Ram of Mendes (figure 9.4).

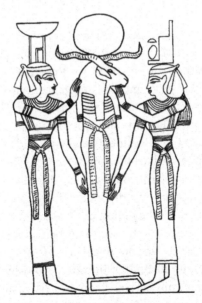

Figure 9.4. The Ram of Mendes, a mummylike god of generation consisting of a fusion of Atum-Re and Osiris. He is shown being sustained by Isis and Nephthys. From the tomb of Ramses II. Adapted from Piankoff, The Tomb of Ramses VI, fig. 5.

February 20 to March 20

Polarity: Feminine, negative

Quality: Mutable (oscillatory activity)

Element: Water (plasticity, mobility, sensitivity)

Planetary Ruler: Jupiter— expansion (the lower aspects: extravagance, impracticality) Neptune—impressionability, sacrifice

Twelfth House: Hidden resources and motivations; clandestine activities, associates, and enemies; confinements

Figure 9.5. Pisces (the Fishes). Systems concepts: differentiation; system; collective behavior; hierarchic levels; reaction-diffusion medium; coupling.

Pisces (the Fishes) ♓

Pisces, represented by two swimming fish and classified as a water sign, evokes the image of a vast celestial river (figure 9.5). In both Egyptian and Mesopotamian mythology, water is a traditional symbol for the ether, so it is natural to equate Pisces with the all-pervading primordial substance. Whereas Aries focuses on the energetic, explosive, dispersive nature of the etheric flux, Pisces emphasizes its substantive and plural nature. The outpouring becomes a great river made up of many intertwining currents, as the Piscean personality is said to embody many levels of complexity and to tend toward nebulousness. Hence Pisces brings in the concept of differentiation. In a cosmological context, it portrays the ether transmutation network, whose multifarious workings are hidden from direct view.

Pisces also clearly portrays the notion of system and the related concept of collective behavior. In a system, as opposed to an aggregation, the parts interact with one another and thereby participate together to form a larger organized whole. Similarly, Pisceans are said to find gratification by losing their personal identity through service to social causes. Their desire to participate in the collective whole is said to conflict with their need to maintain their individual identities, leading to a constant inner struggle.

With its fish and river themes, Pisces encourages us to reflect on how nature forms its multiplicity of hierarchic levels of systems within systems and to ponder how entities at one level interact collectively to constitute a system at the next higher level. Through this metaphor of microscopic dynamic entities collectively forming a substance that functions as a systemic whole, Pisces metaphorically conveys the notion of a reaction-diffusion medium. More specifically, Pisces effectively portrays the concept of a reaction-diffusion ether in which etheron constituents interact with one another to form a unitary collective whole. This contrasts with the mechanical ether of classical physics whose constituents are conceived to be inert and isolated from one another.

The fish and river theme fits quite well with this ether interpretation by showing how individual etherons (the fish) collectively form an integral ether medium (the river). Viewed at a microscopic level, each etheron particle reacts and transforms from one state to another, each such transformation being a discrete event. At a more macroscopic level, these individual transformations merge into one continuous process subject to the law of averages.

One ancient Greek myth provides a clue to the symbolism underlying Aries and Pisces. It tells of a time when Typhon, an exceedingly powerful hundred-headed monster, approached Mount Olympus and caused all the gods to flee in terror. They disguised themselves by changing into various forms. Zeus turned himself into a ram. Aphrodite (Venus) and her son Eros (Cupid) escaped by jumping into the Euphrates and changing into two fish swimming downstream. As a result, the two Piscean fish are sometimes named Aphrodite and Eros.

The frantically fleeing and transforming gods personify transmutative processes. Aries (the fleeing Zeus) depicts the initially unitary ether transmutation flux, and Pisces (Aphrodite and Eros swimming) depicts this flux after it has subdivided into two streams. Typhon personifies the chaos concept. He was one of the destructive Titans born of Gaea (Mother Earth, the ether) and was often associated with Set of Egyptian mythology or equated with Set's breath. Thus, like Set, he symbolizes entropy, the order-destroying tendency of nature that drives process, the fleeing gods.

The V-shaped form of the Pisces constellation displays in a graphic way the two ether transmutation pathways intersecting with one another. This V is sometimes represented as two intersecting streams of water spouting from the mouth of each fish, or as a ribbon or cord tying the two fish together. Alpha Pisces, the brightest star in the constellation that marks the intersection point of the V, was known to the ancient Arabs as Al-rischa (the Cord).

The Piscean glyph (♓) also expresses this theme of intersecting currents; the curved lines would denote the two transmutation pathways (the two swimming fish or two streams of water), and the horizontal slash indicates that these two paths are coupled. As mentioned earlier, an ether reaction network must have two intersecting pathways if it is to spawn a physical universe. In Model G, the two through-puts consist of the reactions $A \longrightarrow G \longrightarrow X \longrightarrow \Omega$ and $B + X \longrightarrow Y + Z$, both of which are interlinked by the reactant X common to both pathways.

Pisces contains in its symbology a rich variety of concepts. In matching this sign with the sequence of Tarot concepts, we find that it represents a synthesis of Arcana 1 and 2. Like Arcanum 1, Pisces connotes dynamism, a trait portrayed by its swimming fish, flowing river currents, and frantic flight of Aphrodite and Eros. But Pisces provides more detail about this flux, namely, that it is multifarious, with the oceanic expanse differentiated into a complexity of hidden currents. Pisces thus represents hidden form much as it is expressed by Arcanum 2. Consequently, like the Magician and High Priestess, Aries and Pisces convey the two basic aspects of *maya:* activity and form.

Gemini (the Twins) ♊

Many cultures around the world have similar representations of the constellation Gemini (figure 9.6). The Assyrians and Babylonians regarded it as "the Twins," as did the Peruvians in their star chart of Salcamayhua. In India, the two were known as "the Horsemen." The Egyptians called them "Horus the Elder" and "Horus the Younger," and also regarded them as two sprouting plants. To the Arabians they were two peacocks. The Eskimos knew this star group as the two vertical door stones of an igloo.[11] The ancient Greeks identified the constellation as Castor and Pollux, twin

May 22 to June 21

Polarity: Masculine, positive

Quality: Mutable (oscillatory activity)

Element: Air (communication, interaction, extension)

Planetary Ruler:
Mercury—communication (the lower aspects: restlessness, indecisiveness)

Third House:
Communications, personal ideas and their expression, short journeys, relatives

Figure 9.6. Gemini (the Twins). Systems concepts: duality; fluctuation.

youths affectionately embracing each other, and named the two brightest stars in the constellation after them. The Chaldeans called these two bright stars "the Two Kids," and the Australian Aborigines knew them as "the Young Men."[12] Interestingly, these various cultures often associated them with the mind and its activities. Such widespread commonalities lead us to suspect that these various civilizations may have inherited their stellar mythologies from some common source at a more distant time in the past.

Among other things, Gemini depicts the concept of duality. This theme is portrayed both by the Twins themselves and by their constellation glyph, which is formed by two vertical bars. This duality theme associates Gemini with Arcanum 2 of the Tarot. In fact, Gemini's glyph is reproduced there in the form of the two temple columns that flank the High Priestess.

Note that the Gemini and Piscean glyphs closely resemble one another. Whereas Pisces is represented by a pair of curved lines (two processes) connected by a single slash, ♓, the curved lines in Gemini are connected by two slashes, ♊. Gemini, then, suggests the concept of double-input coupling, an arrangement that characterizes cross-catalytic reactions. Like the Piscean glyph, it portrays a necessary feature that a reaction system must have if it is to be able to spawn ordered patterns and form a physical universe.

In ancient Greek mythology, the Twins were known as the *Dioscuri*, which means "the striplings of Zeus." They are said to have lived in the ancient Greek city of Sparta, known for its veneration of martial arts. They loved combat. Castor gained fame as a soldier and tamer of horses, and Pollux excelled in boxing. In their many adventures together, these invincible fighters were never separated. In their spare time, they often engaged in playful sparring.

This combat metaphor links the Twins with the systems concept of fluctuation. Ether concentration fluctuations portray a similar combative tendency as they continuously compete with one another for territory. The playfully wrestling Twins are an appropriate metaphor for concentration fluctuations erratically rising and falling in a reacting medium. Just as one twin temporarily gains an upper hand only to be later subdued by his partner, the concentrations of X and Y similarly flicker between two polarities: high Y/low X and low Y/high X. In either polarity, a rise in the concentration of one species is always accompanied by a decline in the concentration of its complement owing to the closed-loop coupling of the X and Y reaction processes.

The personality characteristics associated with Gemini also convey the concept of fluctuation. Astrologers assign Gemini a masculine polarity, which connotes things that are outwardly expressed. Like Pisces, Gemini has a mutable quality that is indicative of oscillation. Both of these traits, outward expression and oscillation, accurately describe the fluctuation process. In addition, people born under Gemini are said to be dualistic, versatile, changeful, unstable, erratic, restless, indecisive, nervous, temperamental, mercurial, and youthful. All of these adjectives describe ether fluctuations as they randomly emerge and subside like waves in a choppy sea.

Gemini people are also said to be extremely communicative. The sign is classified as an air element, which denotes communication, interaction, and extension. Gemini is also associated with the horoscope's third house, which is concerned with communications and short journeys, and is ruled by Mercury, an astrological planet concerned with communication. In particular, Gemini is said to be controlled by Mercury's lower aspect, which imparts restless and indecisive behavior. Fluctuations, in effect, can be regarded as a kind of communication. Just as spoken words are vibrations in the air medium, ether fluctuations are vibrations, but on a microscopic scale. In fact, as mentioned earlier, Hindu tradition states that the ether actually emits a sound as it transmutes (and fluctuates)—the universal OM.

Cancer (the Crab) ♋

Cancer's glyph (♋) reiterates the self-closing reaction-loop idea portrayed in the Gemini ideogram. More specifically, it suggests two mutually transmuting interlinked species. A similar concept is expressed by the Chinese symbol of T'ai-chi T'u (☯), which depicts the mutual transmutation of the female and male polarities, yin and yang, with yin continuously changing into yang and yang into yin. Just as the ancient Chinese sometimes pictured this interactive symbol by two fish swimming head to tail, Cancer's glyph similarly represents the male and female Piscean fish (Eros and Aphrodite) transmuting into one another in circular fashion.

Cancer is a cardinal sign, implying directed activity (figure 9.7). Whereas Aries, the preceding cardinal sign, depicts ether transmutation directed in a straight line, transmutation in Cancer is directed in a closed loop. Hence Cancer depicts the general systems concept of *process looping* or *circular causality*. It portrays the looping reaction process that occurs at our particular juncture in the ether network. Model G and the Brusselator, for example, incorporate two kinds of loop processes, positive feedback and cross-coupling. Both must be present if the transmuting ether is to generate ordered concentration patterns.

It is also appropriate that the authors of astrology chose the Moon to be Cancer's ruling planet. Just as the Moon passes through light and dark phases in its endless succession of cycles, X becomes Y and Y then becomes X, like the cyclic interchange of the Taoist yin and yang. This cycle concept is also found in the personality traits attributed to Cancer, as Cancerians are said to be susceptible to mood swings. Categorized as a water sign, Cancer is appropriately associated with the concepts of plasticity and mobility that suggest process. Moreover, it is fitting that Cancer should be designated a feminine sign, for it depicts the hidden implicit order or form of the processes that underlie our physical universe.

June 22 to July 22

Polarity: Feminine, negative

Quality: Cardinal (directed activity)

Element: Water (plasticity, mobility, sensitivity)

Planetary Ruler: Moon— protectiveness, sustainer of life

Fourth House: Hereditary tendencies, foundations, home and domestic issues, parents, environment, the state of affairs at the close of one's life

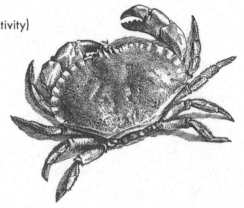

Figure 9.7. Cancer (the Crab). Systems concepts: process looping; circular causality; positive feedback amplification.

One outstanding feature of closed-loop positive-feedback reaction processes is their ability to amplify or nurture minute fluctuations arising in the ether. This nurturing aspect is a prominent feature in Cancer's symbology, as is the concept of protection, the second feature necessary for the growth of fluctuations. For example, astrologers tell us that people born under Cancer are mothering, nurturing, providing, self-preserving, possessive, sensitive, shy, gentle, protecting, covering, and dependable. Cancer is associated with the home, mother, womb, interior of all things, instinctiveness, compassion, and the need for peace. Its ruling planet, the Moon, is associated with both protectiveness and the nurturing tendency. The notion of upbringing is also a theme that underlies the zodiac's fourth house, which oversees hereditary tendencies, foundations, home and domestic issues, parents, and environment.

In summary, Cancer expresses a virgin birth concept similar to that portrayed by the Empress of the Tarot (Arcanum 3) or Isis in the myth of Osiris. The "child" fluctuation that Cancer shelters and nurtures to adulthood is none other than Gemini. The Cancerian symbology suggests that the child being raised is inherently dualistic, emerging as a pair of sparring fraternal twins (*X* and *Y*), with one twin temporarily dominating the other. In Cancer the fluctuation is not yet ready to stand on its own. At this stage it must continue to be protected, like the crab within its protective shell.

Leo (the Lion) ♌

Leo's symbology is most appropriately interpreted as signifying the concept of a critical fluctuation. Creativity and strength, characteristics evident in the list of Leo's attributes (figure 9.8), also recur in the list of associated personality traits, which include power, central will, vitality, dominion, authority, courage, self-confidence, dignity, extroversion, integration, and the ability to organize. Like the critical fluctuation that is about to break the ether's uniform symmetry, Leo portrays the notion of gathered potential, the lightning bolt that is ready to discharge. The idea of the emerging fluctuation is clearly depicted by Leo's arch-shaped ideogram (♌). In a cosmological context, Leo represents the precursor of the first particle of matter, the ether uprising that has the potential to conquer its homogeneous surroundings and bring about the emergence of physical order. This seminal role is consistent with the convention of associating Leo with procreation; Leo is affiliated with the horoscope's fifth house, which is said to determine a person's offspring and generative powers.

In seeking a match for Leo among the Tarot arcana, Arcanum 4 emerges as the natural choice, since the Emperor and the Lion both signify the notion of power. Like the Emperor, the Lion has traditionally been regarded as a symbol for royalty, decorating royal insignia through the ages. The national emblem of the ancient Persians displayed a crouched lion bearing the rising sun on its back. The United Kingdom has a similar insignia as its royal standard. In the Egyptian legend of the Divine Falcon, it is the Leonine One who holds the honor of serving as the keeper of the Nemis Crown, the wig worn only by royalty.

Regulus, the brightest star in Leo, symbolizes the heart of the lion and has long been associated with royalty. Translated from the Latin, *Regulus* means "the Little King." In Arabia this star was known as Malikiyy, "the Kingly One." In ancient Greece it was called "the Star of the King," and Pliny called it Regia, "the Royal One." The ancient Babylonians called it Sharru, "the King," and even earlier the Akkadians of Mesopotamia associated it with Amilgalur, a legendary "King of the Celestial Sphere" who ruled before the Great Flood.[13] The ancient Persians counted Regulus as one of the four "Royal Stars" and distinguished it as their leader, the other three Royal Stars being Aldebaran (the Heart of the Bull), Antares (the Heart of the Scorpion), and Fomalhaut (the Mouth of the Southern Fish).*

The Emperor of the Tarot is the equivalent of the Egyptian warrior god Heru-khuti, the form Horus took when he was ready to do battle with Set to bring physical order into being. Heru-khuti was, in turn, traditionally associated with the lion

---

*All are first-magnitude stars situated close to ecliptic and positioned approximately at right angles to one another. Each is situated in one of the zodiac's fixed signs, with the exception of Fomalhaut, which is situated below Aquarius.

July 23 to August 23

Polarity: Masculine, positive

Quality: Fixed (stability, stored-up potential energy)

Element: Fire (force, activity, creative energy)

Planetary Ruler: Sun—power, will, individuality

Fifth House: Offspring, generative powers, expression of personal creativity, romance

Figure 9.8. Leo (the Lion). Systems concept: the critical fluctuation.

symbol. For example, at Tebt (Tanis) Heru-khuti was worshipped in the form of a crowned lion in the act of trampling its enemies.[14] As the god of procreation, he is also represented in the form of a hawk with a phallus terminating in the head of a lion. In yet another representation he is shown seated on a throne that rests on the back of a lion. All of the above portray Leo as the sovereign warrior who reigns over a local kingdom of order that, with growth, eventually becomes permanently established as it forms the first particle of physical matter.

The Leo constellation, outlined as a crouching lion, suggests the form of the Great Sphinx of Giza or the Aker sphinx of Egyptian mythology through which the Sun was said to pass at its rising and setting. In this context, it is appropriate that Leo's ruling planet is the Sun because Egyptian hieroglyphs portray the rising sun in connection with the horizon lions as a symbol for the dawning of creation.

In summary, the Lion's various associations with Horus, the Sun, and the Aker link Leo to the warrior hero theme of the Osirian creation myth, which portrays the emergence of the first "light" into the universe. The first sextet of astrological principles together describe a nonlinear ether reaction system that is in a fertile supercritical state and in the process of amplifying a spontaneously emerging *X-Y* fluctuation. Up to this point in the zodiac sequence, the ether reaction variables maintain uniform steady-state concentrations, a condition characteristic of empty space. This uniform state is about to be disrupted, however, as the critical fluctuation begins to grow and develop a coherent macroscopic wave pattern, as depicted in the second sextet.

THE SECOND SEXTET:
THE EMERGENCE OF THE PRIMORDIAL WAVE

Scorpio (the Scorpion) ♏

The Scorpion is the revolutionary of the zodiac, the scorner of tradition (figure 9.9). He resists outside influences and all changes that he himself does not originate. Whereas the Hierophant (Arcanum 5) wishes to maintain the status quo of uniformity, Scorpio instead seeks to disrupt, forging the way for the emergence of the new order. In this respect, the Scorpio concept resembles that of Arcanum 16, the Tower of Destruction. In fact, Scorpio has traditionally been regarded as a sign of trouble and death.

For their own purposes, Scorpios inaugurate great changes whose consequences may be either constructive or destructive. These two antagonistic aspects are also reflected in the animal characters chosen to represent this sign. The sign's destructive side is symbolized by the serpent or scorpion and its constructive, creative side by the eagle. These opposites also appear in descriptions of the eighth house, ruled over by Scorpio and said to be concerned with both sex and regeneration as well as with matters of death and legacies. Birth and death are both inevitable aspects of revolutionary transformation, the ascension of the new order occurring simultaneously with the demise of the old.

As a fixed sign, Scorpio signifies the notion of stored-up potential energy. From its appearance, Scorpio's ideogram (♏) suggests that this accumulated power is about to discharge in a lightning-fast flash. The glyph's M-shape portrays the arched body of a snake ready to strike its victim. It is reminiscent of Nehaher, the energy-preserving coiled serpent that encircled Osiris. Where we would expect to find the serpent's head, the Scorpio constellation instead depicts the point of an arrow, the Scorpion's stinger, thereby suggesting that the emerging creative force is focused and penetrating.

We find these various concepts (stored energy, assertion, generative force, and destruction) in the personality traits attributed to this sign. Scorpio is said to be creative, resourceful, firm, determined, self-controlled, self-confident, self-reliant, willful, obstinate, proud, rebellious, resentful, vindictive, and passionate. It is said that Scorpios have strength enough to overcome any adversary and a willingness to tear down in order to rebuild. They are the revolutionaries, the initiators of upheaval, and the power behind the throne.

October 24 to November 22

Polarity: Feminine, negative

Quality: Fixed (stability)

Element: Water (plasticity, mobility, sensitivity)

Planetary Ruler:
Mars—energy (the higher aspects: assertiveness, forcefulness, generative force)

Eighth House:
Death, legacies, sex, regeneration through the use of other people's resources.

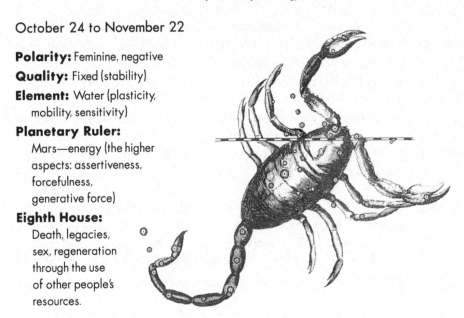

Figure 9.9. Scorpio (the Scorpion). Systems concepts: instability; critical point; hierarchical restructuring.

In systems terminology, Scorpio signifies the moment of instability, the instant when an existing order becomes unstable and begins to crumble, clearing the way for its replacement by a new, more stable order. The Scorpion, like Arcanum 6, signifies the critical point, the moment when the critical fluctuation is about to disrupt the prevailing uniformity of the reaction medium and nucleate the formation of a periodic pattern. In cosmological terms, Scorpio is best portrayed as the stage where matter/energy is about to come into being in the ether.

The concentration fluctuations that randomly arise in the ether are miniature revolts that, with sufficient growth, could destroy the ether's uniform steady state. Because they arise in a disorganized fashion, however, these momentary deviations average out to produce a relatively uniform field of random noise. Like Leo, Scorpio represents the rare occasion in which a fluctuation acquires a sufficiently large size to continue to grow despite the injuries inflicted upon it by competing fluctuations. The Scorpion represents the revolt, the coup d'état, the overthrow of the uniform steady state that creates the new order. In systems terminology, it signifies the concept of *hierarchical restructuring*.

There is a close connection between Scorpio and the Emperor pictured in Arcanum 4 of the Tarot. In the Marseilles Tarot deck, Scorpio is identified with the eagle that props up the Emperor's throne. He is the power behind the throne, the insurrection that elevates the Emperor to rulership. Scorpio, Leo, and the Emperor, then, are one and the same. All signify the emerging critical fluctuation. Whereas Leo embodies the notion of the fluctuation as a sovereign empire, Scorpio emphasizes the consequences of this fluctuation's development, namely the penetration and destruction of the old regime of uniformity and the creation of the new regime of explicit order.

Leo and Scorpio are closely associated in several other ways. Both are concerned with reproduction, although Scorpio is more specifically concerned with the sexual act itself. Ancient reliefs of Heru-khuti, the Egyptian warrior god of procreation, connect him with Leo as well as with Scorpio. Whereas his portrayal sometimes incorporates lion symbols, in one fresco this hawk-headed deity is pictured holding a scorpion in his left hand.[15] In art of ancient Mesopotamia, the lion and bird animal forms sometimes combine in the form of a lion-bird creature, possibly symbolic of Leo and Scorpio. The reverse combination, head of a bird and body of a lion, forms a royal beast called the griffin. A pair of griffins flank the king's throne in the Minoan palace at Knossos on the Island of Crete.

Sagittarius (the Archer) ♐

According to classical mythology, the Sagittarius constellation depicts a centaur named Chiron (figure 9.10), who is armed with a long bow and aiming an arrow at the heart of the Scorpion (Antares). Ancient Greek mythology describes centaurs as a wild race of beings that were gifted with both human cunning and the strength and speed of a horse. When their upper human half was subdued by wine, they would become belligerent and sexually uncontrolled.

Other cultures also associated Sagittarius with speed and aggression. In India, as early as 3000 B.C.E., the constellation was represented in the form of a horse, a symbol of strength, speed, and war. Inscriptions found on Babylonian and Persian monuments laud Sagittarius as "the Strong One" and "the Giant King of War." He was supposed to personify Nergal, their archer god of war and god of the burning midsummer sun. The early Egyptian zodiacs depicted Sagittarius as an arrow.[16]

The racing horse, flying arrow, and battle themes metaphorically express militant forward advance. In addition, astrologers traditionally associate Sagittarius with characteristics that suggest upward advance. Sagittarians are said to be aspiring, freedom loving, independent, friendly, outgoing, direct, optimistic, willing to take a chance, interested in moving into the future, and lacking in discipline. Sagittarius is ruled by the planet Jupiter in its higher aspect, which symbolizes expansion, optimism, and idealism. Sagittarius is also associated with the Ninth House, which is concerned with long-distance travel and expansive thought. The expansion theme is also implied by Sagittarius's classification as a fire sign of masculine polarity, which together signify forceful outward-directed activity.

The expansive growth that Sagittarius presents is very important to any reaction

November 23 to December 21

Polarity: Masculine, positive

Quality: Mutable (oscillatory activity)

Element: Fire (force, activity, creative energy)

Planetary Ruler: Jupiter—expansion (the higher aspects: optimism, idealism)

Ninth House: Higher education, philosophy, religion, long-distance travel

Figure 9.10. Sagittarius (the Archer). Systems concepts: expansive growth; spontaneous symmetry breaking.

process physics that attempts to explain the emergence of an ordered form. Growth allows an initially small ether concentration fluctuation to eventually develop into a macroscopic wave pattern. Since self-generated growth becomes significant only for fluctuations that have attained a critical size, this concept is best placed after Leo and Scorpio to properly explain the zodiac's story of cosmic creation.

In Model G and in the Brusselator, the X and Y reaction variables in a critical fluctuation cannot both increase in concentration simultaneously. Since these two species are part of a self-closing reaction loop that causes them to transform into one another, an increase in one variable (Castor) is necessarily accompanied by a decrease in the complementary variable (Pollux). Growth of a critical fluctuation inevitably causes the medium's uniform concentration to develop a localized hill-valley dichotomy, polarized either as an X-well/Y-hill or X-hill/Y-well.

This subtlety was appreciated by the authors of the zodiac, for in addition to expressing the concept of expansion, Sagittarius also portrays the systems concept of spontaneous symmetry breaking. Astrologers tell us that the Sagittarians are torn between exploring the highest human values and indulging in the lowest sensual desires. In a similar manner, when an X-Y fluctuation grows in size, its X and Y concentrations simultaneously depart from the norm in opposite directions; one soars in magnitude as the other plunges. This bipolar theme is also implied by the dualistic form of the Archer's body, which is half-human and half-horse. Ancient

December 22 to January 20

Polarity: Feminine, negative

Quality: Cardinal (directed activity)

Element: Earth (solidity, stability, nurturing)

Planetary Ruler: Saturn—restraint (the lower aspects: rigidity, miserliness)

Tenth House: Profession, public image, social contribution, parents

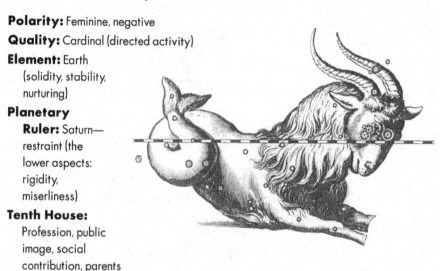

Figure 9.11. Capricorn (the Goatfish). Systems concept: exponential growth.

Egyptian and Assyrian engravings depict the Archer as having two heads, a human head facing forward and an animal head looking back. Sagittarius's designation as a mutable sign again implies duality. In addition, Sagittarius's glyph (♐) expresses both the theme of forward expansion, the arrow, and the theme of dichotomy, the slash across the arrow shaft.

Sagittarius embraces systems principles expressed by both Arcanum 6 (symmetry breaking) and Arcanum 7 (expansive growth). The spirit hovering in the clouds behind the Lovers (Arcanum 6) and the Conqueror, who stands in his forward-rushing horse-drawn war chariot (Arcanum 7), both call to mind the combative zodiacal Archer, thereby implying a close connection to Sagittarius.

Capricorn (the Goatfish) ♑

Capricorn is symbolized by the goatfish, a beast that has the head and upper torso of a goat and the tail of a fish (figure 9.11). The goatfish conveys a bipolar theme, as does the Sagittarian centaur. The personality traits ascribed to Capricorn, however, do not themselves emphasize duality but focus primarily on the theme of exponential growth that was encountered in a preliminary form in Sagittarius. In Capricorn this growth is more directed and rigidified. For example, Capricorns are said to be ambitious, hardworking, disciplined, patient, persistent, dependable, expedient, efficient, confident, realistic, constructive, and practical. They are crea-

September 24 to October 23

Polarity: Masculine, positive

Quality: Cardinal (directed activity)

Element: Air (communication, interaction, extension)

Planetary Ruler:
Venus—attraction (the higher aspects: cooperativeness, sociability)

Seventh House:
Partners, spouse, open adversaries, "others"

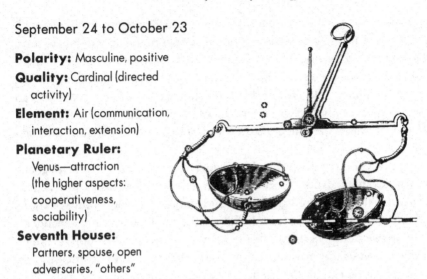

Figure 9.12. Libra (the Scales). Systems concept: balance through process coupling.

tures of routine and lovers of tradition. They willingly take on responsibility, have a strong sense of purpose, and seek prestige, honor, and success before the public. They also have a strong desire for money and a tendency to be materialistic and self-seeking. These themes are also reflected in the attributes associated with this sign, as Capricorn is a cardinal earth sign ruled by Saturn.

On the whole, these traits suggest a relentless upward advance, a theme epitomized by the goat, an animal known for its propensity to climb to great heights. Capricorn portrays the systems concept of exponential growth leading to a state of duality, in which one species (for example, Y) grows exponentially at the expense of its complement (for example, X). This same theme is portrayed by the Charioteer in Arcanum 7.* Exponential growth cannot continue indefinitely in a system of coupled reactions; eventually it will be limited by balancing processes.

Libra (the Scales) ♎

Libra signifies the principle of balance (figure 9.12). Its ideogram (♎) suggests the fulcrum of an equal-arm balance. Alternatively, it represents the counterbalancing of two processes represented by the two adjacent lines—one straight (male), the other curved (female).

Libra does not portray a mechanical type of balancing. Rather, it depicts the

---

*The forward-rushing chariot of Arcanum 7 calls to mind the constellation of Auriga (the Charioteer), which displays a chariot drawn forward by the female goat Capella.

kind of balance that is achieved when processes mutually couple or interact with one another, as in the case where two reaction processes mutually interlink to form a closed loop. In the ether reaction system, the two complementary species X and Y are interlinked by two processes, one that transforms X into Y and another that transforms Y back into X. Because of this bidirectional cross-catalytic coupling, an excessive growth in the product species produced by one reaction will eventually be brought to a halt by the consequent rise in consumption of that species by the partner reaction.

This subtle reaction physics principle was appreciated by the zodiac's designers, for Libra teaches that balance comes about through communication, interaction, and mutual cooperation. Astrologers tell us that Librans seek cooperative relationships and ventures; that they are tactful, diplomatic, manipulative, and dependent; and that they have a strong sense of justice. Libran people are also supposed to be concerned with conjugality, partnership, marriage, union, and "we" consciousness. Libra is concerned more with the relational aspect of a marriage partnership than with the sexual aspect. This theme of balance achieved through mutual communication or exchange essentially describes what takes place in the closed-loop etheron transformations that serve as the generative substrate for our physical universe. Libra makes a natural match to Arcanum 8 of the Tarot (Justice). Not only are the metaphorical principles in each case identical, but Justice is even shown holding aloft Libra's symbol, the equal-arm balance.

Virgo (the Virgin) ♍

With Capricorn limited by Libra's balance principle, neither X nor Y ends up totally dominating its complement; each sibling has its rightful chance to be on top as the X and Y variables cycle to form the emerging steady-state wave pattern. In Virgo, this wave achieves a state of refinement and order as it develops into a mature self-stabilizing wave pattern of precise wavelength. If some disturbance were to perturb it momentarily, the X-Y concentrations forming the wave would immediately proceed to reestablish their proper values.

Virgo portrays this process of refinement and self-correction through her various personality traits (figure 9.13). For example, Virgos are said to be selective, nit-picky, highly critical, analytical, and discriminatory. In their work they desire to reach the highest possible perfection, and they have a fondness for detail. Because they submit their own personalities to this rigor, they are refined in conduct and demeanor, but sometimes can be too self-critical. These traits all suggest the systems concept of *self-stabilization* and the related concept of *order maintenance.*

Virgo is said to rule the sixth house, which is concerned with a person's employment, service to others, and state of health. Employment and service to others helps to maintain the social order, and health refers to a condition of

August 24 to September 23

Polarity: Feminine, negative

Quality: Mutable (oscillatory activity)

Element: Earth (solidity, stability, nurturing)

Planetary Ruler: Mercury— communication (the higher aspects: discrimination, precision)

Sixth House: Jobs, personal service, health

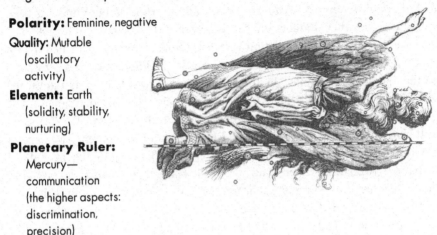

Figure 9.13. Virgo (the Virgin). Systems concepts: order maintenance; self-stabilization; the supercritical region; virgin birth.

optimal physiological or mental order. All of these convey the notion of order maintenance.

Virgo in essence portrays a variation of the balance concept already presented in Libra, since self-stabilization and order-maintenance activities seek to achieve a precise state of balance. It is very appropriate that one ancient Greek myth about the virgin Astraea (Star Virgin) should affiliate Virgo with Libra. According to this myth, Astraea, the goddess of justice, innocence, and purity, was the last of the deities to abandon the Earth during a period when humankind suffered a decline to baser levels. When she left, she was placed in the heavens as Virgo with a pair of scales (Libra) in one hand and a sword in the other.

Virgo is indeed situated adjacent to the Libra constellation; however, Libra is at her feet and there is no sword in her left hand. Nevertheless, the woman portrayed in Arcanum 8 (Justice) fits this description very well. Although Virgo's form-perfecting principle complements the balance principle expressed by Arcanum 8, Virgo is best matched with Arcanum 9, since she and the Sage both express the principle of perfection of form (or knowledge).

For a fluctuation to grow into a wave pattern, the reaction system in its vicinity must be in a supercritical state, that is, the fluctuation must reside in a "fertile" region. The same is true in the open-system ether physics of subquantum kinetics, if an ether fluctuation is to grow into a subatomic particle (see chapter 5). This fertility concept is symbolized in the myth of Osiris by the royal goddess Isis and in the Tarot by Arcanum 3 (the Empress). In astrology, the fertility principle presents itself in the sign of Virgo.* In ancient cultures around the globe, Virgo has been identified as the Mother Goddess and has usually served as a symbol of fertility. The Egyptians associated this constellation with Isis; the Babylonians identified it with Ishtar, Daughter of Heaven and Queen of the Stars. The Euphratean star list names Virgo the "Proclaimer of Life." In India this asterism was known as Kanya, the Maiden, and mother of Krishna. In China she was "the Frigid Maiden." The Peruvians knew Virgo as "Magic Mother" and "Earth Mother." The "Queen's Festival" held in her honor was dedicated to maize and to women. The ancient Greeks and Romans identified Virgo with Persephone, the daughter of the goddess of the harvest, sometimes called Daughter of the Harvest. They also portrayed her as Rhea, daughter of the sky and earth, and mother of Zeus, the creator of the universe. She was portrayed as a beautiful woman wearing a wreath formed from spikes of wheat. Her right hand rested on a stone pillar, while her left hand held a sheaf of wheat, the latter represented by the star Spica.

The zodiac's ether metaphysics implies not only that Virgo is a sexually fertile woman, but also that she is pregnant; her baby is the fluctuation that has grown into the first particle of matter. Both her name, "the Virgin," and her cross-legged posture, also portrayed in her ideogram (♍), suggest that her pregnancy resulted from a virgin conception; the seed that spawned her child arose within her own body through parthenogenesis.

The ancient Egyptians recounted an interesting myth about their fertility goddess, Isis, whom they associated with the Virgo constellation. It states that Isis was fleeing from Set one day and in her haste dropped a sheaf of corn she was carrying. The grain that she accidentally scattered came to form the Milky Way. Another version says she was holding three heads of wheat that subsequently became scattered. Virgo is pictured in the sky scattering this grain with her left hand. The sheaf of corn or wheat in her hand is represented by Spica, the brightest star in Virgo and also one of the brightest stars in the zodiac.

In the context of the galactic-evolution scenario outlined in chapter 5, Isis, the Mother of Creation, signifies the primordial self-created mother star at the center of our galaxy, which has spawned numerous generations of daughter stars. Set, the god of natural catastrophes, signifies the tremendous explosions that repeatedly expel stars and gas from this core object. The grain that Virgo scatters outward represents this expelled material, which has since come to form the stars that make up the Milky Way's spiral arms.

Interestingly, the ancients placed Virgo in a part of the sky that happens to contain an immense concentration of nearby galaxies (see figure 9.14). This collection

---

*Earlier we associated Cancer with Isis and Arcanum 3 as well, in the sense of the related mother concept of child nurturing.

of over ten thousand galaxies makes up the Virgo supercluster that stretches outward over a distance of about 80 million light-years, covering almost half of the area of the sky. Although there are other superclusters in the universe that are as large or larger, this one is unique in that it is the closest; its center lies about 30 to 40 million light-years away. To find a collection of galaxies of comparable size, one must look at least eight times farther away. It is also unique in that the local cluster of galaxies, of which our own galaxy is a member, is part of this supercluster.

The galaxies that compose the Virgo supercluster are mostly concentrated at its center in a cluster that lies about 40 million light-years away. This dense hub, called the Virgo Galaxy Cluster, or "Virgo Cloud," contains about three thousand of the supercluster's more than ten thousand galaxies. It is quite astounding to find that the Virgin, who is stretched out across the supercluster's equator, is gesturing with

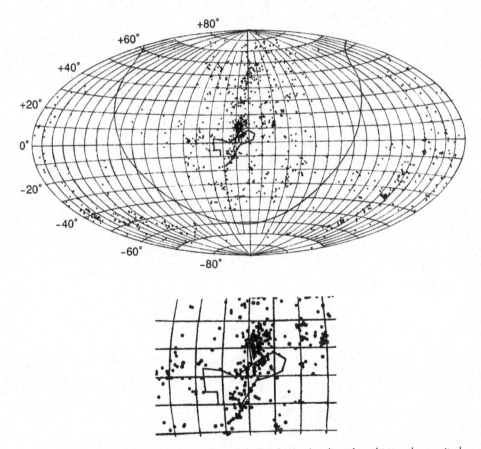

Figure 9.14. This sky map shows the locations of all galaxies brighter than thirteenth magnitude (Shapley and Ames, "Survey of the external galaxies," fig. 4), which is six hundred times dimmer than the naked eye can see. The Virgo supercluster is the elongated concentration of galaxies that extends from north to south. The central region is shown in magnification in the insert shown below. Virgo (figure 9.13) is positioned with her head to the west (right) and feet to the east (left).

her right hand precisely toward this supercluster hub. In fact, the stone pillar, which according to ancient legend supports her right hand, happens to coincide with the supercluster's central polar axis, and her hands are positioned such that they lie along the supergalactic plane, the equatorial plane that extends along the super-cluster's long dimension. Thus the Virgo constellation reenacts the Isis myth using the supercluster metaphor. With her right hand she indicates the supercluster's center, and with her left she seeds thousands of galaxies outward along the superclus-ter's equatorial plane. Quite appropriately, the stream of galaxies forming this equatorial plane passes directly across Virgo's womb.

It is unlikely that the supercluster galaxies were thrown out from a center like the matter forming our galaxy's spiral arms. The clustering of the galaxies is best explained if the ether reactions are more fertile (supercritical) in this part of space as compared with surrounding regions. Nevertheless, by placing Virgo in this important part of the sky, astronomers of old may have been trying to tell us that the same virgin-birth process that spawns matter within our own Milky Way also generates the countless galaxies scattered throughout the universe, and that each distant island of light, like our own Milky Way, consists of billions of stars.

The placement and configuration of the Virgo constellation suggest that the authors of the zodiac had an unusually advanced knowledge of the cosmos com-pared with what is known today. Just to distinguish the morphology of the tiny luminous specks representing the galaxies of the Virgo cluster at the superclus-ter's center, an astronomer would need at least a six-inch-diameter telescope capable of producing a magnification of over fiftyfold. To effectively map out the faint galaxies making up the entire supercluster, a much larger instrument is required, and to recognize that these spiral "nebulae" are actually islands of stars lying millions of light-years outside of our galaxy is by itself a major break-through in scientific thinking. Modern astronomers did not come to this realiza-tion until after the turn of the twentieth century, following the discovery of the cosmological redshift phenomenon.

Aquarius (the Water Bearer) ≈

The ideogram for Aquarius—two wavy lines parallel to one another (≈)—provides a reasonable portrayal of the mirroring X and Y wave patterns that constitute a sub-atomic particle's form. The X and Y peaks that form a subatomic particle's wave pattern are similarly shifted in phase with respect to each other, with one species reaching a maximum where the other reaches a minimum, like the two spirits bound to the turning wheel in Arcanum 10.

Aquarians are said to be very mental people, with a tendency to relate to others through reason and logic, rather than through emotions (figure 9.15). Through this association with thought and knowledge, Aquarius portrays the archetypal concept

January 21 to February 19

Polarity: Masculine, positive

Quality: Fixed (stability)

Element: Air (communication, interaction, extension)

Planetary Ruler: Saturn—restraint (the higher aspects: stability, self-discipline) Uranus—originality, freedom, eccentricity

Eleventh House: Ideals, aspirations, social values, friends

Figure 9.15. Aquarius (the Water Bearer). Systems concepts: explicit order (matter creation); the evolution of order.

of explicit order. Thoughts, for example, constitute our mind's outwardly expressed explicit order, whereas the associative processes taking place in the depths of our unconscious that give birth to thoughts constitute our mind's unexpressed implicit order. The same order-through-fluctuation process that describes the growth of a critical ether fluctuation into a subatomic particle also describes the spontaneous emergence of a creative thought from the mind's unconscious. Applying this thought metaphor of explicit order to the context of cosmology, Aquarius represents the fully created subatomic particle.

The human being is unique in the animal kingdom as a creature with the ability to think creatively and evolve complex thought structures. In fact, the word *man* comes from the ancient Sanskrit *manu*, "to think." So it is fitting that the zodiac has chosen an adult human form to depict this explicit order principle. The Gemini youths represent the precursors to thought, fluctuations arising in the unconscious, and the adult personages of Virgo and Aquarius represent thought when it has entered the conscious domain. Virgo portrays the feminine, intuitive, right-brain mode of thinking that is concerned with the generation of creative thoughts. Aquarius, on the other hand, represents the masculine, logical, left-brain mode of thinking concerned with testing the reasonableness of new ideas and integrating them into the larger whole of the mind's store of learned knowledge. Virgo is the mind's private creative side, while Aquarius is its public social side.

Just as the mind is capable of producing a continuous stream of thoughts, Aquarius signifies the creation of countless numbers of material particles (hydrogen atoms). This continuous stream of matter creation is symbolized by the river of water that emerges from Aquarius's urn and becomes disseminated through space. There is also a social, humanitarian slant to Aquarians. They are said to constantly seek ways to serve the whole of humanity by making active contributions to society. They also tend to take a global long-range perspective oriented toward social progress, constantly seeking to alter present-day circumstances to meet future needs. This social theme is present in the attributes of the eleventh house, which deal with a person's ideals, aspirations, wishes, and social values. By focusing both on ideas and on the much larger knowledge construct that makes up an individual's social environment, Aquarius expresses the long-term evolution of thought, or more broadly, the long-term evolution of an ordered system.

Two opposed sets of Aquarian traits effectively illustrate the Aquarian principle of evolution. On the one hand, Aquarians are said to be opinionated, cold, and detached, inflexible, critical, and demanding in positions of authority. Astrologers attribute these form-crystallizing traits to Aquarius's ruling planet Saturn. On the other hand, Aquarians are also said to be rebels, individualists, liberators, inventors, and generally people with an unabating eagerness to change the established order. These form-shattering traits are attributed to its second ruling planet, Uranus. Through the balanced interplay of these two processes, ordered forms such as subatomic particles or mental thoughts are able to change form and evolve.

Figure 9.16 illustrates this evolutionary process. An ordered state originally stable and closed to change (Saturnian principle) enters a period of instability in which it is open to change (Uranian principle). During this time, positive feedback processes enable an emerging critical fluctuation to grow and eventually drive the system to a new state of order. Once in this new state, the system is once again closed to change. In this new domain, negative-feedback form-stabilizing processes are sufficiently strong that fluctuations are unable to alter the system's state of order. As conditions change with the passage of time, however, a new period of openness or glasnost emerges, and the process repeats. Each abrupt transition constitutes a kind of revolution in which the old order crumbles as the new order emerges to take its place. Systems theorists call these two diametrically opposed processes *morphogenesis* (form creation) and *morphostasis* (form stabilization).[17] Both must alternately operate for evolution to proceed. Without the discipline of morphostasis, a system would evolve into a state of chaos and ultimately become annihilated. Without the occasional liberation brought about by morphogenesis, a system would be unable to adapt and change.

Evolution might be compared to a mountain climber scaling the face of a cliff; the climber consolidates each upward advance by driving crampons into the rock to serve as footholds for his next advance. In this way, evolution proceeds through

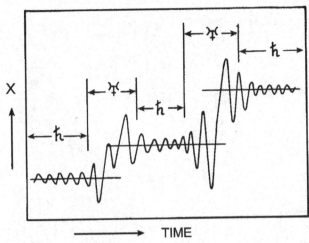

Figure 9.16. The evolution of ordered states through a series of discrete quantum jumps. Alternating stages of Saturnian (♄) and Uranian (♅) principles are indicated.

a series of forward jumps. Systems theorists recognize this stepwise advance to be a basic characteristic of system evolution.[18] Sudden evolutionary transitions are found in a variety of phenomena: in the quantum jumps of physics, the quantum speciations of evolutionary biology, the paradigm shifts of epistemology, the "aha" experiences of psychology, and the revolutions of politics.

In summary, Aquarius depicts the concept of long-term system evolution. In terms of the zodiac's ether metaphysics, Aquarius portrays the ability of an energy quantum to adopt various stable or semistable subatomic particle states (electron, muon, pion, proton, neutron, and so on) depending on its available energy. Thus, with this last sign in the zodiac's decoded sequence, we have moved one step up in nature's system hierarchy; like the sun emerging at dawn from the Aker's leonine underworld bowels, we have emerged from the realm of the ether into the physical world, the realm of humankind (Aquarius).

The ancient Egyptians sometimes imagined the heavens to be a great flood, which they represented in the form of a cow whose star-studded belly formed the sky. This was the mother goddess Hathor, a form of Isis. The relief of Isis-Hathor pouring water from a vase (figure 9.17) calls to mind the constellations of both Taurus and Aquarius, possibly suggesting a close connection between the two. Taurus, the energy source powering the etheric flow, may be identified with Aquarius's jug. In fact, Taurus's ideogram somewhat resembles a jug. The water flowing from Aquarius's jug would signify the etheric flux (Aries and Pisces) that emerges from Taurus. In this regard, it is significant that the stream of water flowing downward from Aquarius's jug in the sky falls directly into the mouth of the Southern Fish, Pisces' southern cousin.

Figure 9.17. Isis-Hathor pouring water into an irrigation canal from which
corn is sprouting (Schwaller de Lubicz, Sacred Science, p. 93). The figure to the left
shows Osiris as the Nile in his underworld cave, guarded by the cosmic serpent
and holding two vases overflowing with water (the symbol of Aquarius).
From Philae, an island in the Nile five miles south of Aswan.

This symbolism suggests that the ordered wave patterns emerging in Aquarius have a double significance. On the one hand, they signify the self-created subatomic particles filling our universe, and on the other hand, they represent the etheron particles that form the transmuting ether. Consequently, besides explaining the origin of physical form, the zodiac's science of creation comes down one level in the cosmic hierarchy to explain the origin of the ether itself. By moving full circle in this fashion, the zodiac suggests that the creation principle it portrays is one that is enacted over and over again to produce the many dimensions of being, physical as well as nonphysical. Like the sacred word *yod-he-vav-he*, the zodiac wheel describes the creation of a succession of systems forming an unending cosmic hierarchy. It is fitting, then, that the sacred symbols INRI on the Rose Cross were interchangeable with the four signs of the sphinx.

With Aquarius, we complete the second sextet of astrological principles, which together describe the emergence and stabilization of the primordial particle wave. Whereas in the first sextet the ether concentrations maintain an unblemished uniformity (space devoid of physical form), the insurgent critical fluctuation in the second sextet disrupts this uniform state and spawns the first particle of matter.

Aquarius enjoys a somewhat isolated position in the decoded zodiac sequence as the only sign to have neither a preceding nor succeeding sign at its side. This isolation suggests a culmination or completion to the symbol sequence depicting the creation process. This unique placement also underscores Aquarius's status as a summary sign, one that subsumes the operation of all the systems principles of the preceding signs and shows this overall creative process to be applicable in several contexts: etheric, physical, and mental. In serving as a summarizer and terminator of the decoded zodiac sign sequence, Aquarius shares much in common with Arcanum 10 of the Tarot. With the signs placed in their proper order, the following correspondences may be established between the first eleven major arcana of the Tarot and the twelve astrological signs:

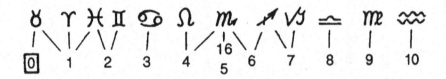

Table 9.2 summarizes the basic steps in this ancient science of systems genesis. Specific Tarot arcana and astrological symbols, listed in the first column, are matched with general concepts, listed in the second column. The last three columns apply these generalizations to explain the origin of matter and energy, of life, and of thought-creation mysteries dealing with the emergence of the three evolution vectors that compose nature's hierarchy of physical systems. Although the ancient science can be generally applied to all of these emergence phenomena, its exposition in ancient creation myths and lore is directed toward explaining the first of these creation mysteries—the origin of physical form.

THE LEGEND OF CASTOR AND POLLUX

One ancient Greek myth about the Gemini twins should dispel any lingering doubts as to astrology's symbolic depiction of a science of order genesis. According to this myth, Castor and Pollux were not identical but fraternal twins. They came from the same mother but had different fathers. Castor and his sister Clytemnestra were fathered by Tyndareus, a mortal. Pollux and his sister Helen, on the other hand, were fathered by the immortal god Zeus but were later adopted by Tyndareus. It was fated that Castor, born of a mortal father, would one day die. In fact, Castor met an untimely death during one of the twins' adventures. The two young men frequently engaged in bitter rivalry with their cousins Idas and Lynceus, who also were twins, and in one of those squabbles Castor was killed. Lamenting his brother's loss, Pollux implored Zeus either to restore Castor to life or to deprive Pollux himself of immortality. Zeus granted his wish by allowing both to share

TABLE 9.2

A Generalized Theory of System Genesis

TAROT ARCANUM & ASTROLOGICAL SIGN	CREATION GENERALIZED	PHYSICAL CREATION	CREATION OF LIFE	CREATION OF THOUGHT
0, ♉	Preexisting substrate composed of individual units.	Ether composed of etherons.	Ocean composed of organic molecules.	Stream of consciousness composed of feeling tones (sensory experience).
1, ♈ ♓	The substrate units engage in transmutative processes.	Etherons irreversibly transmute from one ether state to another.	Genelike and enzymelike polymers are continuously synthesized from nucleic and amino acids.	Feeling tones circulate in the brain's limbic system and continuously trigger new feeling tones from memory.
2, ♊	The substrate is in a state of chaotic fluctuation.	The concentration of each ether state randomly rises and falls, like waves on a turbulent sea.	Sets of genes and enzymes randomly form and dissolve within cell-like proteinoid microspheres.	The intensity of individual feeling tones increases and decreases in a complex manner.
3, 4 ♋ ♌ ♏	Owing to positive feedback in the substrate's transmutative processes, a sufficiently large fluctuation becomes amplified.	Owing to positive feedback in the ether's transmutation, the amplitude of a given ether fluctuation exponentially increases.	As a result of recursive synthesis, a unique set of genes and enzymes duplicate themselves faster than environmental perturbations can fragment them.	Closed-loop cycling of feeling-tones between the thalamus and cortex amplifies the intensity of a specific feeling-tone set.
6, 7, 8, 9, 10, ♐ ♑ ♎ ♍ ♒	A macroscopic ordered pattern or structure spontaneously emerges.	A wavelike concentration pattern emerges in the ether, that is, a subatomic particle forms.	A colony of primitive cellular organisms emerges, each cell capable of synthesizing its own gene-enzyme polymer sets.	A thought emerges to conscious awareness, manifested as an amplified feeling-tone set.

immortality. Each day they were allowed to live and die in an alternate fashion, one of the twins rising to the etheric realm while the other rested under the earth.

Chapter 11 of Homer's *Odyssey* describes the fate of Castor and Pollux as follows: "Both these the life-giving earth holds alive; they having even in the nether world honor from Zeus. Now they are alive alternately, and now again they are dead." Commenting on this passage, mythology researcher Robert Brown states:

> Now I defy any ordinary interpreter either of Homer or of myths generally to explain this precise and very singular statement. It is just one of those sayings so hard to understand, and yet so clear and decided in its terms, in which the real student of mythology recognizes an archaic truth, the primary meaning of which has long been forgotten, whilst the formula descends from age to age.[19]

Nevertheless, armed with an understanding of how open systems generate dissipative structures, we are now in a position to take up Professor Brown's challenge. In this touching tale, the ancient mythmakers have brilliantly captured the essence of how a critical ether fluctuation (the Gemini twins) grows to become a particle of matter (their shared state of alternately living and dying). More generally, it describes how a critical fluctuation leads to periodic behavior in nonlinear open systems.

Castor, an expert horse racer, calls to mind Sagittarius, who is known for his swift speed and fighting prowess. Like Sagittarius, the alternately living and dying Castor and Pollux signify the emerged dualistic state of order. Euphratean cylinders portray the twins as two human figures, one above the other, positioned either head-to-head or feet-to-feet, so that when one is upright, the other is upside down. Similarly, in forming the primordial subatomic particle, X and Y bear a reciprocal relationship, one being high when the other is low.

The myth demonstrates the emergence of the primordial wave pattern in an ingenious way. Immediately following Castor's death, the twins both reside at ground level, the "zero point," the uniform steady state. Through the decree of Zeus they are boosted from this low-amplitude state into the high-amplitude state of being alternately alive in heaven (above zero) and dead in the underworld (below zero). Zeus, the creator of the universe, exemplifies the critical fluctuation that grows in size and overcomes the forces of chaos to establish a coherent state of physical order (see chapter 11). Thus it is very appropriate that Zeus is the one to voice this decree.

The reciprocal oscillation of the twins above and below ground level accurately represents the notion of the oscillatory ordered state. Because of their mutual bond (brotherly love), the Pollux-alive/Castor-dead state (high Y/low X) inevitably evolves into the Castor-alive/Pollux-dead state (low Y/high X), and back again. In so doing, they trace out the way X and Y vary in moving radially outward from the center of the primordial particle. What a romantic way to illustrate the energy potential waveform of a subatomic particle.

Interestingly, the star traditionally named Castor, the brightest star in the Gemini constellation, actually consists of three binary pairs, or six stars in all, the number six appropriately calling to mind the principle of bifurcation (Arcanum 6). But Castor's double nature can be detected only with the aid of a telescope. The renaissance astronomer Cassini telescopically resolved it into two stars for the first time in 1678. These are found to orbit each other about once every four hundred years. Closer scrutiny more recently has revealed that this pair is orbited once every ten thousand years by a very faint red dwarf companion, thereby providing a second level of duality.[20] Moreover, further study with sophisticated spectroscopic instruments revealed that each of these three stellar components is itself a binary star. The separations of the close components in each set present angles that are over ten thousand times smaller than the smallest angle distinguishable to the unaided human eye. Was it just by luck that the Gemini twins were placed in this part of the sky, with this "king of binary stars" being chosen as the constellation's lead star? Somehow, thousands of years before the time of Galileo and Cassini someone must have already charted the heavens using highly sophisticated astronomical equipment.

ASTROLOGY DECODED

An analysis of the sequencing of the polarities and qualities for the astrological signs gives an additional indication of the soundness of our conclusions. When the signs are placed in their original order, the polarities alternate between negative and positive, and the qualities repeat in four cycles of fixed/mutable/cardinal (figure 9.18a). Even if the zodiac circle is begun from a sign other than Taurus, a similar repeating order prevails.

When the signs are arranged to relate their reaction-kinetic physics of cosmic creation, the polarities produce the same pattern, again alternating between negative and positive, while the qualities adopt a new ordered arrangement (figure 9.18b). They now naturally divide the sequence into two sextets, each sextet portraying mirror-image symmetry. That is, the two trinary quality sequences within each sextet appear as mirror images of one another. Thus rearranged, the qualities convey the systems concept of bifurcation, the initial uniformity of fcm | fcm | fcm | fcm now divided into two macroscopic complementary patterns: $[f^c{}_m{}_m{}^cf]$ and $[f_m{}^c{}^c{}_mf]$. These seem to portray the two X-Y concentration alternatives (high Y/low X and low Y/high X) that are responsible for generating the etheric wave patterns that constitute physical form. This is quite appropriate, for the zodiac's encoded science of physical creation describes the emergence of just such a dualistic state.

If the four fixed signs in the decoded sequence are immobilized, any other rearrangement of the intervening signs destroys the prevailing order and ruins either the alternating pattern of the polarities or the mirroring symmetry of the qualities. Removing any restrictions on the ordering of the fixed signs and considering the

Figure 9.18. Astrological polarities and qualities: (a) original order; (b) reaction-kinetic order.

permutations of all twelve signs, there are only three other arrangements of the signs that preserve a comparable order for the polarities and qualities. One arrangement is produced by rotating each sextet around its central axis of symmetry; another by reversing the entire twelve-sign sequence; and yet another by interchanging the two sextets so that sextet II comes before sextet I. These would yield the following alternative sequencings for the fixed signs: Leo-Taurus-Aquarius-Scorpio; Aquarius-Scorpio-Leo-Taurus; and Scorpio-Aquarius-Taurus-Leo. However, none of these gives the proper sequence for the signs of the sphinx, which is traditionally specified as Taurus-Leo-Scorpio-Aquarius. The proper sequencing places the fixed signs in prograde order, the order in which they are transited by the sun, and places Leo and Taurus (the lower part of the sphinx) before Scorpio and Aquarius (the upper part of the sphinx). This is also the same order indicated by the path of the rising sun as it emerges from the underworld through the eastern Aker gate. Thus the sphinx is an essential key to decoding the astrological cipher. Together with the polarity and quality pattern, it restricts the choice of possible sign orderings to just one, the order in which the signs best express their science of creation.

The sphinx also serves as a key to astrology by designating with its body parts the signs that begin and end the two sextet sequences; it points out the entrances and exits to these two sextets. Just as the Aker sphinx of ancient Egyptian mythology formed the entry and exit gates for access to the underworld, the etheric abode of Osiris, the fixed signs of the sphinx form the entry and exit gates to the decoded sextets. In fact, the two-headed Aker, with its sun-entrance head pointing west and sun-exit head pointing east, displays a bilateral symmetry similar to that portrayed by the arrangement of the qualities within each of the two zodiac sign sextets. Thus, the tradition of regarding the sphinx as a gate to the ancient knowledge in itself appears to be an important clue to understanding the manner in which the sphinx signs serve as a key for decoding the zodiac. The postulant priest wishing to become initiated into the Osirian mysteries was required to enter the subterranean passages leading to the Temple of Osiris by passing through a door at the front of the Sphinx. So this sphinx-gate hint was incorporated into the very structure of the initiation ceremony.

In designing messages that must bridge an inherent language/culture barrier, such as communications targeted for extraterrestrial civilizations, scientists usually express their text in terms of a symbolic cipher accompanied by a key and anticryptographic check device. The purpose of the key is to assist the recipient civilization in decoding and interpreting the ciphered message, and the purpose of the check device is to give the recipient an indication when the message has been correctly interpreted. It seems that the sphinx and the polarity-quality designators constitute such a key and check device. In particular, the emergent pattern of polarities and qualities, which has half-a-billion-to-one odds of arising by chance, gives us an indication that we have correctly interpreted astrology's encoded ether science. Furthermore, it informs us that this creation science is not something we simply imagine to be there, but that the zodiac's authors encoded it by conscious design.

THE GALACTIC CREATION CENTER

Among the zodiac sign glyphs, two appear to stand out from the rest, the designators for the signs Scorpio and Sagittarius. These are the only glyphs that incorporate arrows, the universal symbol for direction or location. The arrows represent the Scorpion's stinger and the arrow that the Archer is shooting.

Interestingly, these two signs lie adjacent to one another in the zodiac. When the glyphs are arranged to reflect the way their constellations would appear in the sky, Sagittarius is to the left (east) and Scorpio to the right (west). In portraying this arrangement, Scorpio's glyph must be inverted so as to accurately represent the actual sky appearance of the constellation, since Scorpio's stinger faces left rather than right. Depicted in this fashion, the pointers seem to converge on a certain place. In fact, they are pointing toward that unique point in the heavens around which all stars in our galaxy appear to revolve, the Galactic center. The authors of the zodiac apparently went to considerable trouble to indicate the Galactic center's position, for Sagittarius and Scorpio are the only zodiacal signs to extend so far south of the ecliptic (figure 9.19).

A quick review of the zodiac's creation science shows that the signs Scorpio and Sagittarius fall at the point in the decoded sequence where the critical fluctuation first bifurcates the homogeneous steady state and begins its emergence into physical expression as matter and energy. Since these two signs also point out the loca-

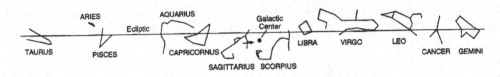

Figure 9.19. The positions of the zodiac signs relative to the ecliptic.

tion of the center of the Galaxy, we are led to conclude that the zodiac cipher indicates this central site as the place where the first particle of matter came into being in our galaxy and where matter continues to come into being as the process of matter creation unfolds.

In fact, a considerable amount of evidence suggests that the centers of galaxies are sites where matter is continually created. This has led some astronomers to conclude that most stars in a galaxy are formed from matter that has been expelled from a galaxy's creative center.[21] Subquantum kinetics predicts just such a creation scenario. So this interpretation is consistent with the creation cosmology that astrology encodes.

One ancient Greek myth states that Sagittarius is aiming his arrow at the heart of the Scorpion. The Scorpion's heart is represented by the red giant star Antares, which is one of the four Royal Stars. At the present time, Sagittarius's arrow points out a trajectory that misses Antares by five degrees of arc. The stars that form this arrow (Gamma and Delta Sagittarii), however, slowly change their position over time, such that the arrow would have been much better aimed in ancient times. In fact, it was aimed precisely on target around 13,865 B.C.E. (± 150). Its flight toward Antares would then have brought it within 0.35° of the Galactic center (that is, within three-fourths of a lunar diameter). This target date may mark an era when the Galactic center was unusually active and hence visible through the thick clouds of dust that now obscure it. A similar date of 13,865 B.C.E. (±20) is meticulously encoded in the zodiac displayed in the Temple of Hathor at Dendera.* If these dates mark the approximate era of the zodiac's origin, the zodiac may be as much as 15,800 years old. Such an age agrees with other estimates, which indicate that the Great Sphinx may date back more than fifteen thousand years (chapter 7).†

---

*This date is indicated by the line that projects due west from the zodiac's polestar, cutting the zodiac's ecliptic near the beginning of the Virgo constellation (ecliptic longitude = 215.9°). This precessional date is determined by reference to the temple's true north axis, whose ecliptic position (longitude = 125.9°) is accurately calibrated by the Sothic calendar date of 7160 B.C.E.

†This zodiac date is also comparable to the 14,700 B.C.E. age that nineteenth-century scholar G. Schlegel estimates for the Chinese duodenary zodiac (G. Schlegel, *Uranographie chinoise*, vol. 2 [Leyden, 1875], p. 796; D. E. Smith, *History of Mathematics*, vol. 1 [New York: Dover, 1951], pp. 18, 23). The Chinese system of astrological symbolism, though, is substantially different from that of Western and Hindu astrology. To date, no creation metaphysics has been discerned in its signs.

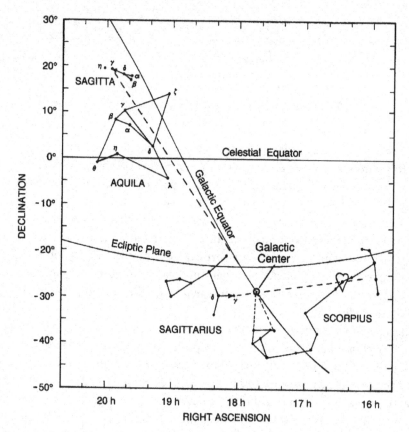

Figure 9.20. The positions of Aquila and Sagitta in relation to Scorpius and Sagittarius.

The Archer aiming his arrow at the Scorpion's heart calls to mind Cupid aiming his arrow at the heart of Vice in Arcanum 6. In fact, like Arcanum 6, Sagittarius and Scorpio represent the principle of spontaneous symmetry breaking. The passage of his arrow through the heart of the Scorpion, the symbolic heart of the Galactic center, indicates this hidden location as the place where explicit order first emerged to form the Milky Way.

At the time of this penetration, the Scorpion presumably transforms into the Eagle, Scorpio's higher form. The constellation Aquila (the Eagle) may represent this resurrected form (see figure 9.20). Aquila is pictured soaring through the heavens clutching in an upraised claw an upward-flying arrow, the constellation Sagitta (the Arrow). Some artists show Aquila actually being pulled along by Sagitta's flight. Sagitta's orientation gives the impression that this is the same arrow that Sagittarius shot through the Galactic center, here flying outward from the center along the galactic plane. This identification of Sagitta with the Archer's heart-piercing arrow finds indirect support in a myth stating that Sagitta is the arrow shot by Cupid.

Ancient Greek mythology specifies Aquila as the Bird of Zeus, the eagle that carried nectar to Zeus when he lay as a child in a cave hidden from his father, Cronus. Later, when Zeus was fully grown, this same eagle carried his weapons as he led his siblings in battle against Cronus and the chaotic giants. With their defeat, Zeus brought the ordered universe into being. Thus Aquila serves as facilitator in the drama of the emergence of order from disorder. We also find an eagle in Arcanum 4 of the Tarot, where it portrays the emergence of explicit order. These associations show that Aquila is closely tied to the creation drama portrayed by the zodiacal constellations.

Aquila seems to be referring to the parthenogenic creation of matter and energy that takes place within the Mother Star at the Galaxy's center. The outward flight of Sagitta and Aquila along the galactic plane suggests that, once created, this matter and energy flies to the Galaxy's periphery. The use of the arrow metaphor and the designation of Aquila as the carrier of Zeus's weapons of war both imply that this matter/energy creation and expulsion process can be quite violent. Perhaps Set symbolizes such a tumultuous birth event when he frightens Isis (Virgo) and causes her to disperse her cosmic grain, thereby forming the stars of the Milky Way.

The ancient authors of the zodiac possessed a relatively sophisticated knowledge of astronomy, for modern astronomers have only recently succeeded in locating the true position of the Galactic center. In the last several decades evidence has begun to surface suggesting that galactic cores not only generate matter and energy at very high rates but at times can violently erupt, spewing out intense volleys of cosmic ray particles and electromagnetic radiation.

Astrology, then, presents a highly evolved physics describing the creation of our universe. In particular, it points to the Galactic center as the site where the first particle of matter arose and indicates that the Galaxy as a whole was formed through a process of central creation and radial dispersal of matter. Virgo's sweeping gesture to the thousands of galaxies in our immediate supercluster suggests that other galaxies form through a similar process. At the same time, the zodiac conveys a second important message, namely, that this ongoing matter/energy creation process can at times become quite violent, resulting in explosive outbursts. Perhaps the authors of the zodiac were attempting to warn us that one such core explosion adversely affected our planet some 15,800 years ago and that the heart of the Galaxy could launch another attack at some unspecified time in the future. The past occurrence of this ancient global catastrophe is examined in detail in the sequel to this book entitled *Earth Under Fire*.

10
SUBATOMIC ATLANTIS

THE MYTH OF ATLANTIS

The story of Atlantis preserved in the dialogues of Plato has long been a source of wonderment for many. This powerful island civilization is said to have flourished during the last ice age and to have had a catastrophic end around 11,600 years ago. The tale, related by Socrates' elder student Critias, was originally told to Solon, ruler of Athens, by priests he met in the Egyptian city of Sais.*

Numerous books have been written about Atlantis, and much time and money has been spent seeking its whereabouts. Some have claimed that it exists as a sunken island in the Atlantic near the eastern coast of Florida. Others have said that it must be located off the coast of Spain. Still others believe that Atlantis was a civilization that once resided on the Aegean volcanic island of Santorini, the greater part of which was destroyed around 1400 B.C.E. by a powerful volcanic explosion. However, this gives a date for Atlantis's demise that is much too recent.

But was Plato's narrative of Atlantis intended to be taken literally? Most probably it was not. This is not to say that an advanced civilization did not once populate the Earth in prehistoric times. Indeed, the priest of Sais spoke ardently of the great accomplishments of the antediluvian Hellenes, a prehistoric civilization from which the ancient Greeks are said to have descended. The legend of the creation of Atlantis, however, is actually an allegorical account of physical creation. It describes how the first particle of matter came into being and how this particle served as a center of creation to spawn all the matter that constitutes our galaxy. Equally important, it seems to describe the occurrence of a major explosion at the center of our galaxy, an event that apparently was responsible for bringing about the downfall of the advanced prehistoric civilization that Plato says once lived upon our planet.

---

*If Plato was in fact an initiate of the Egyptian mysteries, as claimed by Proclus, he may have had the opportunity to check out the details of this myth to ensure that they were accurately recorded.

Any attempt to decipher the symbolic meaning of the Atlantis myth must explain why Plato's dialogues divide the myth into two portions. The first part is in the dialogue *Critias*,[1] which relates the story of how Atlantis was created and gives a detailed description of the physical layout and commercial activities of this metropolis. The second part, given in the *Timaeus* dialogue, describes how the people of Atlantis waged war on the antediluvian Hellene civilization and were finally destroyed in a worldwide flood. It is here that reference is made to the "sinking" of Atlantis.

Of the two narratives, the story in the *Critias* proves to be of particular interest from the perspective of systems science and the ancient ether physics. Like the creation myths analyzed earlier, the legend of Atlantis's creation encodes a sophisticated open-system ether physics that describes how the first particle of matter came into being from the etheric sea. It even encodes a diagram showing how the primordial particle's energy field intensities vary as a function of distance from the particle's center. In some ways, the Atlantean creation myth presents one of the most sophisticated and graphic portrayals of this ancient creation science.

In the creation story found in the *Critias*, we encounter the elder Critias relating to Socrates the following legend of Atlantis's creation. The story begins in the early days of the universe, just after the Olympian gods had defeated the Titans and deposed Cronus and his realm of chaos. They apportioned the kingdoms among themselves by shaking lots. Zeus came to rule the heavens, Poseidon the seas, and Hades the underworld. When they portioned out the earth, Poseidon received for his lot an island in the middle of the ocean. This was Atlantis. It had a diameter of roughly one hundred stadia (one hundred furlongs, or about twenty kilometers) and consisted for the most part of a very fertile plain in the center of which was a mountain of modest height. In this mountain lived two mortals, Evenor and his wife, Leucippe, both offspring of Poseidon. These two mortals had a daughter named Clito.

One day Clito's father and mother died, leaving her alone when she had barely reached womanhood. Poseidon, desiring this maiden, made love with her in her mountain abode. Thereupon, he fortified the surrounding territory by re-forming the ground so that alternating rings of sea and land enclosed the central hill where she dwelt. There were two rings of earth separated by three rings of sea, all concentric with one another so as to form a bull's-eye pattern. In this way, men were prevented from gaining access to the interior of the island where Poseidon lived. He then made two springs well up from underground, one warm and the other cold, and caused the soil to yield an abundance of food plants of all kinds.

Although Plato's dialogues do not specifically state that the Atlantis myth was meant to be a story of cosmic creation, ancient Greek mythology has left hints alluding to such an interpretation. For example, Poseidon, one of the heroes who fought at the side of Zeus to bring order into the universe and thereby create physical form, received this island when the gods were dividing up their newly won empire. So, not only is Atlantis of primordial origin, it is also connected with the first emergence of physical order.

Like the Egyptian, Sumerian, and Greek creation myths, the Atlantis creation myth portrays how a microscopic concentration fluctuation (energy pulse) arising in a chaotic ether grows into a mature subatomic particle. From this ether physics viewpoint, the ocean surrounding Atlantis to the east, west, north, and south symbolizes our ether-filled space. The ocean's featureless surface indicates that this ether is initially devoid of pattern, empty of all material form. The fertile sea-level plain comprising most of Atlantis's landscape represents a supercritical region in this etheric ocean. The vertical displacement of this landscape above or below sea level illustrates how the concentration of one of the ether variables (for example, Y) varies through space. The hill at the island's center depicts a large ether concentration fluctuation (energy potential pulse) that has emerged in this fertile region, locally increasing Y above the uniform steady-state concentration value represented by the surrounding plain.

Clito, the adolescent woman who has been born and raised within this central hill, personifies the fertile G-well in the ether that forms the womb within which explicit order emerges. Poseidon, who desires her and makes love with her at the moment she reaches puberty, represents the critical fluctuation (seed) that emerges within this supercritical region. Poseidon plays a similar order-creating role in the story of Zeus, where, together with Zeus and his four other siblings, he portrays the critical fluctuation that overcomes the titanic forces of chaos (see chapter 11).

Viewed in another way, Poseidon (water) and Clito (land) symbolize the two interacting X and Y ether variables that are cross-coupled into a self-closing transformation loop. Poseidon's breaking of the land contour symmetry to form alternating concentric rings of land and water (elevated land versus depressed land) illustrates how the X and Y ethers depart from their initially uniform steady-state concentrations to form a wavelike concentration pattern configured as a series of concentric shells. Whereas the myth of Mum portrays the symmetry-breaking concept in terms of the separation of the earth (Geb) and sky (Nut), here the concept is portrayed in greater detail using the water and earth as the corresponding ether metaphors.

A cross-sectional contour map of Atlantis's newly emerged landscape drawn to scale from the dimensions given in the *Critias* is shown in figure 10.1. A map showing how Atlantis would appear in an aerial view is presented in figure 10.2. According to the *Critias*:

> The breadth of the largest ring of water, that to which the canal from the sea had been made, was three stadia and a half, and that of the contiguous ring of land the same. Of the second pair, the ring of water had a breadth of two stadia and that of land was once more equal in breadth to the water outside it; the one which

immediately surrounded the central islet was in breadth one stadium; the islet on which the palace stood had a diameter of five stadia.[2]

This ringlike pattern of harbors and dikes described in the *Critias* might seem strange to someone unfamiliar with the science of reaction kinetics. A systems scientist, however, would immediately recognize a localized dissipative-structure wave pattern, similar to the steady-state concentric ring patterns produced in computer simulations of the Brusselator (figure 3.11) or the propagating chemical wave patterns generated by the B-Z reaction (figure 1.1). The subatomic particles spawned by Model G also have this form (see figure 5.4).

As recently as 2002, physicists announced that, using data from particle scattering experiments, they had for the first time accurately determined the shape of the electric field in the core of a proton or neutron. Previously the electric charge density inside a nucleon had been assumed to rise to a sharp peak at the particle's center. But, contrary to expectation, the new data indicates that the particle's electric field rounds off to a plateau as the particle's center is approached and also forms a stationary wavelike field pattern around the particle's core.[3] This recent finding not only confirms a key a priori prediction of subquantum kinetics, but it also validates the particle-wave model of ancient lore which is depicted so clearly in the ancient Greek Atlantean myth of *Critias*. As we ponder figure 10.1 we are numbed by the realization that some civilization thousands of years ago had already known what we ourselves have discovered only now at the dawning of the 21st century with the aid of high-tech particle accelerators.

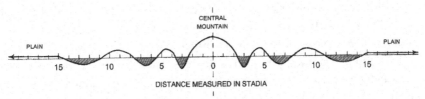

Figure 10.1. A cross-sectional view showing the general contour of Atlantis.

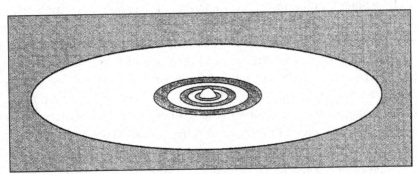

Figure 10.2. A map of Atlantis reconstructed from the description given in Plato's Critias.

Figure 10.3. The cross of Atlantis, an emblem that schematically depicts Atlantis's central mountain, surrounding moats and dikes, and transecting canals. From Muck, Secret of Atlantis, p. 9.

In the context of the ancient ether physics, Atlantis's land contour wave pattern charts the electric field potential that would compose the stationary wave profile of a proton. The myth is accurate even to the detail of describing a wave pattern whose wavelength increases with increasing distance from the pattern's geometric center. Such wavelength-broadening effects are observed in computer simulations of the Brusselator reaction system. The plain extending from the outermost moat to the island's ocean shore would portray the particle's surrounding potential energy field through which it exerts forces on neighboring particles.

Although the *Critias* gives Atlantis's dimensions in stadia, with one stadium measuring about two hundred meters, this presumably was used only as a means for portraying the relative spacing of the moats and dikes. The actual wavelength of this moat-and-dike pattern would be almost a million trillion times smaller, as the so-called Compton wavelength of a proton is about 1.3 trillionths of a millimeter.

Like the other creation myths, the Atlantis myth describes by means of its symbolic allegory the story of how a two-variable Brusselator-like ether reaction system is able to self-organize its constituents into an ordered field pattern. The story of Atlantis, then, is a scientific treatise explaining how the first subatomic particle materialized in the depths of space. The particles of matter forming our galaxy constitute Atlantis's cumulative lineage. The Atlantean contour map and the detailed symbolic structure of the myth itself both challenge us to radically reevaluate the intellectual sophistication of the civilization that produced this legend.

Ancient cults recognized the importance of the Atlantean ring pattern, for certain of its key features appear in the monogram known as the cross of Atlantis (figure 10.3), an emblem found in prehistoric stone circles as well as on sacrificial altars.[4] As mentioned in the discussion of the ankh (chapter 6), the cross has traditionally been the occult symbol for the physical plane. It specifies the place in the ether reaction network where the two primary ether transmutation pathways conjoin and give rise to our physical universe. The Atlantean bull's-eye pattern in combination with the cross represents the emergence of the primordial subatomic particle at this key etheric juncture, an elegant hieroglyphic symbol of physical creation. The sacred symbol of the Rose Cross (figure 8.16) also combines this ring-and-cross design.

As *Critias*'s story of Atlantis continues, Poseidon and Clito beget ten sons, five pairs of twins. The earlier-born of the first pair was named Atlas. He was appointed king and given ownership over his mother's dwelling place and most of the island. Atlantis and the surrounding Atlantic Ocean were named after him.

The Atlas of Atlantis is not the same as the Titan Atlas encountered in the Olympian creation myth; he just happens to have the same name. Very rarely in Greek mythology are different personages designated by the same name; the similarity here may be a literary device purposely introduced to connect the two myths. In the Olympian myth, Zeus assigns Atlas (the Titan) the task of keeping earth and heaven separated from one another. In so doing, Atlas symbolizes the universal power maintaining the material universe in its state of physical existence. In a similar way, Poseidon confers to Atlas (the king) the responsibility of administering this island empire so that its terrain and furbishings are maintained in an orderly state. Both are responsible for sustaining order.

Consequently, by linking the Atlantean and Olympian creation myths through Atlas and Poseidon, the authors of this tale seem to suggest that the Atlantean myth is a continuation or extension of the Olympian creation myth. The firmament or arch that the Titan Atlas maintains aloft in the Olympian myth is the same as Atlantis's central mountain, within whose center Atlas, the king, and his nine brothers are born.

The second half of the Atlantis myth describes how this island empire grew and developed in the years following its creation. The *Critias* reports that under the rulership of Atlas's descendants, Atlantis accumulated considerable wealth:

> They possessed wealth such as had never been amassed by any royal line before them and could not be easily matched by any after, and were equipped with all resources required for their city and dominions at large. Though their entire empire brought them a great external revenue, it was the island itself which furnished the main provision for all purposes of life. In the first place it yielded all products of the miner's industry, solid and fusible alike, including one which is now only a name but was then something more, *orichalch*, which was excavated in various parts of the island, and had then a higher value than any metal except gold. It also bore in its forests a generous supply of all timbers serviceable to the carpenter and builder and main-

tained a sufficiency of animals wild and domesticated; even elephants were plentiful. There was ample pasture for this the largest and most voracious of brutes, no less than for all the other creatures of marsh, lake and river, mountain and plain. Besides all this, the soil bore all aromatic substances still to be found on earth, roots, stalks, canes, gums exuded by flowers and fruits, and they throve on it.[5]

This portrayal of Atlantis as a self-sufficient empire, deriving all of its basic needs from its own soil, very accurately describes the manner in which the transmuting ether sustains the structure of a material particle. In a similar fashion, a subatomic particle is "nourished" by the ether medium, its "soil." The varieties of plant life that flourish in Atlantis's soil are watered by two springs that Poseidon has caused to well up from the ground. These represent the two principal ether transmutation pathways, $A \longrightarrow G \longrightarrow X \longrightarrow \Omega$ and $B + X \longrightarrow Y + Z$, which maintain the subatomic particle's ordered state. In effect, the myth portrays Atlantis as a fertile open system that grows and sustains a variety of living things.

The precious substance orichalch, which is mined on Atlantis and which is crucial to the maintenance of the Atlantean economy, can be compared to the B and G ethers of Model G, which serve as crucial raw materials for the formation of the X and Y ethers, whose wavelike concentration patterns compose the subatomic particle's form. The word orichalch in ancient Greek translates directly as "mountain copper," which has come to signify brass. More specifically, it could be a reference to the copper-rich alloy called red brass (90 percent copper, 10 percent zinc), which has a reddish color similar to orichalch. There is some uncertainty, however, as to whether Plato meant the Atlantean orichalch to be interpreted literally as brass; that is why most mythologists who translate this myth give the word in its original phonetic equivalent. Elsewhere in his dialogue, Plato refers to orichalch as the metal that "glowed like fire," which calls to mind the god Aither (ether), whose name meant "the blazing." So this substance quite possibly was meant as a metaphor for the nourishing ether.

The myth proceeds with the story of Atlantis's rise to power. What was originally an island populated by just Clito, Poseidon, and the ten Atlantean princes has now grown into a metropolis populated by their multitude of descendants. What is originally described as a circular island 100 stadia in diameter later grows to a diameter of 129 stadia at a point in the myth where a canal is cut to connect the outer ringlike harbor with the sea. Still later in the myth this island plain becomes encircled with mountains and grows into a large oblong territory measuring 2,000 by 3,000 stadia (400 by 600 kilometers). By this time the Atlantis population has grown into millions. Just prior to its destruction, Atlantis's dimensions are given as greater than the combined area of ancient "Libya and Asia," implying a diameter of over 5,000 kilometers.

The gradual growth of Atlantis's land area and population portray the process of continuous matter creation through which successive generations of subatomic particles spring into existence in the fertile region surrounding the mother particle. These material offspring eventually accumulate to form a primordial planet that gradually grows into a primordial star, which in time spawns a vast stellar empire, the Galaxy. The sun, moon, planets, and stars would all reside in the suburbs of this Atlantean kingdom. Consequently, as residents of Earth, all of us would be remote descendants of Atlantis.

The *Critias* relates that the kings of Atlantis wisely developed the abundant natural resources of their island, building beautiful temples, royal residences, harbors, and docks. On the central mountain within which Poseidon, Clito, and the princes dwelt, the Atlanteans erected a royal palace. The monarchs of each succeeding generation added to the beauties of this estate, each attempting to surpass the achievements of his predecessors. At the very center of this palace, surrounded by a golden railing, lay the untrodden sanctuary sacred to Clito and Poseidon. This marked the spot where Poseidon and Clito first conceived the race of the Atlantean princes. Here stood an immense golden statue of Poseidon riding in a chariot drawn by six winged horses. This statue reached so high that Poseidon's head touched the temple roof reminiscent of the Titan Atlas supporting heaven. Arcanum 7 (exponential growth) depicts a similar charioteer metaphor, and the six horses call to mind the concept of symmetry breaking (Arcanum 6).

Around the periphery of the outermost of the three circular moats, the Atlanteans built a stone wall fortification, constructing also buildings, towers, and gates. The stone quarried for these constructions was coated with various metals such as copper, tin, and orichalch. They also built a one-hundred-foot-wide bridge to span the three harbors and constructed roads so that they would have access to and from their palace.

Around the entire circumference of the outermost of the two ringlike dikes they built a two-hundred-meter-wide course for horse racing. The horses racing around this circular track might signify a variety of rotational phenomena: subatomic particle spin; electrons orbiting the primordial proton; or on a much larger scale, the rotation of the galaxy's core. If the authors of this myth intended any of these interpretations, we would have to attribute to them a very high level of technical sophistication.

The *Critias* mentions that the Atlanteans excavated a canal from the inner city to the ocean. He describes this work as follows:

> They began on the seaside by cutting a canal to the outermost ring, fifty stadia [ten kilometers] long, three hundred feet broad, and a hundred feet deep; the "ring" could now be entered from the sea by this canal like a port as the opening they had made would admit the largest of vessels. Further, at these bridges they made openings in the rings of land which separated those of water, just sufficient to admit the passage of a single trireme, and covered the openings so that the voyage through

Figure 10.4. A view of subatomic Atlantis, our galaxy's primordial proton.

them became subterranean, for the banks of the rings of earth were considerably elevated above the sea level.[6]

The myth relates that the main canal leading from the sea to the outer harbor was constantly crowded with merchant vessels. Besides goods, these ships also carried large numbers of passengers to and from all quarters of the island. We envision a bustling metropolis continuously engaged in industry, transportation, and commerce of all types. By specifying this ongoing import, export, production, and consumption activity, this legend clearly portrays the open-system character of subatomic Atlantis.

ETHERON FLUXES
AND POTENTIAL FIELDS

In many ways, the flow of citizens and goods into and out of Atlantis's central city resembles the radial flows that arise in a Model G subatomic particle. Consider, for example, a positively charged particle (high *Y*/low *X*/low *G* core concentrations). From computer simulations, we know that *Y*-ons are generated at a faster rate and *X*-ons and *G*-ons at a slower rate inside the particle than outside. Consequently, the particle's *Y*-wave pattern becomes biased upward, and its *X*-wave pattern and *G*-field become biased downward relative to the *Y*, *X*, and *G* levels in the particle's environment. These biasings, in turn, induce a net outward flow of *Y*-ons and net inward flows of X-ons and G-ons (figure 10.5).

Viewed in terms of the Atlantean allegory, Atlantis's central city (the core of a positive subatomic particle) would import G and X etherons and export Y etherons. Imagine a country that, owing to specialization of its production efforts, produces an abundance of computer chips (Y-ons) at the expense of underproducing wheat (X-ons) and milk (G-ons). This country, then, would accumulate a surplus of computer chips (upward-biased Y-field) and a dearth of wheat and milk

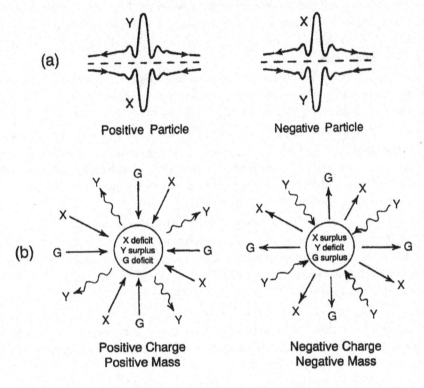

Figure 10.5. (a) Biasing of the X and Y concentrations in a particle of positive (left) and negative (right) charge. (b) Corresponding directions of etheron diffusion.

(downward biased X- and G-fields). It would discharge these production imbalances by exporting computer chips and importing wheat and milk, that is, by producing an outflow of Y-ons and an inflow of X-ons and G-ons. In a negatively charged particle, the X-on, Y-on, and G-on production rate surpluses and deficits would be reversed, as would be the X-on, Y-on, and G-on field potential polarities and the directions of the corresponding etheron fluxes.

These etheron outflows and inflows produce sloping X-, Y-, and G-fields that extend far from the particle's center. This is more easily visualized with the help of the following hydrodynamic analogy. A pool of seawater has a submerged drain at its center that rapidly removes water from the pool. The drain portrays a positively charged particle removing X-ons (seawater) from its vicinity. The outflow creates a valley that induces seawater to flow radially inward at a rate that balances the loss rate. This sloping gradient represents the X component of the particle's electrostatic potential field. To picture the field's Y component imagine a pipe discharging freshwater into a pool of freshwater. The discharge produces a "hill" in the vicinity of the particle with water flowing radially outward down the hill's slope. Superimposing these two pictures, with seawater flowing inward and freshwater flowing outward, we come close to showing how a positively charged particle generates its radial X and Y potential fields.

As the above example illustrates, a particle's long-range energy potential fields necessarily emerge as a direct result of its etheron fluxes. In effect, a particle produces its fields by functioning as a kind of open system. The X-on and Y-on production rate surpluses or deficits in the particle's core constitute its electrostatic charge, and the long-range sloping X and Y potential fields this charge induces constitute the particle's electric field. Particles having no electrostatic charge, such as the neutron or neutrino, produce no radial X or Y fluxes and hence would not generate an electrostatic field. A particle's G-on production rate surplus or deficit constitutes its gravitational mass. A G-on deficit (positive mass) generates a gravity potential well, while a G-on surplus (negative mass) generates a gravity potential hill.*

If today we were to attempt to construct an allegorical model of a subatomic particle dissipative structure, we could do no better than the description so brilliantly encoded in the myth of Atlantis. It is doubtful that the Greeks of Plato's time understood the essence of the sophisticated allegories contained in the Atlantis myth. More likely, like the ancient Egyptians, they served as one more link in a human transmission chain that allowed these entertaining mythical cryptograms to be handed down to modern times.

The idea that a particle's energy potential field could be due to the diffusion of some kind of subquantum medium was put forth in a very hypothetical way in 1964 by the Nobel laureate physicist Richard Feynman and two of his colleagues, R. B. Leighton and M. Sands.[7] In their introductory physics text, *Feynman Lectures*

---

*Why a Model G subatomic particle should incur these surpluses and deficits and generate a charge is explained in my 1985 paper on subquantum kinetics (*International Journal of General Systems* 11:295–328, also discussed in P. A. LaViolette, *Subquantum Kinetics: A Systems Approach to Physics and Cosmology* [Schenectady, N.Y.: Starlane Publications, 1994, 2003]). In accordance with standard physics, magnetic forces can be produced by moving electric potential fields and need not arise from a stationary "magnetic potential" field. Magnetic forces, for example, would arise from a rotational electric potential field, rotational gradients in the X and Y ethers. A subatomic particle's spin magnetic moment is an example of such a magnetic-force-inducing vortex. Rotation of a particle's gravity field, a G-on vortex, would produce a fourth kind of force similar to the magnetic force, but gravitational in nature.

on Physics, they compared the electric potential field around an electron to the concentration profile produced by neutrons diffusing out of the core of a nuclear reactor. They portrayed an electron as a tiny nuclear reactor whose core radiates a flux of subquantum particles called "little *X*-ons"; the concentration of these outwardly diffusing particles, they noted, would drop off inversely with radial distance, just like the electric potential field around an electron. Speculating that the electron's field might actually be an *X*-on concentration profile, they state:

> Could it be that the real world consists of little *X*-ons which can be seen only at very tiny distances? And that in our measurements we are always observing on such a large scale that we can't see these little *X*-ons, and that is why we get the differential equations?
>
> Our currently most complete theory of electrodynamics does indeed have its difficulties at very short distances. So it is possible, in principle, that these equations are smoothed-out versions of something. They appear to be correct at distances down to about 10^{-14} cm, but then they begin to look wrong. It is possible that there is some as yet undiscovered underlying "machinery," and that the details of an underlying complexity are hidden in the smooth-looking equations—as is so in the "smooth" diffusion of neutrons. But no one has yet formulated a successful theory that works that way.[8]

At the time of their writing in the early 1960s, little work had been done on open reaction systems. Had these physicists known about the Model G reaction-diffusion system and its ability to spawn localized field-generating concentration patterns, perhaps they would have given more serious attention to their reactor model analogy. To do so, however, they would have had to take the bold step of rejecting the special theory of relativity and returning to the concept of an ether.

By explaining how a particle's charge and mass originate and how they generate a particle's electric and gravitational potential fields, the open-system physics of ancient mythology explains aspects of the microphysical world left unexplained by modern physics theories. Modern physicists usually reduce charge, mass, and spin to symbols (*q*, *m*, and *s*) and mathematically define them in reference to specific sets of observational data. They do not explain how these properties come into being, nor how they generate a particle's electrostatic or gravitational field.

The energy fields of the ancient ether physics have several other advantages over modern field theory models. First, they avoid the so-called infinite-energy absurdity of contemporary physics. In conventional field theory, a particle's field arises from an infinitely small point, and the energy potential of the field increases without limit toward the particle's center. In the ancient reaction-kinetic physics, on the other hand, the particle's field potential tapers off to a finite value at the particle's center.

The field model that emerges from the ancient physics also resolves the so-called

(a) 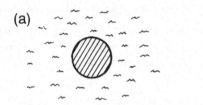   (b)

Figure 10.6. A material subatomic particle represented in relation to its ether ambient, as conceived in (a) the luminiferous ether theory of classical physics, and (b) the ether of ancient times and modern quantum field theory.

field-particle dualism that has long troubled physics. This problem had its roots in the mechanistic luminiferous ether theory devised by physicists of the eighteenth and nineteenth centuries. Physicists in those days sought to describe nature in terms of two very different substances: ether and matter. They hypothesized the existence of an ether primarily as a way of explaining the long-range transmission of light, radiant energy, and forces. Material bodies, on the other hand, were thought to be composed not of ether but of fundamental particles configured as tiny impenetrable spheres. The ether was conceived to surround these spheres as water surrounds immersed stones (figure 10.6a). Moreover, the ether was assumed to be completely frictionless and inert and hence incapable of exerting any kind of force on matter.

This ether-particle dichotomy presented the following problem. An electrically charged particle was supposed to generate and somehow impress an electric field upon the ambient ether, and this in turn was supposed to exert forces upon distant charges and cause them to move. But how could two compositionally distinct entities, matter and ether, act upon one another and, at the same time, be totally isolated from one another and mutually noninteractive? When physicists abandoned the ether concept, they did not rid themselves of this dualism. The same force field equations developed during the era of the luminiferous ether were carried forward, leaving this dualism hanging in the vacuum of space like the grin of an invisible Cheshire cat. Only its name changed; it came to be called the *field-particle dualism*. Fields mediated the interaction of fundamental particles, but paradoxically they did not compose them.

Einstein opposed this fragmented view of nature. He noted that the practice of treating subatomic particles as mass points distinct from their field ambient fragmented the field-continuum of space into a nearly infinite number of pieces. He felt that a workable field theory should require that the field have unbroken continuity throughout all regions of space. In his 1950 magazine article, he stated:

> The combination of the idea of a continuous field with that of material points discontinuous in space appears inconsistent. A consistent field theory requires continuity of all elements of the theory, not only in time but also in space, and in all

points of space. Hence the material particle [as a distinct entity] has no place as a fundamental concept in a field theory.[9]

Einstein sought to devise a unified field theory that could represent physical reality by a continuous field that in turn would account for the laws of electromagnetics as well as for the laws of motion and gravitation. He saw a material particle not as a mass point, but as a limited region in space having a particularly high field strength or energy density, a bunching of the field continuum itself (figure 10.6b).[10] His thoughts later developed into what is today called *quantum field theory*.

In proposing this bunched-field concept, Einstein was borrowing an idea put forth earlier by ether theorists such as Hendrik Lorentz and Gustav Mie, who had proposed that subatomic particles form out of an ether substrate. A similar concept is encountered in the ether physics of ancient times as well as in contemporary subquantum kinetics. The field pattern that forms the subatomic particle farther out becomes the particle's peripheral field; one blends into the other in a continuous manner. Mie, Lorentz, and Einstein, however, did not offer an explanation of how the particle might come into being out of the surrounding ether or field continuum. Nor does modern quantum field theory offer an explanation. On the other hand, both the ancient physics and subquantum kinetics present a feasible theory of matter creation. In overview, it is quite astounding that our ancient predecessors had the insight to devise and encode for posterity a sophisticated physics that in many ways improves on our current physics concepts.

PSYCHIC INVESTIGATIONS OF THE SUBATOMIC REALM

Customarily it has been the practice in physics to base theory upon experimental observation. Observation, however, becomes increasingly fuzzy as the quantum level is approached. Scientists wishing to construct intelligible models of this level find themselves depending to a large extent upon hunches and analogies. Admitting the importance of the intuitive faculties, is there not some advantage to directing the mind to penetrate the barrier of quantum uncertainty and directly "see" what lies below? The clairvoyants Annie Besant and Charles Leadbeater attempted precisely this in 1895. More than thirty years before Heisenberg proclaimed his uncertainty principle, at a time when the atom was still regarded as the most basic unit of matter, these two psychically gifted individuals described the existence of yet smaller units, which they called *Anu* and considered to be the most basic state to which matter could be dissociated.

Their etheric subatomic particles resemble in many ways the open-system subatomic particles illustrated by the Atlantis myth and generated by the Model G reaction system. In their book *Occult Chemistry*, they state:

The fourth dissociation gives the ultimate physical atom on the atomic subplane, the Anu. This may vanish from the plane but it can undergo no further dissociation on it. In this ultimate state of physical matter two types of units, or Anu, have been observed; they are alike in everything save the direction of their whorls and of the force which pours through them. In the one case force pours in from the "outside," from fourth-dimensional space [the astral plane], and passing through the Anu, pours into the physical world. In the second, it pours in from the physical world, and out through the atom into the "outside" again, i.e., vanishes from the physical world. The one is like a spring, from which water bubbles out; the other is like a hole, into which water disappears. We call the Anu from which force comes out *positive* or *male*; those through which it disappears, *negative* or *female*.[11]

The influxes and outfluxes they describe call to mind the *X* or *Y* ether streams flowing into and out of the Model G subatomic particle or the commerce flowing to and from subatomic Atlantis. Both in the psychically "observed" Anu and in the dissipative structures described in ancient mythology and modern systems physics, two particle polarities and ether flow directions are possible. In both cases these fluxes spring into physical existence or disappear from physical existence by transmuting to and from a higher dimensional outside realm. Besant and Leadbeater identify this etheric continuum with the astral plane, or fourth-dimensional space, which is equivalent to the notion of a four-dimensional transmuting ether.

These clairvoyants also concur that these subatomic particles are actually open systems, and that their etheric forms are sustained by a higher dimensional flux:

> The Anu can scarcely be said to be a "thing," though it is the material out of which all things physical are composed. It is formed by the flow of the life-force and vanishes with its ebb . . . if this be artificially stopped for a single Anu, the Anu disappears; there is nothing left. Presumably, were that flow checked but for an instant, the whole physical world would vanish, as a cloud melts away in the empyrean. It is only the persistence of that flow which maintains the physical basis of the universe.[12]

Besant and Leadbeater describe how they are able to discern even greater detail in the structure of these subatomic particles by regulating the etheric flows entering or leaving an individual particle:

> In order to examine the construction of the Anu, a space is artificially made (by pressing back and walling off the matter of space); then, if an opening be made in the wall thus constructed, the surrounding force flows in, and three whorls immediately appear, surrounding the "hole" with their triple spiral of two and a half coils, and returning to their origin by a spiral within the Anu; these are at once followed by seven finer whorls, which following the spiral of the first three on the

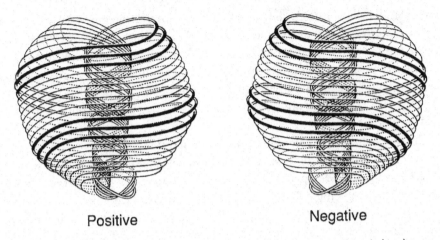

Positive   Negative

Figure 10.7. The structure of a positive and negative etheric subatomic particle (Anu) as seen clairvoyantly by Besant and Leadbeater. Reproduced from Besant and Leadbeater, *Occult Chemistry,* fig. 3.

outer surface and returning to their origin by a spiral within that, flowing in the opposite direction—form a caduceus with the first three. Each of the three coarser whorls, flattened out, makes a closed circle. The forces which flow in them, again, come from "outside," from a fourth-dimensional space.[13]

Their sketches of the two oppositely polarized Anu shown in figure 10.7 are composed of two sets of spiraling, closed-loop etheric flows, one consisting of three whorls, the other consisting of seven finer whorls. These are said to wall off the Anu from the undifferentiated etheric substance that surrounds it. In either polarity, the circulating ether current enters the interior from the particle's top, where it forms a slight depression, and exits from the particle's bottom, where it forms a slight apex. This gives the particle a heartlike shape.*

Vortical structures similar to those drawn by Besant and Leadbeater have been observed at a more macroscopic level in plasma physics experiments. For example, the plasma focus device, a high-current spark discharge device used in fusion experiments, is observed to produce spherical plasma vortices measuring about a half-millimeter across.[14] Each such plasmoid consists of eight or ten electric current plasma filaments twisted into a helical donut-shaped structure that closely

---

*These psychic quantum physics pioneers also describe how subatomic particles can be aligned by an externally imposed electric field. In this aligned state, the heart-shaped depression of one particle receives the etheric flow emitted from the upstream particle and in turn passes this flow out through its own apex into the depression of the next particle, and so on.

resembles the whorl pattern shown in figure 10.7. In the fraction of a microsecond that the plasmoid remains stable, it stores the enormous quantities of energy released in the spark discharge. A similar vortical structure could characterize the centimeter-sized plasma balls formed in artificial lightning experiments as well as the foot-diameter spheres of ball lightning produced by natural lightning strikes, structures that have been observed to remain stable for many seconds before violently exploding.

The flows circulating from the Anu's inner spiral to its outer surface and back again bear some similarity to the diffusive flows of X and Y that circulate within a Model G subatomic dissipative structure. For example, in a positive polarity subatomic particle, the high Y/low X core receives an inward flow of X-ons from its surrounding sheath, which is configured with the opposite low Y/high X polarity This sheath in turn receives from the core an outward diffusive flow of Y-ons. Since the X-ons that flow into the core are there transformed into Y-ons and since the Y-ons that flow into the surrounding sheath are transformed into X-ons, these two complementary core-sheath bidirectional flows combine to form a single closed loop (see figure 5.5). If they were to acquire a little spin, these radial flows would form a whorl pattern, similar to that illustrated for the Anu. The racetrack metaphor in the Atlantis myth might portray this sort of rotational property of subatomic matter.

Whirling etheric currents in the core of a subatomic dissipative structure particle could explain the physical property of subatomic particle spin. The concept of spin was first proposed in 1925 by the physicists Samuel Goudsmit and George Uhlenbeck, who attempted to explain why the lines in the optical spectra of certain atomic elements consisted of pairs of closely spaced lines. Spectral lines are produced when an electron orbiting an atomic nucleus falls from a higher to a lower orbital energy potential level and radiates the lost energy as a light wave. These physicists reasoned that the electron might be continuously spinning on its axis, thereby generating a tiny intrinsic magnetic field pointing in the direction of its spin axis, a field known as the electron magnetic moment. It was hypothesized that the atom's orbital magnetic field (the field generated by the orbital motion of its electrons) would interact with the electron's spin magnetic moment to cause the electron to either slow down or speed up in its orbit, depending upon its magnetic pole orientation. As a result, excited electrons would orbit in either of two sightly different energy states and hence emit photons of slightly different frequency and wavelength when returning to a lower energy orbit. This would explain the closely spaced line pairs in atomic emission-line spectra.

The Goudsmit/Uhlenbeck theory was confirmed in 1927 when Phipps and Taylor discovered that a collimated beam of neutral hydrogen atoms split into two diverging beams upon passing through an externally imposed nonuniform magnetic field.

This behavior could be explained only if the hydrogen electrons possessed spin magnetic moments capable of interacting with the externally imposed field. In later years it was discovered that other subatomic particles also possessed spin.

Although physics books credit Goudsmit and Uhlenbeck for being the first to propose the notion of spin, perhaps this honor should go to Besant and Leadbeater. Thirty years earlier, these two clairvoyants reported observations of three proper motions in the subatomic particles they examined:

> The Anu has—as observed so far—three proper motions, i.e., motions of its own, independent of any imposed upon it from outside. It turns incessantly upon its own axis, spinning like a top; it describes a small circle with its axis, as though the axis of the spinning top moved in a small circle; it has a regular pulsation, a contraction and expansion, like the pulsation of the heart. When a force is brought to bear upon it, it dances up and down, flings itself wildly from side to side, performs the most astonishing and rapid gyrations, but the three fundamental motions incessantly persist.[15]

The first proper motion they mention accurately describes the property of spin. The second motion, the tendency for the particle's spin axis to precess in a circular path, is another important property of subatomic particles, discovered several decades later by the physicist Joseph Larmor. Today this phenomenon is called *Larmor precession,* and the particle's gyration frequency is called the *Larmor frequency.* The diagnostic technique of nuclear magnetic resonance imaging, used extensively in the field of medicine for producing extremely clear images of the body's tissues, makes use of this very same electron precession property. It is indeed quite astounding that these subtle properties were anticipated as early as 1895.

The third property mentioned by Besant and Leadbeater, that of radial pulsation, is today unknown to physicists. This is not to say that it does not actually take place, only that physicists, until now, have had no occasion to suggest its existence. Still, the notion that a subatomic particle continually beats like a heart is quite plausible in the context of the ancient ether physics. It is possible that the radial distance between successive X (or Y) wave crests in a subatomic particle wave pattern might not remain precisely constant over time, but might radially oscillate around an average value. It is also possible to imagine reaction-diffusion waves propagating radially from the particle's core, like the circular waves in the B-Z reaction.

The writings of Besant and Leadbeater provide much food for thought. If indeed the subatomic realm can be accurately observed by means of the psychic faculties, one wonders whether psychically gifted scientists in ancient times may have used similar techniques to gain information about subatomic phenomena and subsequently used such data to construct a sophisticated ether physics.

A JOURNEY TO ATLANTIS

Explorers seeking the underwater whereabouts of Atlantis may be engaging in a futile search, especially if the Atlantean creation myth, like other creation myths, was intended to be interpreted metaphorically, rather than literally. The foregoing analysis suggests that the island metropolis it allegorically refers to does not reside on Earth, but in the heavens at the center of our galaxy. Like the towering city built by Marduk, the city of Atlantis would govern its expanse of stellar creation from this central location. It is here that the first proton would have arisen to spawn the great lineage of matter that today makes up our Milky Way.

Those wishing to travel to Atlantis would have a long journey ahead of them, for the Galactic center lies about twenty-three thousand light-years away. If we were to train a telescope on the Galactic center, we would find that it is not visible from Earth but hidden behind dense veils of cosmic dust that lie in the plane of the galactic disk. To get a clear view, we must image this part of the sky in the radio, infrared, X-ray, or gamma ray spectral regions, since such radiations are not strongly absorbed by the intervening dust. Figure 10.8 shows a radiation intensity contour map of the region within one light-year of the Galaxy's center, made with the twenty-seven antenna VLA radio telescope in Soccoro, New Mexico. The unresolved intensely radiating point source marked Sgr A* (Sagittarius A-star) marks the core's location. Here then lies the long-sought Atlantis, sunken beneath muddy shoals of cosmic dust.

Sgr A* is actually far smaller than this radio image suggests. It is too distant to be resolved even with very-long-baseline radio interferometry, a technique that offers a thousandfold improvement in resolution by linking up radio telescopes at various locations around the world. However, observations with the Chandra X-ray Telescope show that Sgr A* must be smaller than the diameter of Earth's orbit.[16] Although the mass of this object is not known for sure, the velocities of stars orbiting within one hundredth of a light-year of the Galactic center suggest that it contains about 2.6 million solar masses.[17] By comparison, the largest blue supergiant stars are about sixty to eighty times as massive as the Sun.

At infrared wavelengths, this Galactic center source is about 20 million times as luminous as our Sun and many times brighter than the most luminous supergiant star.[18] It emits an intense breeze of ionized hydrogen and helium gas, expelling one solar mass of material every century. The ion wind from the Galactic center is supplemented by a steady breeze of cosmic ray particles, which shoot out at very close to the speed of light.[19] If the majority of the cosmic ray electrons detected in the Earth's vicinity come from this object, then Atlantis's total luminosity might be as high as half a billion times that of the Sun.[20]

Scientists have had difficulty explaining how such an enormous amount of energy could be generated within such a small region. In the past, some

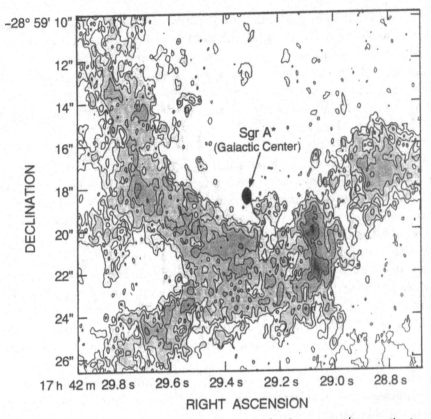

Figure 10.8. A contour map showing the intensity of radio emission (two-centimeter wavelength) coming from within one light-year of the Galaxy's core. The radiation source labeled Sgr A* marks the location of the Galactic center (Zhao et al., "High-resolution VLA images of the Galactic center at 2-cm wavelength with large dynamic range," 48). Photo courtesy of the National Radio Astronomy Observatory.

astronomers have suggested that this object may be a massive black hole that gets its energy by gathering and annihilating gas and dust in its environment. But many disagree with this explanation. For one thing, the energy spectrum of the radiation coming from this region does not match that expected to come from a black hole. It is also doubtful that a black hole could supply the amount of energy coming from Sgr A*. The energy that a black hole radiates is supposed to come from the friction that inward-rushing gases would generate just prior to being consumed. Careful computer simulation studies indicate, however, that nearby gases would be unable to approach close enough to be swallowed. As they become drawn toward the million-solar-mass black hole, their angular momentum forces them to orbit at a considerable distance rather than to fall directly in.[21]

There is little evidence that gas is moving inward toward Sgr A*, as the black hole theory requires. As far out as ten thousand light-years from the Galactic center, gas is instead moving radially outward. Such outflow challenges standard astronomy. For if Sgr A* is losing one solar mass of material per century, unless this material is replenished at its center its entire mass would be dissipated within less than 100 million years. The ancient physics of cosmic creation offers a viable solution by suggesting that this central mother star continuously creates both matter and energy at a prodigious rate (chapter 5).

Even if we successfully crossed the great distance that separates us from galactic Atlantis, we would not get a very close look. By the time we had approached within one hundred astronomical units, three times farther than Pluto's distance from the Sun, the electromagnetic radiation would be so intense that solid substances would spontaneously vaporize. If we could somehow proceed closer to this brilliant display, we would find that Atlantis is enshrouded within a luminous atmosphere formed by the hot gases that are continuously blown off. We would discern an intense bluish white radiance with fuzzy boundaries approximately ten times the diameter of the Sun, with an optical luminosity almost a million times that of the Sun. Atlantis actually would be much more luminous than this, but we would not be seeing the bulk of its radiation, since it would come to us in the form of X rays and ultraviolet radiation as a result of the gases' extremely high temperature ($> 10^5$ Kelvin).

Within Atlantis proper, pressures from the inward pull of gravity would be so great that electrons would be squeezed out of their atomic orbital shells. The conditions would be similar to those theorized to exist in the interiors of white dwarf stars. This "electron degenerate" material at its center would be so dense that one teaspoonful of it on Earth could weigh as much as a thousand tons. On Atlantis the gravitational pull would be so great that this same teaspoonful of material would crush downward with a force of 20 billion tons. At these densities, Atlantis's entire mass of up to a million solar masses would occupy a volume somewhat less than the size of the Sun.

According to the ancient physics, it would be impossible for Atlantis to form into a black hole, since the outward pressure of the tremendous outpouring of energy spontaneously created in its interior would prevent its gravitational collapse. By comparison, conventional physics permits a collapsing body to form a black hole only because it presumes that internal energy generation within the body has ceased. Another reason the ancient physics forbids a black hole from forming has to do with the geometry of the gravity field at the center of each material particle. As discussed earlier, the ancient metaphysics predicts that the gravity field should taper off to a finite value at the center of a subatomic particle, rather than rising to infinity as it does in conventional field theory. Thus at very high matter densities, the gravitational force would decrease rather than increase, and gravitational collapse would come to a halt. Without the infinite-energy absurdities of classical physics, the black hole phenomenon entirely vanishes.

If we were to venture below Atlantis's intensely radiating surface to pay homage to the primordial mother particle that came into being at the dawn of creation, we would encounter great difficulty in locating its whereabouts, for this inner city would be lost in the midst of Atlantis's immense population of densely packed atoms, consisting mostly of elements that had reached the final stages of thermonuclear transmutation, heavy metals such as lead, iridium, platinum, and gold. Heavier elements that would not normally be found on Earth owing to their instability might also be found here (for example, the fabled orichalch). Hydrogen would continuously materialize in Atlantis's interior and would fuse to form helium and other elements. The lighter gases, however, would rapidly float to the surface, where they would be expelled by the pressure of Atlantis's intense radiation. The hydrogen and helium wind observed to emanate from Sgr A* would be this newborn matter journeying away from Atlantis's shores.

Despite its enormous energy flux, astronomers consider our Galactic center to be relatively quiescent at present. At radio wavelengths, this central body emits about ten thousand times less energy than the nucleus of the Andromeda galaxy, our nearest neighboring spiral galaxy. Nevertheless, the centers of galaxies can occasionally undergo violent eruptions that release intense outbursts of cosmic rays. These so-called galactic core explosions rank as the most energetic explosive events known to astronomy. One particularly energetic galaxy, designated CTA 102, is estimated to release each year as much energy as 10 billion of the most energetic kind of supernovae.[22]

Astronomers estimate that approximately one out of every six spiral galaxies like our own has a central object that is presently passing through its active phase. Consequently, it is reasonable to assume that our own galactic core violently erupted sometime in the past. In this context, the description of Atlantis as an empire heavily armed for combat makes sense. The account given in the *Timaeus* of how Atlantis one day launched a vicious attack on the ancient Hellenes could be an allegorical reference to such a galactic explosion event. Available data indicate that the center of our galaxy has been active in the recent past, releasing volleys of cosmic rays that impacted our solar system several times during the last ice age; one particularly intense event occurred not long before the date that Plato gives as the time of this attack.[23]

The change of Atlantis from peaceful civilization to warlike society is first commented on at the end of the account in the *Critias*. The story explains that after many generations of godlike behavior, Atlantis's populace grew wicked and hungry for power. At one point Zeus decides that he must punish them. The myth states:

> Zeus, the god of gods, who governs his kingdom by law, having the eye by which such things are seen, beheld their goodly house in its grievous plight and was

minded to lay a judgement on them, that the discipline might bring them back to tune. So he gathered all the gods in his most honorable residence . . . that stands at the world's center and overlooks all that has part in becoming, and when he had gathered them there, he said . . . [24]

We may never know what Zeus actually said, for the narrative breaks off here in midcourse. It is interesting, though, that this very same sentence reports that Zeus (the creator) gathered the gods together at the center of creation, hence at the center of the Galaxy. Perhaps he executed his disciplinary action during this unrecorded pronouncement. Zeus, who symbolizes the principle of spontaneous matter/energy creation, may have unleashed within Atlantis a powerful outburst of energy, a galactic core explosion. Following this cathartic release, Atlantis would have returned to its former quiescent state, and thereupon come "back to tune" in accordance with Zeus's plan.*

---

*The ancient Maya similarly regarded the center of the Milky Way as a cosmic center of celestial creation. They identified the Creation Place of heaven with the "Crossroads," the location in Sagittarius where the Milky Way equator intercepts the ecliptic, the path followed by the Sun and planets. This intersection point, which is situated just six degrees northeast of the true Galactic center, was considered by the Chortí Maya to be the "heart of heaven and earth." The dark rift of light-obscuring cosmic dust that extends through this part of the Milky Way was identified by the Maya as the vagina of a cosmic Great Mother deity which gives birth to the Sun (J. M. Jenkins, *Maya Cosmogenesis 2012* [Santa Fe, N.M.: Bear and Co., 1998], pp. 106, 117–118, 176).

11

MYTHS FROM THE ANCIENT EAST AND MEDITERRANEAN

THE SUMERIAN MYTH OF CREATION

One of the earliest Near Eastern creation myths comes to us from ancient Sumeria. Although the tablets that record it date from the third millennium B.C.E., it probably has a much earlier prehistoric origin. Although this Sumerian myth is brief and lacking in detail, its resemblance to the Egyptian story of Atum suggests its relevance here. The creation myth may be summarized as follows:[1]

ठ   (a) Initially there was the primeval sea.

♈ ♉ (b) The primeval sea begot the cosmic mountain, which consisted of heaven (the god An) and earth (the goddess Ki), who existed in an unseparated state, An-Ki.

♋ ♌ (c) An and Ki united to produce the air god Enlil.

♏ ♐ (d) Enlil, the god who "brings up the seed of the land from the earth," then separated his father and mother, heaven from earth, "the great above" from "the great below."

♍ ♒ (e) Enlil then carried off his mother, Ki, and together they created the universe.

Like Nun in Egyptian mythology, the primeval sea of the Sumerians may be understood to represent metaphorically the primordial ether. The birth of the cosmic mountain from this sea resembles the rising of the primeval hill (or primeval mound) from Nun in the story of Atum, except that in this case the mountain is already differentiated into male and female aspects at the time of its birth. Like the primeval hill, the birth of the cosmic mountain may be understood to signify the emergence of the primal energy source in the ether. This etheric sea comprises an inherent duality, represented by the gods An and Ki, of whom the myth presents little detail. Let us assume, however, that like Shu and Tefnut in the myth of Atum,

245

they represent etheric fluxes. In that case, the united An and Ki may be identified with the cross-coupled ether reactions that mutually convert X into Y and Y into X. Their child, Enlil, may be identified with the critical X-Y fluctuation that is spawned from this positive-feedback interaction.

The part of the myth where An and Ki separate from each other resembles the separation of Geb and Nut, except the genders are here reversed such that heaven (An) is masculine and earth (Ki) is feminine. Like Geb and Nut, An and Ki signify the X and Y ether complements of the heterogeneous transmuting ether. Their separation illustrates how a growing critical X-Y fluctuation (Enlil) ultimately causes the X and Y ethers to deviate in opposite directions from their former concentration values, Ki (X) decreasing as An (Y) increases. When fully separated, An, Ki, and Enlil together compose the first self-created particle of physical matter.

The symbolic roles of Enlil and Ki appear to shift at this point in the myth. Enlil, who earlier represented the critical fluctuation, now represents the newly formed particle of matter, whereas Ki represents the fertile G-well supercritical region that surrounds it. Both work together synergistically to spawn all matter. In other words, ether fluctuations emerging within their fertile region would rapidly grow in size and develop into additional particles of matter. Just as the massive primordial mother star in a galaxy's core is surrounded by the stars that it spawned, so too is Enlil conceived to occupy a central position amid his creation, for one passage states that the "little ones" (the stars) are scattered about him like grain.

Both Egyptian and Sumerian myths similarly invoke the air metaphor in connection with the act of separation, Enlil being described as "the air god" and Shu being assisted in his separation effort by birdlike air spirits. This serves as an appropriate image for conveying the emergence of physical creation, since the growth and expansion of the critical fluctuation may be thought of as a kind of upward flight.

The story of Rangi and Papa of the Maori people of New Zealand bears an intriguing resemblance to the Sumerian creation myth. At the dawn of creation, Rangi, the sky god, and Papa, the earth goddess, are torn from their love embrace by their son, Tane, the god of forests, birds, and insects. Reminiscent of the story of Osiris, Ta-whiri-matea, the king of darkness and chaos, opposes the act of creation. It is tempting to speculate on why this myth from the South Pacific bears such a close similarity to creation myths originating thousands of miles away in the Mediterranean region. Did these peoples have contact long ago? Or could they have received this information from some common source in very ancient times?

THE BABYLONIAN CREATION MYTH

The Babylonian creation myth *Enuma elish* ("When on high") has survived the ravages of time in the form of seven clay tablets inscribed around the second millennium B.C.E.[2] Yet the myth is believed to have much earlier origins, possibly dating back to the time of the Sumerians. It encodes a creation metaphysics equally complex as that found in the Egyptian myths. Like the myth of Osiris, it tells a story of a struggle and ultimate victory of the forces of order over chaos, a conflict that culminates with the creation of the universe. The epic's key contestants are Marduk, the champion of order, and Tiamat, the great mother goddess, who personifies chaos.

The Birth of the Gods and Death of Apsu

The story begins by describing the initial state of the universe:

> *When on high the heaven had not been named,*
> *Firm ground below had not been called by name,*
> *There was nought but primordial Apsu, their begetter,*
> *And Mother Tiamat, who bore them all,*
> *Their water commingling in a single body*

Apsu was the primeval freshwater ocean and Tiamat the saltwater ocean. Together they represented the original cosmic substance from which all things were eventually made. During this early time they abided in peace and quiet. But then from within Apsu and Tiamat two other gods came into being, the brother-and-sister pair Lahmu and Lahamu. These two in turn brought forth the brother and sister gods Anshar and Kishar, who in turn spawned Anu, who became the god of heaven. Anu later fathered Ea, a god of unusual wisdom and strength who came to rule over the subterranean fresh waters. In this fashion the divine family proliferated.

They were a young and boisterous lot, always throwing parties, quarreling, and raising much clamor. This caused considerable distress to Apsu and Tiamat, who desired peace and quiet. Finally Apsu could bear it no longer and went to Tiamat, urging her to do away with them. She refused, saying that she could not kill her own children. But Apsu, persisting in his desire, planned to undertake the task himself. The gods became worried when the news reached them; however, Ea, the wisest among them, came to their rescue. While Apsu was asleep one night, Ea placed him under a spell and then killed him. Afterward Ea donned Apsu's robe and crown.

Apsu and Tiamat signify the two fundamental ether transmutation pathways. Their original state of intermingled repose portrays the ether in its initial state of unstructured uniformity. The gods that they bear as their children are the X-Y fluctuations that emerge at the juncture where these two transmutation streams mutually cross and interact. The gods' boisterous contentiousness describes the disorder of the fluctuating ether, in which the "noisier" fluctuations would be those that have a greater size or amplitude. Individually they are tiny pulses of order, but collectively they produce an incoherent disorder.

Apsu and Tiamat personify that disorder. Their desire to maintain peace and quiet reflects the tendency for fluctuations to maintain the etheron concentrations in a spatially uniform "peaceful" state by mutually eroding one another's growth, the principle of Arcanum 5. It is only the particularly noisy, rebellious gods that Apsu and Tiamat find disturbing, for such greatly deviant fluctuations could eventually destroy the ether's established state of uniformity and create a state of explicit order. These parent gods attempt to prevent the emergence of nonuniformity by planning ways of doing away with the gods—destroying the fluctuations. This paraphrases the second law of thermodynamics. By killing Apsu and saving the gods from their impending destruction, Ea brands himself as a rebel, a champion of order.

The Birth and Coronation of Marduk

Ea and his wife, Damkina, celebrated the demise of Apsu by building a beautiful cottage. They dwelt there and to them was born Marduk. Marduk was quite a sight to behold. He had four eyes, four ears, and breathed fire from his throat. He was the king of kings, unsurpassed for his wisdom and strength, taller and stronger than any divine creature. But he began to make a nuisance of himself and to enrage the other gods. Some went to Tiamat and, reminding her that Ea had killed her husband, persuaded her to do away with Marduk, Ea, and his family.

Tiamat began by calling a conference of the rebel gods to devise a plan for carrying out the slaughter. She then gave birth to eleven kinds of serpent monsters and ferocious dragons in whose veins ran poison instead of blood. These she placed in the high command of the god Kingu. She gave Kingu dominion over all the gods and entrusted him to bear the tablet of destinies, with all of its magic powers.

Ea did not hear of the impending danger until Tiamat was almost ready to launch her assault. Ea saw that he was no match for the eleven terrible monsters, and he retreated in fear and dismay. His father, Anu, attempted to pacify Tiamat, but his efforts failed. It was apparent that Tiamat had to be subdued by physical force. Ea's grandfather Anshar and the council of gods were in despair, for it was now clear that Tiamat intended to slay all the gods, not just Marduk and his immediate family. Yet none of the gods dared to face Kingu and his terrible monsters.

At this point the happy thought occurred to Anshar that Marduk might succeed. Although he was ready and willing to do battle, Marduk demanded that if he was to be their champion and fight Tiamat himself, he must be given supreme and undisputed authority among the gods. So a messenger was sent to Lahmu and Lahamu, who agreed to call a council of all the gods. At their banquet assembly it was decided that Marduk should be made king of the gods and ruler of the whole universe, and that he should have the final say in all things. After the feast, the gods built an impressive throne on which Marduk then sat to receive sovereignty. To test his powers they placed before him a garment. At his command the fabric was destroyed; again at his uttered word the fabric was miraculously restored. Upon beholding the efficacy of his word, the gods rejoiced. They invested him with the royal insignia, the scepter, the throne, and the royal robe.

The quelling of Apsu's impending attack and the construction of the cottage signify the emergence of a protected region in the ether that is temporarily free of the order-destroying effects of chaos. In the shelter of this cottage Ea and Damkina are able to raise the hero Marduk. Like Horus, who is suckled to maturity in the Nile marshes, Marduk grows big and strong by receiving divine power by suckling the breasts of goddesses (the Cancer principle). Like Horus, he signifies the fluctuation that, with proper nourishment, eventually matures to the stage of criticality. His reputation of being independent and rebellious singles him out as a clear threat to Tiamat's regime of peaceful homogeneity.

The ceremony acknowledging Marduk's supremacy and making him king of the gods allegorically portrays the moment when the fluctuation has grown to a critical size. In pledging him their allegiance, the gods, Marduk's confederate fluctuations, agree to work in unison to create a state of coherent order. By working in harmony, rather than against one another, they are able to recruit the energy-amplifying resources of the surrounding ether toward the common objective of emergence. The test in which Marduk demonstrates his ability to destroy and recreate a garment by the power of logos symbolizes the principle of transmutative regeneration of form, for it is by such means that a fluctuation is able to survive the process of erosion. As king of the gods, Marduk is the equivalent of Horus in his form of Heru-khuti. Like Heru-khuti and like the Emperor (Arcanum 4), Marduk sits on a throne holding a scepter in his right hand.

The Battle

Marduk then departed to prepare for battle. He grasped a club in his right hand and carried a bow and arrow at his side. He also made a net, which the winds carried for him. He filled his body with a blazing flame and mounted his storm chariot, drawn by four terrifying creatures. As he ventured forth, he raised an inundating rain and caused lightning to precede him.

Kingu and his forces were thrown into a state of confusion at the sight of Marduk in his dazzling splendor. But Tiamat, unperturbed, stood her ground and let out a loud roar. Ignoring this, Marduk proceeded to denounce her for her wicked schemes and challenged her to a duel. At these words Tiamat went into a frenzy. Crying out furiously, she accepted the challenge. The two then proceeded to engage in combat. Marduk immediately spread out his net and entrapped her. Then, as Tiamat opened her mouth to devour him, he called up an evil wind, which blew so fiercely that her jaws could not close. As the raging winds distended her body, Marduk shot an arrow through her open mouth. It penetrated through her belly and struck her heart, and thus she was killed. She fell to the ground, and Marduk stood victoriously upon her carcass.

When her followers saw that their leader was dead, they attempted to flee. But none escaped. Marduk took them prisoner and disarmed them, rendering harmless the eleven dragons. He took from Kingu the tablet of destinies, affixed his own seal on it to establish his ownership, and fastened it to his breast. He then returned to Tiamat. With his mace he crushed her skull, splitting her huge body in two parts like a shellfish. With one half of her corpse he formed the sky, and with the other half he made the earth. He then established Anu as the god of heaven, Enlil as the god of the air and earth, and Ea as the god of the waters below. Afterward, he constructed stations in the sky for the great gods, forming their likenesses as constellations. Then he formed the days of the year, assigning three constellations for each of the twelve months. He built gates in the east and west for the sun to enter and depart and caused the moon to shine forth. Finally, with the help of the great gods Marduk created humans out of the blood of Kingu.

The duel between Marduk and Tiamat represents the classic struggle in which order emerges from disorder. Their contest is symbolically identical to the battle fought between Horus and Set. Marduk's battle regalia (mace in right hand and bow and arrow at side) strikingly resemble those worn by Horus (see figure 11.1). The chariot that Marduk rides into battle portrays the idea of exponential growth, the same metaphor encountered in Arcanum 7 of the Tarot. The killing of Tiamat (chaos) marks the turning point in which the growing fluctuation dominates the eroding forces of disorder. After this critical transition, nothing can stop Marduk's continued growth to supremacy. The transference of the tablets of destiny from Kingu to Marduk signifies the irreversibility of the event that has just taken place. Now destiny rests with the forces of order.

By describing Marduk as having two faces (four eyes and four ears), the myth portrays the bivalent nature of the critical fluctuation. Interestingly, Sagittarius, the astrological sign that specifically conveys the notion of symmetry breaking, was sometimes depicted in Babylonian art as a double-headed personage, having a human head facing forward and an animal head looking back. Following Marduk's victory over the forces of disorder, this incipient bivalent order becomes expressed at a macroscopic level with the "division" of the chaotic ether, represented as the primeval saltwater ocean. Marduk's division of Tiamat's body to form the sky above and the earth below illustrates how the growing critical fluctuation induces the X and Y ethers to separate and form the central arching field of the primordial proton. The shellfish metaphor appropriately illustrates the symmetrical arched divergence of X and Y.

Once the forces of disorder are conquered and Tiamat is divided, the main portion of the drama of creation is completed; the primordial self-created particle of

Figure 11.1. This Egyptian fresco showing Horus setting out to engage Set in battle might equally well depict Marduk setting out to confront Tiamat.

matter (Marduk) has emerged and is now able to form the stars, sun, and moon. That is, the primordial proton nucleates progeny particles whose descendants eventually spawn a galaxy full of stars and planets.

The Erection of the Ziggurat

In honor of Marduk's victory, the gods built the city of Babylon and constructed within it a massive staged pyramid called a ziggurat (derived from *zagaru,* "to be tall"). This is the famous Tower of Babel referred to in early Hebrew legends (Genesis 2:1–9). At the summit of the tower they built Marduk a great temple, where he was to live and where they would pay him homage. Then, after taking part in a joyful banquet, the gods solemnly assembled to confer upon Marduk fifty titles in appreciation of all that he had done.

The city of Babylon is an allegorical representation of the Milky Way. The temple housing Marduk, fashioned atop the Tower of Babel built at the city's center, signifies the massive mother star that resides at our galaxy's center and marks the location where the first particle of matter came into being. Like the pyramids of ancient Egypt, symbolizing the primeval mound and palace of the resurrected Osiris, the heavenward-reaching ziggurats of ancient Babylonia would have symbolized this awe-inspiring event of physical creation accomplished by Marduk.

Although this myth speaks of one primordial creation event that led to the formation of the earth, sun, moon, planets, and stars of our galaxy, there are many galaxies like ours strewn throughout the cosmos. Each of these "island universes" would be a site where a heroic battle was once fought and won by its own Horus or Marduk, with his own valiant story to tell.

Lacking an understanding of the metaphysics encoded in the *Enuma elish*, some contemporary myth historians have questioned why the myth would describe extended disputes and battles prior to the emergence of physical creation. Not realizing their purpose, they have discounted the bulk of the *Enuma elish* as being simply an entertaining tale, devoid of any cosmological significance. For example, in his book *The Babylonian Genesis*, Alexander Heidel comments:

> *Enuma elish* is not primarily a creation story at all. If we were to put together all the lines which treat of creation . . . they would cover not even two of the seven tablets but only about as much space as is devoted to Marduk's fifty names in Tablets VI and VII. The brief and meager account of Marduk's acts of creation is in sharp contrast to the circumstantial description of his birth and growth, his preparations for battle, his conquest of Tiamat and her host, and the elaborate and pompous proclamation and explanation of his fifty names. If the creation of the universe were the prime purpose of the epic, much more emphasis should have been placed on this point.[3]

In accordance with the contemporary Western cosmological view that existence itself came into being simultaneously with the emergence of the physical universe, Heidel assumes that the act of creation takes place at the very end of the epic, where Marduk forms the sky, earth, stars, sun, moon, and human race. Realizing that the myth portrays an elaborate reaction-kinetic metaphysics, however, we find to the contrary that the entire myth deals with the process of creation, and in a profound scientific manner.

Our own myths color and distort our interpretation of the thinking of past cultures. As one notable philosopher of science, Paul Feyerabend, aptly states, "the apparent primitivism of many myths is just the reflection of the primitive astronomical, biological, etc., knowledge of their collectors and translators."[4] The modern view that this and other creation stories have been devised by people far more primitive than ourselves inevitably leads scholars of mythology to misinterpret these works and underestimate the intellectual prowess of the peoples who originally devised them. But by first acquiring a knowledge of how open systems function to create order out of chaos, we can unmask the archaic wisdom in these allegorical stories. The ancients now become our teachers, and we their students.

THE OLYMPIAN CREATION MYTH

One of the earliest accounts of the ancient Greek story of creation comes from Hesiod's poem *Theogony* (eighth century B.C.E.).[5] Like the cosmogonies of the ancient Egyptians and Babylonians, the ancient Greek myth uses the language of metaphor to portray, through the actions of its gods, a highly sophisticated science of order creation. The myth begins by describing the first entities to inhabit the cosmos, who signify an originally turbulent state of existence. The second part of the myth, which describes the birth of Zeus and his five siblings and how they came to create the physical world, appears to explain through metaphor how the first subatomic particle emerged from this chaotic substrate.

The Birth of the First Gods

As an answer to how the "shining stars and the wide heavens above came into being," the *Theogony* begins by explaining the genealogy of the gods (figure 11.2). It relates that in the beginning there arose Chaos, a formless mass of infinite extent.* Out of Chaos, in turn, came the first five deities: Gaea (Earth), Tartarus (a dark pit in the depths of hell), Eros (Love), Nix (Night), and Erebus (the darkness of Tartarus). The union of Nix and Erebus brought forth Hemera (Day) and Aither (Ether).† Gaea on her own brought forth Ouranos (Heaven), "equal to herself to cover her on all sides and to be a home forever for the blessed gods."[6]

Like Tiamat of the Babylonian myth, Gaea gave birth to quite a raucous bunch. All were creatures of gigantic size and strength. First came the Titans, or "elder gods," as they are sometimes called. There were twelve in all, six male and six female. They personified the mighty convulsions of the physical world, such as earthquakes, hurricanes, and volcanic eruptions. Next, Gaea and Ouranos spawned three immense one-eyed monsters, known as the Cyclopes, whose name translates as "wheel-eyed." These three represented the terrors of thunder, bolt lightning, and sheet lightning. Finally she bore three other children, which emerged as huge monsters having one hundred hands and fifty heads.

The kingdom of Gaea and Ouranos, like that of Tiamat and Apsu, depicts the ether in its initial featureless state of spatial uniformity, the condition prevailing prior to the emergence of matter. Gaea's unruly children personify the concentration

---

*In ancient Greek, chaos (χαος) did not mean "disorder," as in the English meaning of the word. It instead signified "shapeless mass" or "infinite space." Nevertheless, the grandchildren of Chaos (the Titans, Cyclopes, and Hundred-Handed Ones) do metaphorically convey the notion of a disorganized, disorderly state of preexistence.

†On occasion Aither was identified with Ouranos. And in another variation of the Greek cosmogony, Gaea is descended not from Chaos but from the union of Aither, Hemera, and Eros.

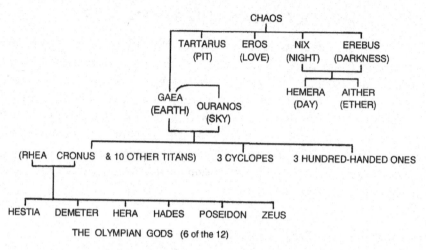

Figure 11.2. Genealogy of the Greek cosmogony.

fluctuations that chaotically arise throughout the ether. Each of these "monsters," if grown sufficiently large, could potentially seed an ordered pattern in the ether. It is no wonder, then, that Ouranos strives to suppress the emergence of these imminent creation centers, for their growth would blemish the uniform state of chaos that he seeks to maintain throughout his regime. Just as Apsu and Tiamat seek to restore peace and quiet in their kingdom by attempting to destroy their noisy children, Ouranos seeks to maintain the status quo of uniform disorder by suppressing the freedom of his rebellious progeny. Myth relates that immediately after Gaea bore the Cyclopes and the rebellious Hundred-Handed Ones, the fearful Ouranos imprisoned them deep under the earth in the pit of Tartarus. Like Tiamat and the Hierophant of the Tarot (Arcanum 5), Ouranos personifies the thermodynamic principle of entropy. Actually, fluctuations bring about their own suppression as a result of their own mutually competitive actions. As each monster-child fluctuation attempts to develop its potential to grow and seed explicit order, it necessarily interferes with its siblings' efforts to do the same. So the tyrannical actions of Ouranos really symbolize the order-suppressing effects of the collective chaos produced by his rebellious children.

To continue the story, Gaea became enraged with the way Ouranos was treating her children. So she pleaded with the Titans to help her. Only Cronus (Saturn), the youngest among them, was bold enough. One night while Ouranos slept, Cronus mortally wounded his father by castrating him with a flint sickle, whereupon Cronus took command of the universe. But in so doing he immediately reinstated his father's policy of repression, and once again the Hundred-Handed Ones and the Cyclopes were confined to Tartarus.

Cronus does not fit the conventional hero mold, as do Horus and Marduk. His overthrow of Ouranos was essentially a planned attack on an unsuspecting victim, a strategy more akin to the way Set slaughtered Osiris. Furthermore, since Cronus reinstated his father's tyrannical suppression of the monster fluctuations, like Ouranos, he personifies the principle of entropy. Despite the "change of throne," the ether continues to maintain a uniform state of chaos. The decisive act of creation, the crucial challenge to entropy, must await the birth of Cronus's son Zeus. It is Zeus's victory over Cronus that finally brings the physical universe into being.

Note that Cronus's name, Κρόνος in ancient Greek, sounds very much like *chronos* (χρόνος), which in Greek means "time," a metaphor of flux. This time metaphor is well suited to Cronus, whose destructive actions portray entropy (order-destroying process), time and entropy being closely related unidirectional processes. Recall that the *Mahabarata* assigns to the Hindu destroyer god Shiva the epithet Kala, meaning "time," suggesting a similar link between time and entropy. Interestingly, after he takes control of the kingdom, Cronus marries his sister Rhea, whose name in ancient Greek translates as "flux." The flux symbolism of Rhea and Cronus is discussed in Plato's *Cratylus* dialogue, which relates the following conversation between Socrates and one of his students:

SOCRATES. My good friend, I have discovered a hive of wisdom.
HERMOGENES. Of what nature?
SOCRATES. Well, rather ridiculous, and yet plausible.
HERMOGENES. How plausible?
SOCRATES. I fancy to myself Heraclitus repeating wise traditions of antiquity as old as the days of Cronus and Rhea, and of which Homer also spoke.
HERMOGENES. How do you mean?
SOCRATES. Heraclitus is supposed to say that all things are in motion and nothing at rest; he compares them to the stream of a river, and says that you cannot go into the same water twice.
HERMOGENES. That is true.
SOCRATES. Well, then, how can we avoid inferring that he who gave the names of Cronus and Rhea to the ancestors of the gods agreed pretty much in the doctrine of Heraclitus? Is the giving of the names of streams to both of them purely accidental?[7]

Like Shu and Tefnut in the myth of Atum, or Osiris and Isis in the story of Osiris, the inherently dynamic Cronus and Rhea may be identified with the two pathways of etheric flux: $A \longrightarrow \Omega$ and $B + X \longrightarrow Y + Z$, at whose intersection the physical universe arises. Alternatively, they may represent the cross-catalytic reaction loop that transforms X into Y and Y back into X. For, later in the myth, Rhea and Cronus spawn Zeus and his siblings, who together form a combative force (critical fluctuation) that brings physical order into being.

The Story of Zeus

Soon after Cronus came to power, taking as wife his sister Rhea, he learned through prophecy that one of his children was destined one day to dethrone him. In an attempt to prevent this, each year he swallowed the newly born whom Rhea bore him. Five children met their fate in this fashion: first Hestia, then Demeter and Hera, then Hades, and then Poseidon. All had become imprisoned within Cronus's bowels.

But when Rhea had her sixth child, her third son, she foiled Cronus. She bore Zeus in the dead of night and afterward had him secretly hidden away in a cave in the mountains of Crete. Instead of the child, Rhea gave her husband a large stone wrapped in swaddling clothes, which he unwittingly swallowed. Nursed by three nymphs in the protection of his cave, Zeus gradually grew in size and strength. Later, when he had grown to manhood, he sought out his mother Rhea. With her help he mixed a vile potion, which he secretly added to Cronus's drink. The potion made Cronus's stomach so upset that he disgorged the five children and the stone.

There followed a war in which Zeus and his five siblings fought against Cronus and his brother Titans. Early in this confrontation, Zeus freed the three multi-handed monsters and the three Cyclopes. With their help, his ranks now twelve strong, he managed to conquer the twelve Titans and dethrone Cronus. Zeus punished his conquered enemies by binding them in chains in Tartarus, with the exception of Atlas, who was instead condemned to bear forever upon his shoulders the great pillar that holds apart earth and heaven. With the forces of disorder and destruction finally conquered, Zeus and his brothers and sisters became the new rulers of the universe. They made their home on Mount Olympus and came to be known as the Olympian gods.

There are strong parallels between the characters of Zeus and Horus. In both legends, the mother hides her newly born child to prevent its destruction at the hands of a tyrannical ruler. Later, after having grown to maturity, these young warriors challenge their oppressors in battle, and in the end emerge victorious. Parallels can also be found with the Babylonian creation epic. Like these other myths, the story of Zeus metaphorically depicts the emergence of order out of disorder. Zeus, as Rhea's newborn, symbolizes a newborn fluctuation emerging in the ether. Just as Zeus grows in size and strength by suckling nymphs, traditional symbols of fertility, so too an emergent fluctuation grows in intensity when situated within a fertile supercritical region. Moreover, just as Zeus must be hidden from Cronus (entropy) in the shelter of a cave, so too an emerging fluctuation must have the good fortune of being protected from competing fluctuations.

At maturity Zeus represents the fluctuation that has achieved critical size, sufficiently large that competing fluctuations, Cronus and his sibling Titans, are no longer able to eradicate it. Zeus, however, does not conquer his enemies on his own; he is helped by his brothers and sisters as well as by the Cyclopes and

Hundred-Handed Ones. Properly organized, their energies constitute a formidable order-creating insurgence (a giant fluctuation). United into a coherent front, this band of rebels ultimately defeats Cronus and the Titans.

Zeus's victory over the forces of destruction, like that of Horus and Marduk, marks the joyful moment when the critical fluctuation disrupts the initially homogeneous, featureless state of the ether to create the first subatomic particle. Being Rhea's sixth child, Zeus numerologically signifies the concept of bifurcation and spontaneous symmetry breaking. Consequently, he represents the fluctuation that is ultimately destined to induce explicit order. His use of a thunderbolt as a weapon to strike down Cronus calls to mind Arcanum 16 of the Tarot, which pictures a lightning bolt striking a tower to depict the sudden overthrow of the established state.

The separation of heaven and earth is mentioned near the end of the creation myth in connection with Atlas. As in the ancient Egyptian and Sumerian creation myths, earth and heaven may be identified with the X and Y ether variables, whose separation represents the divergence of the X and Y concentrations from their homogeneous steady-state values. The localized, archlike nonuniformity that Atlas forms, then, would constitute the charged core of the self-emerged primordial proton. Just as Atlas is condemned to maintain heaven and earth in a separated state for eternity, so too does matter, once it has come into being, continue to maintain itself in a stable state.

Ancient Greek initiates must have been aware that the story of Zeus, employed symbolism similar to that of the Egyptian creation myths, for they identified Atlas with Shu, the Egyptian god who separated Geb and Nut (earth and heaven) at the dawn of creation. Likewise, they identified Hermes (Mercury), the son of Zeus, and Maia, Atlas's daughter, with Thoth, the Egyptian god of wisdom traditionally considered the original teacher of the sacred Osirian mysteries.

THE STORY OF ADONIS

The Phoenician story about Adonis and his two lovers, Aphrodite and Persephone, also presents aspects of this ancient creation science.[8] According to this story, the Queen of Assyria once proclaimed her daughter Smyrna to be more beautiful even than Aphrodite, the goddess of love. Offended, Aphrodite cast a mischievous spell that induced the maiden to sleep with her father while he lay drunk. When the king later discovered the truth of what had happened, he seized a sword and chased the pregnant Smyrna from his palace. Overtaking her at the crest of a hill, he raised his sword. But, at that moment, Aphrodite transformed Smyrna into a myrrh tree. The king's descending sword then split Smyrna in halves, and out tumbled the infant Adonis.

Repenting her mischief and concerned for the child's safety, Aphrodite concealed Adonis in a chest, which she entrusted to Persephone, queen of the dead and daughter of Demeter, the goddess of the harvest. Curious, Persephone opened the chest and found Adonis inside. She found him so attractive that she raised him in her underworld palace and made him her lover. Hearing of this, Aphrodite at once journeyed down into the underworld (Tartarus) to claim Adonis for herself, whereupon a quarrel ensued. The court of the gods settled the dispute by deciding that both goddesses had equal claims on Adonis. They decreed that Adonis was to spend one third of the year with Persephone, one third with Aphrodite, and the remaining third with himself to give him a respite from the insatiable amorous demands of the two goddesses.

But Aphrodite did not play fair. Wearing a magic girdle, she enticed Adonis to disobey the court order by giving her his own share of the year and part of the share allotted to Persephone. Annoyed by the life he led, Adonis went off alone to hunt in the forests of Lebanon, where he met an untimely death. One of Aphrodite's jealous ex-lovers (Aries or Apollo) had disguised himself as a wild boar and gored Adonis to death before Aphrodite's eyes. Adonis's soul subsequently descended into the underworld, whereupon Aphrodite tearfully pleaded with Zeus that Adonis should not spend the entire year with Persephone. Moved by her grief, Zeus permitted the resurrection of Adonis, that he might pass the spring and summer seasons in the skies with Aphrodite and the fall and winter seasons in the underworld with Persephone.

Like the myth of Castor and Pollux, the story of Adonis graphically illustrates the periodic reciprocity of the X and Y ethers in forming a subatomic particle. The child Adonis, who emerges from the chest and is raised in Persephone's underworld palace, symbolizes the growing ether fluctuation that is destined to create explicit order. Persephone, who in classical times was associated with the Virgo constellation, here personifies the nourishing ether. Persephone and Aphrodite, who both desire Adonis, symbolize the two alternative polarities that the emerging fluctuation could adopt in bifurcating the ether's prevailing state of uniform concentration. A similar choice between alternatives is depicted in Arcanum 6 of the Tarot (the Lovers, or the Two Ways). This bifurcation theme is hinted at earlier in the myth when Adonis is born through his mother's bisectional death. Adonis's dominant partnership with Aphrodite results in his untimely death, indicating that this particular partnership polarity (high X/low Y) is inherently unstable. It is comparable to the Lover of Arcanum 6 choosing the path of ostentation to his left. Thus the Adonis myth indicates that one of the two possible states is unstable, thereby portraying the sophisticated cosmological concept of matter/antimatter asymmetry.

The subsequent descent of Adonis's soul into the underworld invokes the opposite polarity of Persephone-Adonis, low X/high Y, which has a more enduring fate. That is, at Zeus's decree, Adonis is allowed to spend the harsher seasons in the underworld with Persephone (low X/high Y) and the milder seasons in heaven with Aphrodite (high X/low Y). In other words, the X and Y ether concentrations cyclically alternate, tracing out the wavelike energy field that disseminates outward through space to form the emergent subatomic particle.

The theme of Adonis being removed from the tree trunk and placed within a chest finds a parallel in the myth of Osiris, where the chest containing Osiris's body is cut out from the trunk of a fragrant tree. Moreover, according to one version of the myth, Smyrna's father was King Phoenix of Byblos, the same Phoenician seaport where Osiris's chest washed ashore. In fact, in ancient times, feasts were celebrated at Byblos in honor of Adonis, an event known as Adonias. Byblos once carried out extensive trade with ancient Egypt, furnishing it with papyrus for making scrolls. The town apparently acquired its name from its product since *byblos* in ancient Greek means "book." It is quite possible that Egyptian priests transplanted the Adonis myth to Phoenicia, which could explain some similarities to the story of Osiris.

THE I CHING

According to ancient Chinese thought, all physical things are maintained in existence through the ceaseless interplay of two fundamental cosmic principles, yin (darkness) and yang (light). Subsequently, yin also came to connote passivity, coldness, weakness, softness, the negative, and the feminine, while yang came to connote activity, heat, strength, hardness, the positive, and the masculine. Yin and yang were understood to be inherently dynamic, each aspect being in the process of transmuting into its complement.

The exposition of the yin-yang metaphysics appears primarily in the Appendices and Commentaries to the I Ching, the Book of Changes. It presents a system for constructing hieroglyph-like symbols by combining broken lines (yin) and unbroken lines (yang) in sets of three to form what are called trigrams. By making various pairings of the eight possible trigrams (figure 11.3), it is possible to form sixty-four different hexagrams, each of which represents a cosmic archetype or pattern observed to recur in nature. The original text of the Book of Changes describes the symbolic meaning assigned to each of these hexagrams. Together, these symbols describe the constant changes that everywhere take place.

Because of its great antiquity and wisdom, the Book of Changes has been held in great veneration by the Chinese, in a manner comparable to the Vedic scriptures in India or the Bible in the West. Although it is not known for sure when this classic work originated, the symbols and their explanations have been attributed to the

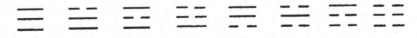

Figure 11.3. The eight trigrams.

legendary emperor Fu Hsi, who lived between 2953 and 2838 B.C.E. This would place the origin of this work about half a millennium prior to the inscription of the Pyramid Texts. The Appendices and Commentaries, however, are more recent additions attributed to Confucius (551–479 B.C.E.).

Like the myth of P'an-Ku, the Book of Changes relates that physical form emerged as a result of the cyclical mutual transformation of yin and yang principles. It envisioned this process of continual change as a unitary and indivisible process termed the Tao: "One Yang and one Yin: this is called the *Tao*."[9] *Tao* translates as "the way" (of nature) or, in other words, the manner in which things come into being and continue to manifest. Lao Tzu, the sixth-century B.C.E. philosopher who is credited with the founding of Taoism, in his book *Tao Te Ching* wrote the following about the Tao:

There was something undefined and complete, coming into existence before Heaven and Earth. How still it was and formless, standing alone and undergoing no change, reaching everywhere and in no danger of being exhausted! It may be regarded as the Mother of all things. I do not know its name, and (in the absence of a better word) I give it the designation of the Tao. . . .

All-pervading is the Great Tao! It may be found on the left hand and on the right. All things depend on it for their production. . . .

We look at it, and we do not see it, and we call it "the Equable." We listen to it, and we do not hear it, and we name it "the Inaudible." We try to grasp it, and we name it "the Subtle." With these three qualities, it cannot be made the subject of description and hence we blend them together and obtain the One.[10]

Whereas Lao Tzu considered the Tao as beyond description, Confucianists sought a more tangible portrayal of the idea in terms of the *ch'i* ether concept. *Ch'i* which translates literally as "gas," metaphysically signified the vital breath or vital energy that animates the cosmos. It was conceived of as a subtle primal substance characterized by continuous activity. The neo-Confucian philosopher Chang Tsai traced the idea back to the Book of Changes wherein the Tao is described as an "emanation," a term similar in meaning to *ch'i*. The word *ch'i* was first used in a philosophical context by Mencius (380–289 B.C.E.), who spoke of an "all-embracing force" *(hao jan chih ch'i)*.[11] The cosmologist Tung Chung-shu (176–104 B.C.E.) also used the idea when he taught that yin and yang were two types of *ch'i*. These were conceived to undergo reciprocal oscillation. Wang Ch'ung (80 C.E.) described this action as follows: "The yang having reached its climax retreats in favor of the yin; the yin having reached its climax retreats in favor of the yang."[12]

One cannot help noticing the close resemblance of this metaphysics to that of subquantum kinetics. The Tao may be identified with the transmuting ether, and the yin and yang principles with the X and Y ether states of Model G. As with yin and yang,

Figure 11.4. Diagrams depicting the mutual transmutation of yin and yang: the original diagram showing two fish (left) and its later schematization as the t'ai-chi-t'u (right).

so too the X and Y ethers continually transmute into one another. Just as these co-transmuting ethers give rise to physical waves (matter and energy) formed by the opposed divergence of X and Y from their prevailing mean concentration, so too yin and yang oscillate in magnitude relative to one another. Moreover, just as these emergent X-Y waves involve moderate and equable concentration excursions above and below a common mean value, so too the I Ching emphasizes the ideal of seeking a central equilibrium and cautions that extremes in either direction, toward either yin or yang, produce opposite reactions of equal magnitude. The Appendices apply this Epicurean doctrine of the mean to daily human affairs, advising that people should seek a golden mean between extremes, neither too much nor too little. This same important concept is expressed by Arcanum 9 of the Tarot (the Sage) and by the zodiac sign of Virgo.

The recursive transformation of yin into yang and yang into yin was represented in ancient times by two closely entwined fish, one dark and one light, swimming in a circle (figure 11.4, left). The symbol calls to mind the two Piscean fish as well as the glyph for Cancer. In the eleventh century C.E. the Chinese philosopher Chou Tun-yi devised a more schematic version of this ideogram called the *t'ai-chi-t'u*, the "Diagram of the Supreme Ultimate" (figure 11.4, right). The fish eyes were made into circles and the outlines of the bodies were each formed by drawing partial circular arcs. Chou Tun-yi explained the *t'ai-chi* diagram as follows:

> The Supreme Ultimate through Movement produces the Yang. This Movement, having reached its limit, is followed by Quiescence, and by this Quiescence, it produces the Yin. When Quiescence has reached its limit, there is a return to Movement. Thus Movement and Quiescence, in alternation, become each the source of the other.[13]

The swimming fish and *t'ai-chi* diagrams graphically portray the yin and yang forces as they engage in complementary oscillation relative to each other. Through growth, the less prominent species ultimately attains the state of superior abundance once held by its waning predecessor. The dot of light within the region of maximum darkness, like the dot of darkness within the region of maximum light, indicates that the minor species is never totally eradicated, even when the more predominant complement species is at a maximum. Without these residuals, the ongoing yin-yang transmutation would cease, and with it, so would the Tao, whereupon all of physical existence would dissolve. The two dots in the *t'ai-chi* diagram, then, serve as reminders of the delicate mutually interdependent relationship that yin and yang always maintain.

This depicts a universal truth about all oscillating reaction systems, whether the oscillations be of ether concentration, chemical concentration, or predator-prey population. What transpires in the one state intimately affects the complementary state. If either state were to become totally consumed, the polar opposite state would also disappear and the oscillations would abruptly end.

These bipolar diagrams portray in integrated fashion both the implicit and explicit orders of physical existence. The head-to-tail arrangement of yin and yang illustrates the circular transmutation that characterizes the ether's implicit order. The two dots within each symbol imply the reciprocal bivalent oscillation that generates explicit order—physical form. In fact, the fish diagram shown in figure 11.4 is essentially a polar coordinate plot of two reciprocally oscillating waves. If a radius is extended from the center of the circle (between the two fish) and is swept counterclockwise, the length of its dark and light segments would vary in such a way that, when plotted in Cartesian coordinates, they would produce two reciprocally oscillating sine waves like those graphed in figure 3.9. This insignia encodes in its form key aspects of a highly sophisticated reaction kinetic science.

PART 3

Changing the Paradigm

12
ETHER OR VACUUM?

THE MICHELSON-MORLEY EXPERIMENT

The downfall of the nineteenth-century mechanical ether theory came about as a result of laboratory experiments conducted during the latter part of the nineteenth century. The initial purpose of these experiments was to determine the speed of the Earth's motion through the ether. It was known that the Earth travels in its orbit around the Sun at a speed of about thirty kilometers per second and that the solar system moves at an even greater velocity relative to other stars in the Galaxy.* So physicists expected that the ether should be drifting past the Earth at a substantial velocity and that this drift speed should slightly alter the speed of a light beam seen by an Earth observer. Since light travels at nearly 300,000 kilometers per second, Earth's motion would be expected to change this measured velocity by only a small fraction of a percent. If we visualize this ether drift as a river current and a light beam as a person rowing a boat downstream in the same direction as the ether drift, then an Earth observer (standing on the riverbank) would see the boat traveling slightly faster, whereas if the boat were moving upstream against the ether drift current, the observer would see it traveling slightly slower.

Unfortunately, no one knew of a way to measure the one-way velocity of light to see if light traveled slightly faster in one direction compared with another. Nevertheless, two American physicists, Albert Michelson and Edward Morley, devised a way to determine the direction and velocity of the ether drift simply by comparing the round-trip travel times of light waves directed along two mutually perpendicular directions. They reasoned that a light beam traveling over to a mirror and back again perpendicular to the ether drift should take longer to complete its round-trip journey, as compared with a beam making a similar round trip along

---

*Modern observations indicate that the solar system orbits around the Galaxy's center at a speed of around 250 kilometers per second.

264

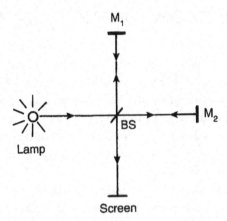

Figure 12.1. A schematic representation of Michelson and Morley's interferometer experiment. BS is a beam splitter; M_1 and M_2 are mirrors.

a path parallel to the ether drift;* the magnitude of this time difference was expected to provide a measure of the Earth's relative velocity.

In 1887 these two physicists constructed a light interferometer to check this prediction. A simplified drawing of their apparatus is shown in figure 12.1. By means of a set of mirrors, two light beams from a common sodium vapor lamp were made to traverse mutually perpendicular round-trip paths and combine into a single beam. The wave fronts of the two beams, when combined, were expected to interfere with one another to produce a pattern of light and dark fringes. Different transit times were expected for light beams oriented perpendicular to one another. The time difference between beams would be at a maximum when the apparatus was oriented so that one beam was parallel and the other perpendicular to the ether drift direction and would be at a minimum when the apparatus was turned so that both beams were oriented at 45° to the ether drift direction. As the apparatus was rotated, this direction-dependent change should have changed the relative phasing of the beams and shifted the light and dark fringes making up the light wave interference pattern.

To their surprise, and to the surprise of the rest of the physics community, no fringe movement was observed, regardless of how the apparatus was oriented. The experiment was repeated many times by many investigators, and still the expected phase shift was not found. Their results delivered a lethal blow to the mechanical ether theory. Physicists still accepted that an ether must exist, but they were unable to explain its inner workings as they once had done. They could no longer reasonably describe wave propagation through this medium in mechanical terms.

---

*The mechanical ether theory required that a light beam directed parallel to the ether drift should travel slower by a factor of $1/\sqrt{1-v^2/c^2}$ where v is the Earth's velocity through the ether and c is the light beam's velocity in the ether rest frame.

THE ELECTROMAGNETIC ETHER THEORY

As a result of these developments, a new nonmechanical, electromagnetic view of nature began to emerge in the 1890s. One of the key people shaping this view was the Dutch physicist Hendrik Lorentz. His attempts to account for the results of the Michelson-Morley experiment led him to devise a radically new theory of matter. In 1904 he proposed that the subatomic particles making up material bodies were not billiard-ball-like spheres distinct from the ether, but resilient wavelike excitations formed in the ether itself. Thus he conceived matter, like energy waves, to be basically electromagnetic in nature.[1]

According to his theory, as a body accelerated to the speed of light its constituent particles would become increasingly foreshortened in the direction of motion as well as increasingly massive. He demonstrated that if matter behaved in this way, then the arms of Michelson and Morley's interferometer would contract in the ether drift direction by an amount sufficient to produce a null result in the Michelson-Morley experiment. With the Lorentz contraction, a light signal would take the same amount of time to cover its round-trip journey through the ether in whatever direction it traveled.

The equations of motion that Lorentz formulated, the forerunner of today's "Lorentz transformations," also imply that a moving clock should slow down and its clock mechanism should physically move more and more slowly as its speed increases toward the speed of light.[2] Yet Lorentz did not come to recognize clock retardation as a real physical effect until several years later.[3] As a result, physics books often credit Einstein for first pointing out this phenomenon; however, Einstein interpreted the effect quite differently from Lorentz. His special theory of relativity, published in 1905, attributed the slowing of the clock to the slowing of time itself. This came to be called the time dilation effect.* Although Lorentz's clock-retardation interpretation is not as well known, it is still considered just as valid.

Gustav Mie further developed Lorentz's electromagnetic concept of matter by suggesting that fundamental particles such as electrons were simply places in the ether where the electric and magnetic field intensities achieved particularly high values, where the normal equations of electrodynamics no longer applied and a new type of nonlinear behavior emerged giving rise to matter.[4] Lorentz, Mie, and the many other proponents of the electromagnetic ether concept, therefore, envisioned the ether as the only reality. All physical phenomena, material particles as well as energy waves, were understood to be superficial excitations in the universal

---

*Einstein is also often given credit for proposing in his 1905 paper that matter is convertible to energy. This concept, however, had already been in existence for several years within the electromagnetic ether formulation. It was implied in the works of Jules-Henri Poincaré published in 1900, and later, in 1904, it was stated more clearly by the British physicist Frederick Soddy (D. Turner and R. Hazelett, *The Einstein Myth and the Ives Papers: A Counter-Revolution in Physics* [Old Greenwich, Conn.: Devin-Adair, 1979], pp. 182–87).

ether. Thus the ether/particle dichotomy that had once plagued physics had now become repaired. The ancient concept of an omnipresent ether serving as the fundamental stuff of the universe was once again reinstated.

About a decade before Lorentz and Mie proposed their ether-wave theory of subatomic particles, a nineteenth-century Chinese physicist, T'an Ssu-t'ung (1865–1898), developed a similar ether theory of matter based instead on the neo-Confucian idea of a transmuting ether. Like the ancient ether physics, his theory assumed the existence of a bipolar ether consisting of mutually transmuting yin and yang states. He proposed that the fundamental particles of matter are formed through a "condensation" of this ether substance. He explained the transmutation process of his ether in considerable detail, proposing that the ether as a whole maintains a dynamic steady state of continuous renewal, as its individual components engage in "the minute process of production and destruction." In other words, at a "microscopic" level, ether transmutation would consist of individual etherons entering and leaving a given etheron state, whereas at a more macroscopic level, these countless events would blend together to produce a continuous transformation process:

> If we look at the past, the process of production and destruction has never had a beginning. If we look at the future, the process will never come to an end. And if we look at the present, it is constantly going on. . . . Once when Confucius was standing by a stream he observed: "All is transient, like this! Unceasing day and night!" The [change of] day and night is the principle of that stream. . . . There is neither singleness nor duality, neither interruption nor continuity. The cycle of production is followed by that of destruction; as soon as there is this destruction, it gives way to production. The alterations between production and destruction become minute and still more minute, until they can be made no further minute; they become hidden and still more hidden, until they can be made no further hidden. In this way they merge into a oneness in which there is neither production nor destruction. But though there is this state, that which causes the minute process of production and destruction certainly cannot be easily concealed.[5]

T'an developed his ether theory in an era when physical science was dominated by the mechanical ether theories of the West. In an effort to integrate his theory into Western terminology, he abandoned the traditional term *ch'i* and instead coined the term *yi-t'ai*, a transliteration of the word "ether." It is interesting to speculate what physics would be like today if the past three decades of research on oscillating reaction-diffusion systems had been carried out just one century earlier. Had it been further developed and elaborated using modern reaction-kinetics concepts, T'an's theory could have evolved into a version very similar to the twentieth-century theory of subquantum kinetics.

FROM ETHER TO VACUUM

The unitary ether of Lorentz and Mie apparently did not secure a very strong hold on Western physics. In 1905, just when the description of physical phenomena in terms of an electromagnetic ether was gathering momentum, Einstein published his special theory of relativity. He noted that the results of the Michelson-Morley experiment could be explained in a simple fashion if one accepted the premise that the one-way velocity of light would always be equal to a constant value regardless of the observer's speed through space.

This interpretation, of course, refuted the classical notion that the metrics of space and time are absolutes. Einstein theorized instead that the distance between two points in space or duration of a given event is an "elastic" quantity that can take on an infinite number of possible values depending on how fast an observer happens to be traveling relative to the object or transpiring event. Making space-time dependent on the observer's speed could make life very complicated. For example, suppose a person is in a crowd and simultaneously observed by a hundred people moving in a hundred different directions. If relativity theory is correct, these annoying observers should be causing that person to exist simultaneously in a hundred different space-time frames, each with its own unique space-time metric. Is one of these many space-time clones the real person? Or is his or her consciousness somehow distributed among all of them simultaneously, only to rebound into a singular state once the crowd stops observing? Indeed, this schizophrenic concept of reality is difficult to reconcile with any commonsense view of the world.

Einstein's theory led to other equally perplexing contradictions, such as the twin clock paradox [6] and the light-source-velocity paradox,[7] neither of which had troubled the electromagnetic ether theory, with its single measure of time and space. But the majority of physicists, accustomed to thinking about electromagnetic phenomena solely in mathematical terms and to working with field equations divorced from any kind of concrete conceptual grounding, were not bothered by these paradoxes. They were willing to accept relativity on the basis of its mathematical elegance and to overlook its counterintuitive implications. By 1910 special relativity began to be widely accepted.

A choice to adopt relativity was a choice to deny the existence of the ether. Relativity theory with its infinite space-time frames was incompatible with the ether concept, which involves the existence of just one metric of space and one metric of time uniformly valid for all reference frames.

Shortly after Einstein proposed his theory, relativity encountered a serious challenge. In 1913 the French physicist Georges Sagnac conducted an experiment in which he mounted a light source on a turntable. He used a half-silvered mirror to divide its beam into two beams and, by means of a system of mirrors, made these two light beams travel in opposite directions around the perimeter of the turntable.

He then recombined the two beams to produce a light interference pattern and found that clockwise rotation of the turntable caused the fringes of the interference pattern to shift by an amount proportional to the turntable's speed. This indicated that rotation of the turntable had caused the counterclockwise traveling light beam to complete its circuit in less time than the clockwise traveling beam. He considered this as direct evidence that light travels in an ether. According to Sagnac:

> The observed interference effect is clearly the optical whirling effect due to the movement of the system in relation to the ether and directly manifests the existence of the ether, supporting necessarily the light waves of Huygens and of Fresnel.[8]

Sagnac's discovery later led to significant advances in guidance system technology. The ring-laser gyroscopes that daily guide passenger jets such as the Boeing 757 and 767 through the skies operate on this very same principle.[9]

Although Sagnac's experiment initially threw relativity theory into a state of turmoil, it was not long before relativists proposed a way to explain its results. Paul Langevin in 1921 claimed that the experiment's results would be "neutralized" if its calculations were adjusted to take into account relativity's time-dilation effect, which he assumed would apply to the locally revolving reference frame of the apparatus.

Herbert Ives, a prominent American inventor and Bell Laboratories physicist, published a paper in 1938 demonstrating Langevin's "local time" argument to be incorrect.[10] Therefore, Sagnac's interpretation remained valid and special relativity stood disproven, at least for rotating frames of reference. Nevertheless, few physicists read Ives's astute rebuttal. In the years following Langevin's paper, the ether concept gradually faded into the background, lingering only as a philosophical abstraction. With special relativity, physicists could focus their attention just on the field equations and forget about the ether. It was a relatively small step for them to deny the ether's existence entirely. Contrary to what some textbooks say, the ether notion was not abandoned because of any experimental disproof; it just went out of style.

With this shift in thinking, physics entered the era of the vacuum. In so doing, it excluded the possibility of a nonphysical realm of existence. If there was a spirit world beyond the material, it had to be potentially observable and quantifiable, just like the rest of the physical world. Those adopting this modern worldview had the choice of either viewing God as a physical entity composed of energy fields drifting in space, or of simply denying his existence altogether.

In 1951 Ives exposed a crucial flaw in Einstein's theory. Applying Jules-Henri Poincaré's principle of relativity to the results of the Michelson-Morley experiment, he demonstrated that the one-way velocity of light, as defined by Einstein for a relatively moving frame, is *not* equal to a constant c as Einstein had claimed. Rather, what remains constant from one reference frame to another is a very complex mathematical function that includes readings of rods and clocks and terms describing their method of use. Apparently Einstein's result is obtainable only by using time and space quantities that are not measurable by normal physical means. Irritated by the physics community's complacency with special relativity's insecure observational foundations, Ives commented:

> The assignment of a definite value to an unknown velocity [the one-way velocity of light] by fiat, without recourse to measuring instruments, is not a true physical operation, it is more properly described as a ritual. . . . The "principle" of the constancy of the velocity of light is not merely "understandable," it is *not* supported by "objective matters of fact."
>
> With the abandonment of the "principle" of the constancy of the velocity of light, the geometries which have been based on it, with their fusion of space and time, must be denied their claim to be a true description of the physical world.[11]

Ives continued his battle against relativity throughout the 1940s and early 1950s. He published a series of papers demonstrating that the electromagnetic ether theory accounted for the results of all experiments normally cited in support of special relativity.[12] His elucidation of Lorentz's theory has today come to be known as the *rod-contraction-clock-retardation ether theory*.[13] His efforts and those of others, however, did not sway the scientific community away from their relativistic outlook.

THE ETHER RETURNS

Einstein's special theory of relativity specifically requires that the one-way velocity of light be a constant. If that turns out not to be so, special relativity falls. The Michelson-Morley experiment, however, demonstrated only that the two-way, over-and-back, average velocity of light was constant. It did not necessarily prove that the one-way velocity of light in any direction was also constant. Consequently, special relativity is founded on a tentative extrapolation that goes far beyond the experimental results of the Michelson-Morley experiment.

Although no one has succeeded in accurately measuring the one-way velocity of light, in 1987 Ernest Silvertooth published the results of an experiment that clearly showed that the wavelength of light varies with the direction of light propagation.[14] His finding provided evidence that the one-way velocity of light also varies with direction. While the Sagnac experiment showed that special relativity does not apply for rotating frames of reference, Silvertooth's experiment indicated that special relativity also does not apply for straight-line motion. Silvertooth assembled a special kind of laser interferometer apparatus to carry out his wavelength measurements, shown in figure 12.2. His apparatus included a complex array of adjustable mirrors and beam splitters that caused two oppositely directed

Figure 12.2. Apparatus used by Ernest Silvertooth to determine the speed of the Earth's motion through the ether. Photo courtesy of E. Silvertooth.

laser beams to interfere and produce a standing wave pattern of regularly spaced bright and dark bands or fringes (see figure 12.3). He was then able to determine the spacing of these fringes, using a specially built television camera tube with a transparent light-sensing surface. Since the effective thickness of his detector's light-sensing surface was less than 10 percent of the laser light wavelength, he could very accurately determine the positions of consecutive bright fringes. He found that the fringes achieved their closest spacing of approximately one-fourth of a millimeter (one-hundredth of an inch) when the opposed laser beams were pointing along a direction aligned with the constellation of Leo. When the path of

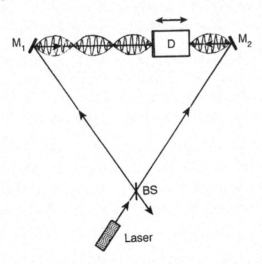

Figure 12.3. A simplified representation of Silvertooth's wavelength-measuring apparatus. Light from a laser source is divided into two beams by beam splitter (BS). A mirror system (M_1 and M_2) causes the two beams to pass in opposite directions through photoelectric detector D.

the opposed laser beams was rotated away from that heading, the fringes spread apart to greater distances. He concluded that this unique direction in which the fringe pattern attained a minimum spacing marked the direction of the Earth's motion through the ether.

By measuring the minimum fringe spacing, Silvertooth determined that the Earth moves through the ether toward Leo at a speed of about 378 (±19) kilometers per second. Several years later he built a substantially improved version of his earlier apparatus and obtained a similar result.[15] By comparison, astronomers have found that the solar system is moving toward the southern part of Leo at a speed of about 365 (±18) kilometers per second relative to the surrounding 3 Kelvin cosmic microwave background radiation field.[16] This matches Silvertooth's result within the accuracy of the respective measurements, implying that the microwave background radiation is stationary with respect to the local ether rest frame.

Other evidence against relativity and in support of an ether comes from electrodynamic experiments carried out by the Greek physicist Panagiotis Pappas[17] and the American physicist Peter Graneau.[18] These indicate that the relativistic Lorentz force law, which physicists use to describe how moving charged particles generate a magnetic field, is not universally valid and should be replaced instead by the more correct nonrelativistic cardinal force law of André Ampère. According to Ampère's force law, all electrodynamic interactions take place relative to a preferred absolute reference frame, namely that of the ether rest frame.

The findings of Silvertooth, Sagnac, Ives, Pappas, and others bring the relativistic era to an end. Relativistic concepts such as "space-time" and "warped space," must be replaced by Newton's concept of absolute space and absolute time, and the ether concept must come forward once again to fill the vacuum of the past century. With the undermining of special relativity, however, general relativity is rendered invalid as well, and with it goes the expanding-universe theory. Consequently, the entire edifice of modern relativistic cosmology has begun to crumble.

To what conceptual model can physicists now turn? They cannot return to the nineteenth-century mechanical ether theory, since that fails to explain the results of the Michelson-Morley experiment. The mathematically abstruse electromagnetic ether theory of Lorentz and the rod-contraction-clock-retardation ether theory of Ives explain the Michelson-Morley experiment results, but they are not grounded in a conceptual model. In seeking a conceptual model for the ether, perhaps scientists would do well to turn to the notion of a transmuting ether, such as that described in ancient myth and lore or more explicitly explored through Model G of subquantum kinetics.

THE TWENTIETH-CENTURY CREATION MYTHOS

MODERN COSMOLOGY PRESENTS A PICTURE of the origin of things that strongly contrasts with the conceptions of our ancient predecessors. Today's cosmologists assert that all matter and energy in the universe came into being as a result of a tremendously energetic explosion, or "Big Bang," and that prior to this creation event nothing existed. Not only was there no matter nor energy, but supposedly no space nor time either. For general relativity proclaims that space and time are intimately intertwined with matter and energy and that all came into being simultaneously.* Astronomical data, however, favors the ancient view of a static, gradually created universe as opposed to the expanding big bang universe advocated by the majority of twentieth-century cosmologists. Before considering the details of this challenge to contemporary cosmology, it is helpful to review the historical events that brought the big bang theory into being in the first place.

THE MAKING OF THE MYTH

In the early years of astronomy the visible universe was believed to be considerably smaller. The Ptolemaic model, which dominated medieval cosmology, fixed the outer boundary of the cosmos just beyond the orbit of Saturn. This containment shell was conceived to consist of a star-studded crystalline sphere positioned so that the Earth lay at its center. Following the development of the telescope, early astronomers such as Copernicus and Galileo discovered that the Sun, not the Earth,

---

*According to general relativity, material bodies exert their gravitational pull on neighboring masses by warping the geometry of space around themselves. The theory does not explain how material bodies warp space; it simply states that they do and proclaims that matter and space are intimately interconnected as an integral unit, "matter/space." Since matter and energy are interchangeable, relativists further consider "matter/energy/space" to be a single unit. In addition, special relativity regards space and time to be intertwined as "space-time." In combination, special and general relativity theory propose that the physical universe is bound up together with its space and time dimensions to compose a single entity: "matter/energy/space/time."

lay at the center of the solar system and that our Sun was just one star out of many scattered throughout space. As a result, the Ptolemaic model was eventually discarded in favor of one that visualized a large collection of stars with the solar system positioned somewhere near its middle. In 1790 this stellar universe was estimated to have a radius of about eight thousand light-years, quite small by current standards.

Astronomers had yet to accept the notion of the existence of galaxies other than our own Milky Way. What are now recognized to be other galaxies were then thought to be nebulae, large clouds of luminous gas, located not far from the Sun. Nevertheless, there were some who felt otherwise. Immanuel Kant, for example, proposed in 1755 that many of these nebulae might be "island universes," systems of stars comparable in size to our own galaxy. The English philosopher Thomas Wright put forth a similar view five years earlier in his book entitled *An Original Theory of the Universe*. Wright suggested that our galaxy, which he referred to as the "Creation" or the "known Universe," is one of many, many galaxies scattered throughout space. He wrote: "As the visible Creation is supposed to be full of sidereal Systems and planetary Worlds, so on, in like manner, the endless Immensity is an unlimited Plenum of Creations not unlike the known Universe."[1]

The majority of the scientific community, however, continued to regard the universe as being of finite extent and consisting of a single galaxy of stars, of which our Sun was a member. Whether the nebulae were distant galaxies or just clouds of gas within the Milky Way did not become seriously considered until the early part of the twentieth century. The first confirmatory evidence came in 1924. Using the hundred-inch telescope then recently installed on Mount Wilson in southern California, Edwin Hubble reported that he could discern the presence of individual stars in some of the more prominent nebulae. Moreover, he was able to identify some of these as Cepheid variables: stars that pulsate in brightness with a well-defined cyclic period ranging anywhere from a few days to many months. He compared the luminosities of Cepheids in these nebulae to those found in our own galaxy, and knowing the distances to the latter, he was able to estimate the distances to these nebulae. The Andromeda galaxy (figure 5.12), our nearest neighboring spiral galaxy, was the first to have its distance announced. Hubble estimated that it lay 900,000 light-years away, well beyond the outer boundary of the Milky Way. With this announcement, the perceived universe suddenly increased thirtyfold.

In 1924, following Hubble's announcement, the perceived size of the universe expanded at a tremendous pace. By 1929 Hubble had estimated distances to galaxies lying as far away as 25 million light-years. By 1931 he and his co-worker Milton Humason had pushed the outer limits to 100 million light-years. At no other time in the history of astronomy did scientists' concept of the extent of the universe expand at such an enormous rate (see figure 13.1). It is no wonder that during these very same years the expanding-universe hypothesis successfully secured a foothold in modern cosmology.

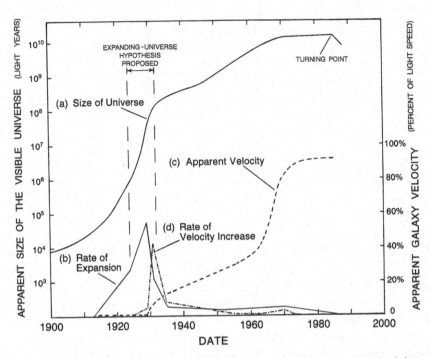

Figure 13.1. Profiles illustrating: (a) how the perceived size of the universe has increased during the twentieth century; (b) how the rate of this size increase has varied; (c) how the value of the maximum observed redshift (or "recessional velocity") has increased; and (d) how the rate of this redshift increase has varied.

The notion that the universe might be expanding came about as the result of the misinterpretation of wavelength shifts seen in light spectrograms of distant galaxies. The first spectrograms of individual stars were made in the 1870s. These were produced by using a prism or diffraction grating to disperse starlight according to its wavelength (or color) and photographing this spectral smear.* Astronomers found that the spectrum of each star has a recognizable pattern or "fingerprint": a series of bright and dark lines (figure 13.2) whose particular arrangement depends on the temperature of the star and the specific elements present in its atmosphere.

By observing whether a star's spectral fingerprint is shifted to higher or lower wavelengths (toward the blue or red ends of the spectrum), scientists found they

---

*When sunlight or starlight is passed through a prism, it becomes dispersed into a rainbow of colors, a light spectrum. Each color denotes a particular light wavelength; the red end of the spectrum is formed by long wavelength (low frequency and low energy) photons and the violet end of the spectrum by short wavelength (high frequency and high energy) photons. The spectrum of a typical star extends well beyond the visible frequencies from radio waves at the low frequency end to gamma rays at the high frequency end.

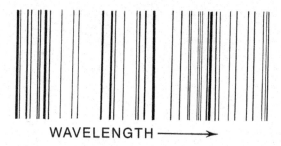

WAVELENGTH ⟶

Figure 13.2. A hypothetical stellar spectrum. Each spectral line represents a specific frequency or wavelength of light. In proceeding from left to right, wavelength increases and photons become redder.

could determine whether the star was approaching or receding relative to the Earth. The light from an approaching star appears bluer than normal (indicating a higher frequency and shorter wavelength), whereas the light from a receding star appears redder (indicating a lower frequency or longer wavelength). A similar phenomenon happens with sound. If an approaching train blows its horn, the sound will be heard at a higher than normal pitch, since the sound waves are slightly compressed by the train's forward motion. Then, as the train passes, the horn's pitch falls, as the sound waves are now expanded by the train's receding motion. The faster the train's speed, the greater is the change of pitch. This phenomenon is known as the Doppler effect, in honor of the German physicist who once studied its properties.

Just as it is possible to determine a train's speed from the pitch of its horn, astronomers realized that they could determine the speed of a star's approaching or receding motion by measuring the amount of the blueward or redward shift in its light spectrum. By means of this technique, they were able to observe the cyclical shifts in the spectra of stars orbiting in a binary star system and thereby determine the orbital speeds of each star. Vesto Slipher, an astronomer at Lowell Observatory in Flagstaff, Arizona, began preparations in 1909 to use this technique to determine the rotational velocities of the spiral nebulae, which at the time were thought to reside relatively close to the Sun. He took his first spectrogram in 1912 by training his telescope on the Andromeda nebula. As expected, he found that stars to one side of the nebula appeared to be approaching relative to stars on the nebula's opposite side, thus indicating that they were orbiting the nebula's center. To his surprise, however, he found that light from Andromeda was on the average shifted to the blue end of the spectrum, amounting to a wavelength shift of about 0.1 percent. Since spectral shifts were customarily interpreted as evidence that the light-emitting source was in motion, he naturally interpreted this overall spectral line displacement as evidence that Andromeda was approaching us at a velocity of three hundred kilometers per second—somewhat greater than velocities typically observed for stars in the Milky Way.*

During the following two years, Slipher measured the spectra of thirteen nebulae in all. But his results began to look a bit peculiar. All but two of the nebulae in his sample showed spectral shifts toward the red, with the fainter objects exhibiting higher redshifts. Moreover, one of these nebulae had a redshift that projected a velocity of one thousand kilometers per second, considerably in excess of velocities typical of stars. Slipher reasoned that such high speeds would not be so unusual if the spiral nebulae actually lay outside the Milky Way and were in rapid motion relative to our galaxy. So as early as 1917 he reported that his findings seemed to support the idea that these nebulae were actually vast systems of faraway stars—an idea further supported by Edwin Hubble's 1924 measurements of the distances to Andromeda and several other prominent nebulae.

By 1925 Slipher had accumulated spectral shift measurements for forty galaxies, and together with the measurements made by other astronomers, the total came to forty-five. Of all these, only two exhibited blueshifts; the others exhibited redshifts that, if interpreted as Doppler effects, indicated recessional velocities of up to eighteen hundred kilometers per second. At that time no other mechanism was known that could cause spectral shifting. So the notion that galaxies were moving relative to our own at very high velocities eventually became firmly entrenched despite its astounding implications.

During the earlier years of the galactic redshift discoveries, several papers were published that came to provide a theoretical basis for the expanding-universe hypothesis. All of these theories employed variations of Einstein's general relativity equations. Einstein was the first to forge this area of investigation. In 1917, two years after publication of his general theory of relativity, he wrote a paper that applied his gravitational field equations to the universe as a whole. His paper theorized how galaxies would gravitationally affect neighboring galaxies through their alleged ability to curve space. Thus began a field of study that has since come to be known as *relativistic cosmology.*

Einstein's general relativity equations, however, predicted that space should be either expanding or contracting. They would not produce a universe that remained stationary. Believing at the time that the universe was static, Einstein introduced into his equations a stabilizing term that he called the *cosmological constant,* whose sole purpose was to ensure that his equations would produce a stationary space-time continuum.

The Russian mathematician Alexander Friedmann took a different approach, however. In 1922 and 1924 he published solutions to Einstein's general relativity

---

*This speed of approach later was revised downward to only fifty kilometers per second by subtracting the Sun's orbital velocity about the Galactic center.

field equations making no attempt to force a static-universe solution. His equations predicted a variety of possible dynamic behaviors, the type of behavior depending on what one assumed was the average density of matter in the universe. Cosmologists usually speak of a parameter called omega (Ω), which represents the ratio of the assumed matter density to the *critical* density required to bring the universe's expansion to a halt.* If the average matter density is assumed to be less than the critical density, the gravitational attraction force exerted by all the matter in the universe would be unable to completely contain the expansion, and as a result the universe would expand forever. This type of universe is termed *unbounded,* or *open.* On the other hand, if the average matter density is assumed to be greater than the critical density, the gravitational force would be sufficient to halt the universe's expansion and cause it to subsequently contract. Such a universe is termed *bounded,* or *closed.* Friedmann's equations predict differing types of space-time geometry for each of these three types of universes. The open universe would have a hyperbolic geometry (parallel lines projected to great distances would gradually diverge from one another); and a closed universe would have a spherical geometry (parallel lines projected to great distances would gradually converge). A universe having a matter density equal to the critical density would have a *flat* or *Euclidean* geometry.

Although Friedmann's papers did not make much of an impact on the Western astronomical community at the time, his equations later formed the foundation for the modern big bang theory. These equations, however, must be very finely tuned to ensure that the universe starts its expansion with a matter density exceedingly close to the critical value of $\Omega = 1$. Calculations show that if the initial matter density deviates by as little as one part in 10^{40} from this value (that is, 0.0000000000000000000000000000000000000001), in less than a second the universe's space geometry either curves wide open (leaving space essentially empty) or curves tightly closed, collapsing into a black hole.[2]

Not having prior knowledge of Friedmann's work, the Belgian cosmologist George Lemaître put forth very similar ideas in 1927. Lemaître published solutions to the relativity equations that, like Friedmann's models, required space to be expanding, or alternately expanding and contracting. He cited the galactic redshift discoveries of the previous decade as evidence that the universe should presently be in an expanding mode. In this same 1927 paper, he also originated the notion of the "Big Bang," although he did not refer to it by that name. He noted that if space was indeed expanding, then all matter in the universe must at one time have originated from a single point of very high density. Lemaître did not postulate a primordial fireball for the creation event. Rather, he imagined a superdense primeval atom,

---

*Some instead use the symbol q_0, where $q_0 = \Omega/2$.

thirty times the size of our Sun, which supposedly fragmented explosively to produce the multitude of atoms that eventually came to form the stars and galaxies of our universe. He theorized that this fragmentation gave the universe the initial push that caused its expansion.

Three years earlier, in 1924, the German astronomer Carl Wirtz had demonstrated that the galaxies having the smallest apparent diameters exhibit the highest redshifts. This led Wirtz to conclude that redshift (or recessional velocity) increases with distance and that the more distant galaxies are moving away from us at greater velocities. This concept fit precisely the predictions of the expansion model. Then in 1929 Edwin Hubble announced a finding that supported Wirtz's redshift-diameter relation. Hubble, who had just finished determining distances to thirty-three galaxies, discovered that galaxy redshift indeed increased in direct proportion to distance (see figure 13.3). Provided that redshift is interpreted as a recessional velocity, Hubble's data indicated that for every million light-years of distance, galaxies appeared to be receding faster by an additional 150 kilometers per second. Consequently, Hubble suggested that his data might be evidence that the universe is expanding. His graph later became known as the Hubble diagram; the redshift-distance relation it plotted came to be called Hubble's law; and the slope of this redshift-distance relation came to be termed the Hubble constant, written as H_0.

As larger and larger telescopes were built and as photographic techniques were continuously improved, spectra of increasingly distant galaxies could be recorded. Although the unusually high redshift velocities ran counter to commonsense

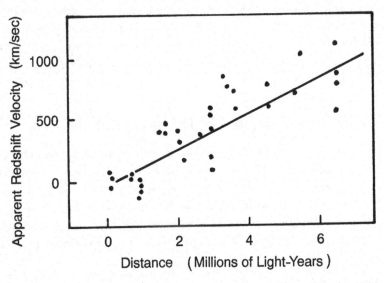

Figure 13.3. The redshift-distance diagram that Edwin Hubble published in 1929 (in *Proceedings*, pp. 168–73).

expectations, the expanding-universe scheme had now found a legitimate form of expression in the relativity cosmologies of Lemaître, de Sitter, and Friedmann. Having secured a foothold, the expanding-universe scheme was destined to rule cosmology for many decades to come. By 1931 recessional velocities had been found to reach as high as 20,000 kilometers per second (7 percent of the speed of light), by 1935 as high as 40,000 kilometers per second (13 percent of the speed of light), and by the late 1950s the speed record had passed 100,000 kilometers per second, over one-third of the speed of light! With the discovery of quasars* in 1963, redshifts reached as high as 0.5 (a 50 percent change in wavelength), implying a velocity of 41 percent of light speed. As additional quasars were discovered, the quasar redshift maximum rose, reaching 4.73 by 1990 and 6.3 by 2001. A 630 percent increase in wavelength implies a recessional velocity of 96 percent of light speed. Figure 13.1 shows how this progressive increase of inferred recessional velocity (curve c) compares with the increase in the perceived size of the universe during this period (curve a).

When Lemaître had proposed his big bang theory, the maximum observed redshift was over 100,000 times smaller than the current maximum, and the known universe was over one thousand times smaller. If today's redshifts and distances had been discovered all at once sixty years ago, would astronomers have been as inclined to interpret them as evidence that the universe is expanding?

It is known that if a frog is placed into a pot of boiling water, it will react by immediately jumping out. But if instead it is placed in a pot of lukewarm water that is heated gradually, the frog will unwittingly remain there and become thoroughly cooked. Once accustomed to thinking of redshifts in terms of recessional velocities, expanding-universe cosmologists found it difficult to jump out of their pan of water. With the inferred velocities now pressing close to the speed of light, an unbiased onlooker might suspect that the expanding-universe hypothesis has been in hot water for some time.

TIRED LIGHT, A BETTER ALTERNATIVE

Although the big bang theory and expanding-universe hypothesis have dominated cosmological thought for the greater portion of the twentieth century, not all scientists accepted these concepts. For example, in October 1929, just seven months after Hubble published his results on the redshift-distance relation, Fritz Zwicky proposed quite a different interpretation of Hubble's findings.[3] This Swiss-born Cal Tech physicist suggested that galaxies and space were cosmologically static and that

---

*Quasars are galaxies whose cores are in a highly energized state. The intense emission radiated from their cores allows them to be easily detected at great distances.

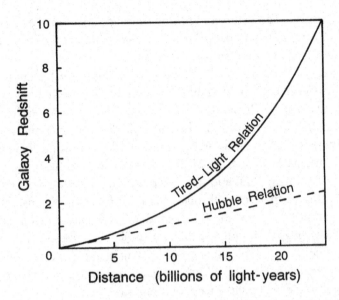

Figure 13.4. The logarithmic increase of redshift with distance in the "tired-light" model. Hubble's linear relation is shown for comparison. The profiles assume a redshifting of 10 percent per billion light-years.

the redshift was instead due to light photons gradually losing their energy during their long journey through space. Since a photon's energy and wavelength are inversely related, a reduction in photon energy translates into a wavelength lengthening, or redshift. Zwicky speculated that this energy loss could be caused by some kind of frictional drag associated with a photon's passage through the intergalactic gravitational field, with the lost energy reappearing perhaps in the form of secondary photons of very low energy. If the parent photon were to lose energy in a uniform manner throughout its journey, the magnitude of its redshift would be expected to increase progressively with distance in a logarithmic fashion, as shown in figure 13.4. For travel distances of less than 1 billion light-years, this redshift-distance relation would appear linear, just as Hubble observed.

The ether physics of ancient mythology and its modern counterpart, subquantum kinetics, also predict an energy-loss tired-light effect in intergalactic space, but one in which the lost energy actually vanishes from physical existence. Unlike Zwicky's gravitational-drag redshift, this tired-light effect would be energy nonconserving; that is, input physical energy would not equal output physical energy. But this seeming violation of the first law does not pose a problem in a universe that is presumed to function as an open system.

Zwicky's theory was motivated solely in an attempt to explain Hubble's redshift data, whereas the subquantum kinetics tired-light effect emerged as a prediction of Model G, which itself was developed primarily to describe the structure and origin of subatomic particles and fields. So not only does subquantum kinetics offer a complete replacement of the big bang theory, but its tired-light prediction also does not require making any new ad hoc assumptions.

Actually, Zwicky was not the first to propose a tired-light effect. Eight years earlier, in 1921, the German physicist Walther von Nernst proposed a tired-light cosmology which, like that of subquantum kinetics, was energy nonconserving. Nernst pointed out that in a universe of unlimited age, whether it be stationary or freely expanding, the temperature of interstellar space should be continually increasing, owing to its accumulation of stellar radiant energy. Noting that the temperature of space has instead remained quite low, he proposed that light photons must lose energy to the ether as they travel through space. In his book on cosmology, he wrote: "Luminous ether . . . thought to be a conductor capable of assuming energy, a fact which may only be disputed with great difficulty, possesses the ability to absorb radiant energy even if only in extremely small quantities. One might imagine that this absorption would redistribute the irradiated energy over the long term, thus returning it to the zero point energy of luminous ether."[4]

Interestingly, Nernst's prediction came one year before Alexander Friedmann published his expanding-universe equations, and six years before George Lemaître put forth his big bang theory. Nernst wrote his book at a time when it was still not clear whether the spiral nebulae were galaxies or whether their modest redshifts might be due to something other than peculiar motion. Consequently, his photon energy-loss effect may be regarded as a bonafide a priori prediction, predating Hubble's discovery of the cosmological redshift-distance relation. By all rights, Nernst's tired-light theory should have caught astronomers' interest immediately following Hubble's 1924 announcement. Unfortunately, Nernst's book was not well known. Astronomers did not become aware of his prediction until 1938, when he wrote a scientific paper calling attention to it.[5]

As more distant regions of space were probed and the redshift magnitudes continued to increase, other astronomers also began to have their doubts as to whether the spectral shifts were really produced by motional effects. For example, when Hubble and Humason published their 1931 results, which projected galaxy recessional velocities as high as 7 percent of the speed of light, they began to refer to these as "apparent" velocities, with the idea that the phenomenon might be due to some cause other than the Doppler effect.

In 1935, with apparent velocities reaching 13 percent of the speed of light, Edwin Hubble and Richard Tolman coauthored a paper comparing the tired-light and expanding-universe hypotheses on the basis of galaxy number count data. They suggested that some mechanism other than expansion might be responsible for producing the cosmological redshifts, although they did not entirely rule out expansion as a possibility.[6] A year later, armed with a much better set of galaxy number count data, Hubble wrote a follow-up paper that came out decidedly in favor of the tired-light model. In the abstract of his paper, he wrote: "If red shifts are not velocity shifts, the apparent distribution agrees with that in an Einstein static model of the universe."[7] In other words, his data agreed with a stationary Euclidean universe in which the redshifts were due to some "unknown effect" causing photons to lose energy as they traveled through space.

He noted that the main objection to the assumption that redshifts are not velocity shifts is the absence of any other satisfactory interpretation of the redshifts. The tired-light effect hypothesized by Nernst and Zwicky had never been demonstrated in the laboratory—nor could it ever hope to be, since the required energy change would amount to only one part in 10^{26} for every meter that the photon traveled. This is many orders of magnitude smaller than what is measurable even with today's best instruments. Hubble went on to comment on the incompatibility of his data with the expanding-universe hypothesis:

> If red shifts are velocity shifts which measure the rate of expansion, the expanding models are definitely inconsistent with the observations unless a large positive curvature [small, closed universe] is postulated. The maximum value of the present radius of curvature would be of the order of 470 million light-years; and the mean density of the general order of 10^{-26} grams per cubic centimeter. The high density suggests that the expanding models are a forced interpretation of the observational results.[8]

In 1938 Walther von Nernst praised Hubble's conclusions, noting that his own energy-loss hypothesis had anticipated the cosmological redshift discovery as early as 1921: "On the basis of this simple formula, we think we have replaced the fairly unreliable theory of the exploding universe with a much simpler concept of vast importance, which also accounts for redshifts in the most distant objects. . . . And it is highly significant that Hubble, one of the discoverers of redshifts, should consider the model of the expanding universe to be unreliable."[9] Despite Hubble's recognition as the discoverer of the redshift-distance relation, however, he was unable to lure many cosmologists away from the now-orthodox Doppler shift interpretation.

A COSMOLOGY OF CONTRADICTIONS

Although most cosmologists in the 1930s believed the universe to be expanding, not all were convinced that Lemaître's big bang theory was correct. The main problem concerned the young age that the theory predicted for the universe. The big bang theory requires that the universe should have an age equal to $1/H_0$, the inverse of the Hubble constant. Since H_0 was then believed to equal 150 kilometers per second per million light-years of distance, this value yielded an age for the universe of only 2 billion years. This fell embarrassingly short of the 4.5 billion years estimated for the age of the Earth's crust.

The modern version of the big bang theory was proposed in 1948 when two U.S. nuclear physicists, George Gamow and Ralph Alpher, replaced Lemaître's fragmenting superdense atom with a hot dense expanding fireball consisting of electromagnetic radiation and highly energetic subatomic particles. Their theory was inspired by the then recent detonation of the first atomic bomb. Like Lemaître's version, however, this "hot" big bang theory predicted an unrealistically young age for the universe. As a result, astronomers began to give more serious attention to another expanding-universe theory published that same year—the *steady-state theory* developed by Hermann Bondi, Tommy Gold, and Fred Hoyle. This proposed that the universe has been expanding for an indefinitely long period of time and that some unspecified process causes individual atoms of hydrogen to be continuously created throughout all parts of space. But the steady-state theory was later discarded when it was found to make a poor match to redshift data.

The 1952 discovery by astronomer Walter Baade that the Andromeda galaxy actually lay two and a half times farther away than Hubble had estimated brought about an upward revision of the entire extragalactic distance scale that, in turn, proportionately reduced the value of the Hubble constant and raised the predicted age for the universe to 7 billion years. Over the course of the following two decades the estimated size and age of the universe increased with additional revisions of the extragalactic distance scale. By the 1960s age estimates for the universe ranged from about 10 to 20 billion years, depending on the particular value adopted for the Hubble constant.

During the 1960s and 1970s the big bang theory developed considerable support and came to dominate the field of cosmology. In the mid to late 1980s, however, it began to run into difficulty once again. One paper published by the cosmologist R. Tully summarized conclusive evidence that the correct value of the Hubble constant was around 30 kilometers per second per million light-years of distance.[10] The age of the universe then approached 10 billion years, with some astronomers even quoting ages as young as 7 billion years.[11] Measurements later made with the Hubble Space Telescope of the distance to stars in moderately distant galaxies indicated a Hubble constant value of 22 ± 3 kilometers per second per million light-years, predicting a somewhat older age of about $13\frac{1}{2}$ billion years for the big bang universe.[12]

Such young ages leave insufficient time for some stars to form and evolve. For example, some of the oldest stars in the Galaxy are believed to reside in globular clusters, dense clumps of stars that form a symmetrical halo about the Galactic nucleus. Spectral analysis of these stars indicates that they are more than 15 billion years old, which means they were created at least 2 billion years before the Big Bang.

The existence of galaxies at very high redshifts also causes problems for models predicting a recent big bang explosion. One survey for high-redshift quasars turned up a quasar galaxy having a redshift of 6.3.[13] If we adopt a big bang universe age of

13½ billion years, then the big bang theory would have to maintain that light from this distant quasar had been emitted 13.1 billion years ago, or just 400 million years after the time of the Big Bang.* By comparison, the most successful galaxy formation model that big bang theorists are able to propose requires a galaxy gestation period of at least three-fourths of a billion years. But the redshift threshold is being pushed out even further by tentative sightings of even higher redshift galaxies. For example, in 2000, one group of astronomers discovered that one Hubble Space Telescope deep field view contained an unusual infrared object that had an unconfirmed redshift of 12![14] Galaxies at such high redshifts are so faint that our present technology of telescopic observation must be pressed to its limits to get reliable redshift readings. If this redshift of 12 is later confirmed, the big bang theory would be forced to embarrassingly claim that this galaxy was existing at a time when the universe was only 160 million years old and that it had formed out of intergalactic gas five times faster than even the most liberal galaxy formation model predicts.

The discovery of an elliptical galaxy at a high redshift of 3.4 also poses a substantial age problem for the big bang theory. Elliptical galaxies are reddish in color, indicating that they are populated by "old population" stars conventionally believed to have ages of 12 billion years or more. Yet according to the big bang cosmology, a galaxy at a 3.4 redshift would have an age of just over 1 billion years. Big bang cosmologists, then, must explain how a galaxy that is about 1 billion years old could contain stars that are over 12 billion years old.

If we live in a static, tired-light universe, as the ancient cosmology and subquantum kinetics predict, then we could expect to see star-populated galaxies with redshifts perhaps as high as 30, forcing the big bang theory into an even more tenuous position. For a galaxy at a redshift of 30 would have to be receding at 99.8 percent of the speed of light and would supposedly have existed, fully developed, when the cosmic fireball was only 20 million years old.

It is doubtful that the big bang theory can account for the formation of even the closest of galaxies, our own included. If the matter density is too low, slight concentration nonuniformities in the expanding fireball will not exert sufficient gravitational attraction to gather themselves into protogalactic gas clouds. For such matter clumping to take place, the big bang theory requires the universe to be at or above its critical density of three hydrogen atoms per cubic meter. In other words, it requires a geometrically closed universe, one that should eventually stop expanding

---

*This value is calculated using the relativistic formula $d = [(z + 1)^2 - 1/(z + 1)^2 + 1] (c/H_0)$, where z is redshift, c is the velocity of light in kilometers per second, and H_0 is the Hubble constant in kilometers per second per million light-years. The tired-light redshift model predicts somewhat greater distances for galaxies than those predicted by the expanding-universe cosmology. For example, a galaxy having a redshift of 6.3 would be located about 26 billion light-years away, or about 100 percent farther away than the big bang theory would predict. This assumes that photons lose about 7.4 percent of their energy for every billion years of travel time.

and later begin contracting. An assessment of the amount of visible matter seen to compose galaxies, however, indicates a universal density which is about fifty times less than this critical value (about one hydrogen atom every sixteen cubic meters). Moreover, this estimate is most likely high, since it is based on observations made in our own sector of the universe, which is quite well populated with galaxies. When the universe's immense matter-free voids are taken into account, the matter density is found to be as little as 0.02 percent of the critical value.[15]

Big bang theorists attempted to circumvent this matter density problem by suggesting that much of the required matter was present, but just not visible. They suggested that this so-called dark matter might exist in the form of black holes, massive "cosmic strings," or nonluminous dust particles hiding in the midst of galaxy clusters. But the evidence to support their dark-matter hypothesis simply was not there. The laws of physics dictate that the speeds of galaxies in a given cluster depend on the total gravitational mass present in that cluster. The larger the gravitational mass of that cluster, the greater the velocities of its galaxies. The mathematical relation setting forth this correspondence is known as the Virial theorem. The astronomers Mauri Valtonen and Gene Byrd used this approach to carefully assess the gravitational masses of several clusters and found no evidence of dark matter.[16] The velocities they measured were normal, based on estimates of the clusters' visible mass.

Supercluster complexes such as that shown in figure 13.5 present the greatest difficulty to big bang theorists. Observations indicate that individual galaxies in these complexes do not move faster than about one thousand kilometers per second. At this speed, it would take more than 80 billion years for an initially homogeneous distribution of galaxies to aggregate into such large structures,[17] making them at least eight times older than the big bang universe. Moreover, there is also the problem of coming up with an attractive force that would be sufficiently strong to pull such large structures together. Earlier we saw that the mass density of the universe is several orders of magnitude smaller than what the big bang theory requires in order to allow it to explain the agglomeration of galaxies. The disparity between theory and observation becomes even greater when big bang theorists try to explain the formation of these supercluster complexes. And even if dark matter were to exist, it would be unable to explain structures even one-tenth the size of the Great Wall.[18]

Structures even more immense than supercluster complexes present even bigger problems for big bang theorists. An international team of astronomers has discovered that galaxies are grouped into a more or less regularly spaced pattern that stretches over a distance of 5 billion light-years in either direction, covering a large fraction of the observable universe (figure 13.6).[19] Groups of galaxies are spaced apart by about 600 million light-years, based on a Hubble constant of 22 kilometers per second per million light-years. For an initially homogeneous distribution of galaxies to form any given rung on this titanic "ladder" would require over 100

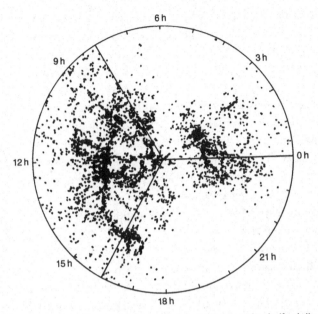

Figure 13.5 A map showing the locations of galaxies lying within half a billion light-years of our galaxy, which is positioned at the center of the circle. The nearest supercluster complex, named the "Great Wall," consists of an immense sheet of galaxies measuring about 600 million light-years in length, 200 million light-years in width, and 20 million light-years in thickness. Adapted from Geller and Huchra, "Mapping the universe."

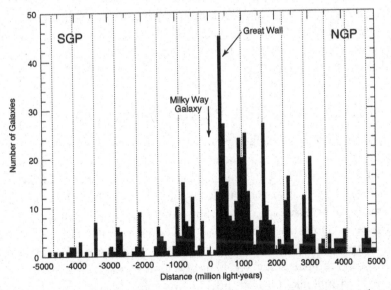

Figure 13.6. A cosmic dissipative structure. A plot of the number of galaxies versus their distance from the Earth in two opposing directions exhibits a regular pattern. Regions of peak density are separated by approximately 600 million light-years. Diagram courtesy of David Koo.

billion years. Moreover, the big bang theory is clearly at a loss to explain how such a large structure could maintain a regular pattern over more than 5 billion light-years' distance and over a comparable period of light-travel time, during which the universe and its relative spatial distances would have supposedly grown some 60 percent larger.

The static-universe continuous-creation cosmology of ancient times does not suffer from this time constraint problem. The ancient cosmology allows galaxies trillions of years to develop and adopt their present grouping configurations. Its method of generating new matter from existing matter and young dwarf galaxies from older spirals tends to foster a nonuniform distribution of galaxies, thereby explaining the formation of aggregates such as galaxy clusters.

Superclusters and supercluster complexes could simply be regions where the ether was particularly fertile and prone to nucleate matter. Reaction-diffusion processes might form vast patterns of galaxy groups that stretch out over billions of light-years, which suggests that the concentration of an ether species determining criticality (fertility) varies regularly through space with a wavelength of hundreds of millions of light-years. In a Model G–type ether, this would happen if the B ether had a very large yet finite diffusion coefficient, allowing it to develop an astronomically large spatially periodic concentration pattern. Recall that the ether's state of criticality is determined by the values of both B and G. On distance scales smaller than several tens of millions of light-years, where B would appear to have a uniform distribution, particle materialization would primarily be determined by variations in G. Over larger distances, however, the long-range influence of B would become important and result in the emergence of cosmologically large matter-creation patterns. In an open, reaction-diffusion ether of such indescribably large dimensions, it would not be surprising to find dissipative structures as large as that shown in figure 13.6.

IS THE UNIVERSE REALLY EXPANDING?

The inadequacy of the big bang expanding-universe cosmology becomes most apparent when its cosmological predictions are checked against actual astronomical data. For example, Hubble became convinced of the superiority of the tired-light cosmology after he had compared the predictions of both theories on the galaxy number count cosmological test. Since that time, additional cosmological tests have been developed, and the quality of the observational data used in these tests has improved. In one study, which appeared in *Astrophysical Journal* in 1986, I simultaneously compared the big bang, expanding-universe, and tired-light, static-universe cosmologies on four different cosmological tests and showed that on *all* tests the tired-light cosmology made a better fit to the data.[20] The results of this study received international attention,[21] and for fifteen years big bang theorists did not challenge its conclusions. But in 2001 two papers were published that claimed to have overturned the tired-light theory on the basis of new evidence.[22] A careful examination of those supposed negative findings, however, shows that just the reverse is true: the tired-light model instead fares better than the expanding-universe model. The refutation of these challenges to the tired-light theory is presented in the revised edition of *Subquantum Kinetics*.[23]

The revised edition of *Subquantum Kinetics* also contains a new comparison between the tired-light and expanding-universe cosmologies using more up-to-date data. Reviewing these data, which are reproduced as figures 13.7 through 13.11, we find that, as before, the tired-light model consistently makes a better fit with the data. Once again, we are led to conclude that the universe is in fact not expanding and that the cosmological redshift instead results from light waves diminishing their energy as they travel through space.

Prior to the publication of the 1986 *Astrophysical Journal* paper, big bang cosmologists were unaware that the tired-light model made a superior fit with test data. They simply had not attempted to plot the tired-light model to see how it compared with their favored expanding-universe cosmology because they were already convinced that the expanding-universe theory was correct. Whenever their cosmology did not fit their data, they attributed this not to an inadequacy of the expansion hypothesis but to a failure to adjust their model to account for certain assumed details, such as the supposed gravitational curvature of space, the supposed dimming of galaxies due to stellar evolution, or the supposed gradual contraction in the size of galaxy clusters. Consequently, rather than using cosmological tests to validate or refute their hypothesis, big bang cosmologists have used the tests to make ad hoc adjustments to their model so as to make it properly fit the data (see table 13.1, number 5).

Like the epicycle theorists of medieval times, modern cosmologists have fallen into the trap of attempting to repair the deficiencies of their cosmology by complicating it with more and more ad hoc assumptions instead of abandoning it in favor of a better alternative. But the situation for the expanding-universe theory is far worse than that of the epicycle theory. Not only do cosmologists make an inordinately large number of assumptions on its behalf, but often these assumptions work against one another, throwing the big bang theory as a whole into a confused state. Generally, when cosmologists make these ad hoc adjustments to eliminate their theory's discrepancy on one cosmological test, they avoid considering how this assumption affects their model's fit on other cosmological tests. More often than not, their adjustments end up worsening the fit on those other tests. They are able to make the big bang model look inordinately good by focusing their study on just one cosmological test, adjusting the expanding-universe model to obtain a good fit to the data of that test, and avoiding plotting competing cosmologies such as the tired-light model.

Consider, for example, the Hubble diagram test shown in figure 13.7, which is similar to the one Hubble used to plot his first redshift-distance relation. The lower trend line, marked as $q_0 = 0$, plots the big bang prediction for a geometrically open Friedmann expanding universe having a mass density consistent with estimates of the observed amount of visible matter in the universe. However, the model makes a poor fit to the Hubble diagram data, erring on the dim side at high redshifts. In the past, big bang theorists have attempted to eliminate this discrepancy by fitting the data with a curved-space Friedmann model which assumes a mass density ratio, Ω, that

---

TABLE 13.1

A Comparison of the Assumptions Made by Two Rival Cosmologies

EXPANDING-UNIVERSE COSMOLOGY (TWENTIETH-CENTURY COSMOLOGY)	TIRED-LIGHT COSMOLOGY (ANCIENT, TWENTY-FIRST-CENTURY COSMOLOGY)
1. Space is non-Euclidean. Amount of curvature depends on how much "missing mass" the cosmologist decides to assume is present.	1. Space is Euclidean.
2. Space is expanding. Distant galaxies are receding from one another at very high velocities.	2. Space is static. Galaxies drift relative to absolute space at low velocities.
3. All inertial frames of reference are relative and equally valid.	3. There is a unique absolute inertial frame.
4. Photon wavelength remains constant with time. Photon energy is strictly conserved.	4. Photon wavelength changes with time. Photons lose energy in intergalactic space at an average rate of 7.4 percent per billion years.*
5. Discrepancies between observed data and expanding universe model predictions are assumed to arise from the following evolutionary effects, all occurring at just the right rate: a) Galaxy luminosity is decreasing. b) Galaxy clusters are collapsing. c) Lobe size of radio galaxies and quasars are decreasing. d) The number density of visible galaxies is decreasing. e) Radio galaxy lobe brightness is increasing. f) The number density of radio galaxies is decreasing.	5. Evolutionary effects are minimal.

---

*The tired-light phenomenon is a rigid prediction of the subquantum kinetics model. Only the specific value of the energy loss rate is assumed.

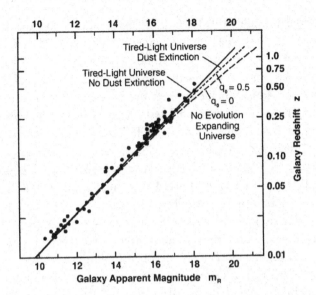

Figure 13.7. The Hubble diagram test compares galaxy redshift (vertical scale) to distance (horizontal scale). Compared with the open expanding-universe model ($q_0 = 0.5$ or $\Omega = 1$), the static-universe tired-light model lies closer to the data trend. Distance is judged by the apparent brightness of the galaxies. The data points are taken from Kristian, Sandage and Westphal, "The extension of the Hubble diagram," Paper II, Fig. 4.

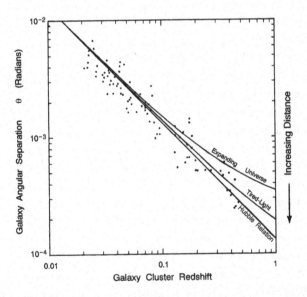

Figure 13.8. This test compares the angular size of a galaxy cluster, used to judge galaxy distance (vertical scale), to the average redshift of the cluster's galaxies (horizontal scale). Again, the tired-light model makes a better fit. Decreasing cluster angular size denotes increasing distance. The data points are taken from Hickson and Adams, "Evidence for cluster evolution," table 1.

is 3.4 times the critical density ($q_0 = 1.7$). This would suppose that the universe contains hundreds of times more mass than telescopic observations indicate, hidden in the form of dark matter (table 13.1, ad hoc assumption 1).

While this assumption moves the expanding-universe model to the left on the Hubble diagram, into coincidence with the data trend line, it worsens the model's fit on the angular-size versus redshift test (figure 13.8) which plots the angular sizes of galaxy clusters as a function of redshift. That is, with the assumption that the geometry of space is positively curved, distant objects such as galaxy clusters would appear larger than expected. So, the curve representing the expanding-universe model prediction would move upward at its high-redshift end, thereby worsening the fit to the data on those tests. Similarly, in figure 13.9, which plots the angular separation of radio lobe separation in radio galaxies, the expanding universe prediction would again move upward away from the data trend. Thus what seems to improve the fit on one test deteriorates the fit on others.

As we saw earlier, the Valtonen-Byrd galaxy velocity study indicates that dark matter is essentially absent within galaxy clusters. Also, astronomical observation of how the space concentration of galaxies varies as a function of distance indicates that the universe is Euclidean out to the farthest limits of observation, in agreement with tired-light cosmology predictions. Thus, in addition to being counterproductive on cosmological tests, the big bang theory's assumption that the universe is spatially curved by the presence of large amounts of dark matter also appears to be contradicted by observation.

Since our big bang cosmologists would not want to make matters any worse on the angular-size versus redshift test, they could instead try to eliminate the discrepancy on the Hubble diagram test by assuming that galaxies were formerly brighter than they are today (table 13.1, ad hoc evolutionary assumption 5a). They would fudge the Hubble diagram data (figure 13.7) by brightening the high-redshift end of the expanding-universe prediction so that it would move sufficiently to the left to close the gap between theory and observation. However, this same amount of evolution is insufficient to allow the expanding-universe theory to make a consistent fit to data on the Tolman surface brightness test (figure 13.10). To fit that data set, at a redshift of $z \sim 0.9$, galaxies must be assumed to be over four times brighter due to evolution than what must be assumed to get them to fit the expanding-universe model on the Hubble diagram test.

Luminosity evolution does not help much to close the gap for the expanding-universe prediction on the galaxy number count versus magnitude test (figure 13.11). On this test, the data trend at dimmer magnitudes lies well above the $q_0 = 0.5$ ($\Omega = 1$) expanding-universe model prediction (dashed line). The accelerating-universe model prediction, which is plotted as the dotted line, fares better but still lies somewhat below the data trend. The discrepancy may be eliminated by introducing yet another evolutionary correction, one that assumes that the number of

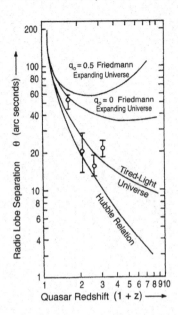

Figure 13.9. This test compares the angular separation of radio lobes in double-lobe radio quasars to quasar redshift plotted as 1 + z. The tired-light model is seen to make a far superior fit to the data. Adapted from Ubachukwu and Onuora, "Radio source orientation and the cosmological interpretation of the angular size-redshift relation for double-lobed quasars," fig. 2.

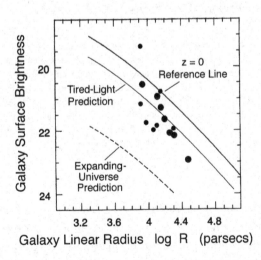

Fig. 13.10. The Tolman surface brightness test. Galaxy surface brightness in magnitudes is plotted against the log of the galaxy's radius in parsecs using R-band data for cluster 1604+4321 (z = 0.92). The no-evolution, tired-light prediction with the added assumption of a small amount of intergalactic light absorption (solid line) makes a good fit. The no-evolution, $q_0 = 0.5$ expanding-universe prediction (dashed line) requires the introduction of a large evolutionary correction in order to fit the data. The plotted data is adapted from Lubin and Sandage, "The Tolman surface brightness test for the reality of the expansion," Paper IV, fig. 1; see LaViolette, *Subquantum Kinetics,* for an explanation.

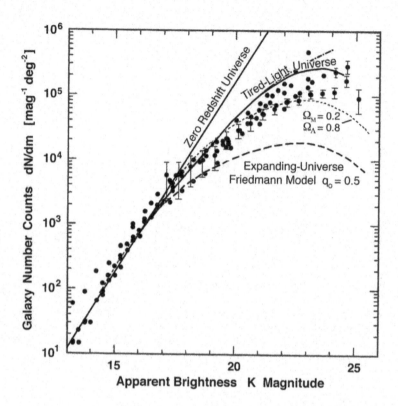

Fig. 13.11. This test compares the number of galaxies within an incremental volume of space at a given distance from Earth (vertical scale) to the distance that volume increment is from Earth (horizontal scale). As in the Hubble diagram test, galaxy apparent magnitude is used as the distance indicator. Going from left to right, brightness decreases and distance increases. The tired-light model (solid line) is seen to make a better fit. Hubble used a similar comparison in his 1936 study that pointed out the inadequacies of the expanding-universe cosmology. The diagram is taken from LaViolette, *Subquantum Kinetics*, second edition; the data is from Totani, "Near-infrared faint galaxies in the Subaru Deep Field," fig. 1.

galaxies in a given volume of space has been decreasing (for example, through galaxy mergers) in just the right way to make the expanding-universe model fit the data (table 13.1, ad hoc evolutionary assumption 5d). The tired-light model, on the other hand, makes a good fit to the data trends on all of these tests without requiring any assumption about excessive luminosity evolution.

We have not covered all the possible conflicts that could arise from the process of making adjustments to cosmological test data, but this should give a reasonably good idea of some difficulties; see *Subquantum Kinetics* for a more in-depth analysis. The important lesson from this is that cosmologists have not made themselves accountable for the ad hoc corrections they make to their data. To properly test the expanding-universe and tired-light models, both cosmologies must be compared side by side on several different cosmological tests, preferably without making any ad hoc adjustments to the data. When such a comparison is made, it is clear which model is superior: the tired-light cosmology predicted by the ancient physics of creation. Observations made with the Space Telescope upgrades are expected to greatly increase the data base available for testing competing cosmological models. Since the tired-light static-universe and big bang expanding-universe cosmologies make diverging predictions at high redshifts, an expanded data set should even more decisively favor the tired-light model.

Challenges to the Tired-Light Model Addressed

The Tolman Test: The data for figure 13.11 is taken from the Tolman surface brightness test published by Lubin and Sandage.[22] To make the expanding universe model prediction fit their data trend, they introduced the assumption that galaxies have undergone substantial luminosity evolution, causing them to have much higher intrinsic surface brightnesses in the past. That is, they assumed that in the past a typical galaxy's intrinsic brightness per unit area was much greater than it is today and of just the right amount so as to allow the expanding universe model to fit the observed data trend. They then imposed this same luminosity evolution assumption on the tired-light model and noted that this drove the tired-light model far away from the data trend. They then concluded that since the expanding universe model made a better fit to the data trend (with this luminosity evolution correction), the tired-light model was by comparison inadequate.

Their conclusion might have been correct if in fact they knew with certainty that stellar luminosity should evolve in the manner they assumed. However, the luminosity evolution model they use is quite flexible and allows the cosmologist wide latitude in choosing how much galaxy luminosity change he should assume. For example, a galaxy at a redshift of $z = 0.9$ could be anywhere from 1 to 4 times brighter than current galaxies, clearly offering a very flexible choice. Consequently, by introducing the assumption of luminosity evolution, big bang cosmologists in effect are adjusting their model to fit the data; they are not testing the validity of the expanding universe cosmology. Moreover they are incorrect in applying to the subquantum kinetics tired-light cosmology the same amount of luminosity evolution that they arbitrarily assume for the big bang cosmology. For subquantum kinetics departs radically from standard stellar evolution theory. In subquantum kinetics, galaxies are expected to gradually become brighter, not dimmer, with the passage of time, and this change is expected to take place much more gradually than the dimming assumed to take place in standard astrophysics. Consequently, over the redshift range covered by existing data, it is best to assume a no-evolution tired-light model. When we include a small brightness correction to the data to compensate for intergalactic light absorption (not accounted for in the Lubin-Sandage paper), the tired-light model is found to fit the data set quite well without the need for assuming any major change in galaxy surface brightness.

Moreover Lubin and Sandage did not consider how the expanding universe model would fit on other cosmology tests. That is, they did not take into account that the amount of luminosity evolution they were assuming on the Tolman surface brightness test would overcorrect the expanding universe cosmology on the Hubble diagram test and produce a poor fit to that data set. Hence they failed to acknowledge that the expanding universe model made an inconsistent fit to both cosmology tests. The no-evolution tired-light model, on the other hand, makes a consistent fit to the data.

The Claim of Supernova Time Dilation: The other study that attempted to disprove the tired-light model claimed that distant supernovae took longer to occur, as judged by the time it took them to fade from maximum brightness.[23] The authors of that paper then attributed this inferred stretching of time to a time dilation effect caused by the galaxy's assumed relativistic recession from us due to the expansion of the universe. That is, as a galaxy's speed of recession approaches increasingly close to the speed of light, the special theory of relativity predicts that time should accordingly slow down in that galaxy's reference frame. The tired-light alternative, on the other hand, predicts no time dilation effect since it assumes that the universe remains stationary.

However, there is good reason to believe that the time dilation effect they claim to have found is actually a misinterpretation of observation. A study of their supernova data reveals that it is flux limited, meaning that at large distances (high redshift) the more prevalent, less luminous supernovae are too dim to detect, leaving only the more rarely occurring bright supernovae to be detected, supernovae that are far more luminous and far longer lasting than those typically seen in our own galactic neighborhood. But intrinsically more luminous supernovae are known to last longer. So, a failure to take account of this so-called Malmquist bias would lead astronomers to unwittingly conclude that the supernovae were similar to those seen in our neighborhood, but taking longer to occur due to their favored time dilation mechanism. For a more thorough discussion of this see Subquantum Kinetics, chapter 7.

Furthermore, a similar time dilation phenomenon is not seen in gamma ray burst data. If the universe were expanding and if time were substantially dilated in distant galaxies, distant gamma ray burst light curves should be seen to be lengthened, just as is claimed for the supernova light curves. However, this is not observed. Gamma bursts so distant that their light has been redshifted into the X-ray spectral range are seen to last just as long as more nearby gamma ray bursts; see "X-ray 'gamma-ray bursts.'" *Sky and Telescope*, August 2002, p. 20.

DYING ECHOES?

Contemporary cosmologists traditionally call upon three main pieces of evidence to support their claim for a Big Bang: (1) the existence of the cosmological redshift; (2) the existence of the cosmic microwave background radiation; and (3) the abundance of the light elements deuterium, helium, and lithium relative to hydrogen. The previous section shows that the first of these supporting pillars has completely crumbled. The microwave background radiation, the second pillar, is argued to be a relic from the Big Bang's incandescent fireball, which would be observed at a much lower temperature today owing to the expansion of the universe. But this interpretation is also not without its problems.

The microwave background was observed for the first time in 1955 by E. Le Roux, a radio astronomy doctoral student at the Ecole Normale Supérieure in Paris. Using a World War II surplus radar antenna, he found that all parts of the heavens were radiating a microwave noise with a temperature of 3 ± 2 Kelvin.[24] Unfortunately, his work went unnoticed. But nine years later, Bell Lab researchers Arno Penzias and Robert Wilson made an independent discovery of this cosmic noise using a microwave antenna horn with a helium-cooled detector. Their calculations indicated a radiation temperature of about 3.5 ± 1 K. Today we know the temperature of the microwave background to be more accurately 2.73 ± 0.01 K, a measurement made in 1989 by the Cosmic Background Explorer satellite (COBE). One key feature of this cosmic microwave emission is that it has a *blackbody* radiation spectrum (figure 13.12) that is similar to an object that has reached a state of thermal equilibrium with its surroundings. The COBE satellite found that the 2.73 K spectrum deviates from a perfect blackbody spectrum by less than 1 percent.

Most astronomy textbooks credit big bang theorists as the first to foresee the existence of the microwave background, citing a 1948 paper written by Ralph

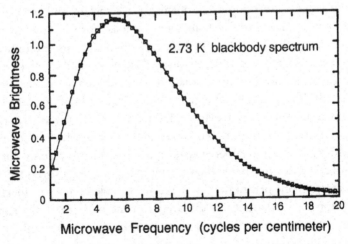

Figure 13.12. The blackbody spectrum of the cosmic microwave background constructed from radiation intensity measurements made at various frequencies by the COBE satellite.

Alpher and Robert Herman. Nevertheless, in 1933, fifteen years before Alpher and Herman's prediction, the German physicist E. Regener predicted the existence of a 2.8 K microwave background produced from the warming of interstellar dust particles by high-energy cosmic rays.[25] Surprisingly, Regener's estimate erred from the true value by less than 3 percent. The cosmologist Walther von Nernst cited Regener's microwave background prediction in a 1938 paper that discussed a tired-light interpretation of the cosmological redshift.[26]

Although Regener correctly predicted the temperature of the cosmic microwave radiation, his assumption that the emission originates from dust particles is incorrect, since that would require microwave radiation to be concentrated toward our galaxy's equatorial plane, where the dust concentration is highest. Instead, the microwave background is observed to be fairly uniformly distributed across the sky. Nevertheless, as we shall see shortly, he may have been correct in suggesting that the microwave emission derives its energy from the ambient cosmic ray particle flux.

In competition with tired-light theorists, big bang theorists Alpher and Herman predicted in 1948 a microwave temperature of about 5 K, which they revised upward to 28 K in the three years that followed. Their temperature estimate was tenfold too high. This translates into a ten-thousand-fold error in estimating the energy density of this background radiation, which varies as the fourth power of temperature. So tired-light theorists were not only the first to predict the existence of the microwave background but also those who predicted it with the greatest accuracy.

Nevertheless, big bang theorists were more organized and lost no time in claiming the newly discovered cosmic noise for their own. In the very same journal issue in which Penzias and Wilson announced their microwave findings, four Princeton cosmologists published a companion paper linking this discovery to the big bang theory, even though the original temperature predictions were substantially higher than the observed value.

For a long time afterward, the 2.73 K microwave background was passed off as proof of the correctness of the big bang theory. But as more data on the microwave background were gathered, little evidence appeared of any connection between the alleged big bang fireball and this microwave radiation. The uniform manner in which the microwave background radiation is distributed across the sky implied that the fireball should have been extremely uniform and that matter should also be uniformly distributed in space. Instead, the universe is seen to be very clumpy. Matter is gathered in the form of gas clouds and galaxies, which in turn are gathered into clusters, then into superclusters, and so on. Just to account for the existence of galaxies, the big bang theory required that the microwave radiation intensity vary from one part of the sky to another by at least one part in a thousand. To account for the vast structures discovered in the mid-1980s, the supercluster complexes and immense periodic structures stretching across the universe, even greater nonuniformities would have been needed.

But such nonuniformities were not found.[27] To date, the most accurate observations of the microwave field have been made with the COBE satellite. These indicate that when the Earth's motion relative to the microwave background is taken account of, there are intensity variations of less than one part in 100,000, a hundred times smaller than the big bang theory's most modest prediction. When the COBE scientists first announced discovering these "ripples" in April 1992, they proudly asserted that they had finally proven the existence of the Big Bang. The news media blindly echoed their claims, and soon even theologians were purporting the ripples to be evidence of the biblical act of creation. Yet if anything, the COBE measurements had definitively disproven the big bang theory by showing that the microwave background was far too smooth to account for the universe's clumpiness.

An alternate theory proposed by the cosmologist Eric Lerner suggests that the 2.73 K microwave blackbody radiation arises from a fog of ionized filaments theorized to fill intergalactic space. In 1990 Lerner found evidence for the existence of such a plasma filament fog through his discovery that radio and microwave signals from distant radio galaxies are attenuated to 15 percent of their original strength after traveling a distance of just 100 million light-years.[28] As a result of repeated absorption and reradiation of these radio galaxy microwaves, this plasma fog would be expected to emit a thermal blackbody spectrum. On the other hand, intergalactic cosmic rays, rather than radio galaxy microwaves, might be the primary energizers of these filaments. In fact, the filaments would radiate at the observed 2.73 K temperature if they were to absorb 4 percent of the ambient cosmic ray energy flux passing through them. Given the well-established existence of cosmic rays and the recently established existence of an intergalactic radio-absorbing plasma, we are naturally led to conclude that an isotropic blackbody microwave radiation field must exist and that it should have a temperature close to what is observed. No Big Bang is required. Regener, it seems, may have been right all along about cosmic rays producing a microwave background radiation—except instead of heating dust particles, as he had suggested, the cosmic rays would be heating intergalactic plasma filaments.

As for the third pillar of observational evidence that is claimed to support the big bang theory, the abundances of the light elements, observational analysis places the cosmic helium abundance at approximately 22 to 23 percent, the deuterium-to-hydrogen ratio at approximately seventeen parts per million, and the ratio of lithium-7 to hydrogen at approximately one part in 10 billion. Big bang theorists claim that most of this helium, deuterium, and lithium was synthesized from hydrogen during the first minutes of the Big Bang. But big bang models fail to predict the correct ratio of these elements.[29] If the big bang fireball model is adjusted to predict the correct helium abundance, it ends up predicting a deuterium abundance that is approximately ten times too high and a lithium abundance that is approximately three times too high.

In the absence of the fictitious Big Bang, stars are the most plausible sites of element synthesis. Elements would be synthesized in nuclear fusion reactions taking place inside the more massive stars and in cosmic ray collisions taking place in the stars' outer atmospheres. The most rapid rates of synthesis would occur in and around the mother star that forms the galaxy's core, an object that also would have the highest matter-creation rate in the galaxy. In the continuous-creation cosmology described in ancient myth, there is plenty of time available for all the elements to be produced. As we begin the twenty-first century and watch cracks start to form in our modern myth of creation, the cosmological wisdom preserved by our ancient ancestors looks increasingly attractive.

A Brief History of Cosmology

150-1600 c.e.	The Ptolemaic system dominates Western cosmology. The outer bound of the universe is defined by a crystalline sphere positioned just beyond the orbit of Saturn.
1543	Copernicus publishes a book suggesting that the Sun and not the Earth is at the center of the solar system.
1750	Thomas Wright envisions the Galaxy as a disk of stars.
1755	Immanuel Kant suggests that the nebulae are distant galaxies, "island universes" like our own.
1760–70	The distance from the Earth to the Sun is determined.
1790	William Herschel estimates the universe to extend out ~8,000 light-years from the Sun.
1838–40	The distance to three nearby stars is determined; distances range from 4 to 11 light-years (ly).
1912	Vesto Slipher measures a blueshift in the spectrum of the Andromeda galaxy; $v = 300$ kilometers per second (km/sec) = 0.1% of the velocity of light (c).
1914	Slipher measures the spectra of 13 galaxies, 11 of which exhibit redshifts reaching as high as 0.3% ($v_{max} = 1000$ km/sec = 0.3% c).
1915–16	Einstein publishes his general theory of relativity.
1917	Einstein and de Sitter each publish solutions to the general relativity field equations that describe a static universe.
1918–19	Shapley discovers the center of the galaxy and determines that the Sun lies near the galaxy's outer rim.
1919	Observations of a solar eclipse confirm a key prediction of general relativity.
1920	A famous debate is held between Heber Curtis and Harlow Shapley as to whether or not the spiral nebulae lie outside of the Milky Way. The extragalactic interpretation presented by Curtis wins out.
1921	Walther von Nernst proposes that photons gradually lose energy as they travel through space, thereby predicting the existence of a tired-light redshift.
1922	Alexander Friedmann publishes his solutions to the cosmological equations of general relativity describing an expanding universe.
1924	Carl Wirtz shows that the apparent diameters of nebulae seemed to be correlated with their redshifts.
1924	Edwin Hubble measures the distance to the Andromeda galaxy to be 900,000 ly.

1925	Slipher and other astronomers make spectral shift measurements on 45 galaxies, 43 of which have redshifts corresponding to velocities as great as 1,800 km/sec.
1927	Georges Lemaître publishes his big bang theory.
1929	Hubble publishes his redshift-distance relation for 33 galaxies, showing that redshift increases linearly with distance (Hubble's law) out to a distance of 7 million ly.
1929	Milton Humason discovers a galaxy at a redshift corresponding to a velocity of 3,800 km/sec (1.2% of light speed), and Hubble infers that it lies at a distance of 25 million ly, consistent with his redshift-distance relation.
1929	Fritz Zwicky publishes his paper proposing the tired-light model.
1931	Hubble and Humason publish data showing that redshift increases linearly with distance out to a distance of 100 million ly (v_{max} = 20,000 km/sec = 7% of light speed).
1933	E. Regener predicts the existence of a 2.8 K microwave back-ground energized from cosmic rays.
1935	Milton Humason catalogs galaxies with redshifts of up to 30% believed to be as distant as 250 million ly (v_{max} = 40,000 km/sec = 13% of light speed).
1935	Hubble and Tolman propose that the cosmological redshifts might be better explained in terms of a tired-light mechanism, rather than in terms of recession.
1948	George Gamow and Ralph Alpher lay out the foundations for the modern version of the big bang theory.
1952	Walter Baade revises the extragalactic distance scale, making the universe $2^1/_2$ times larger (d_{max} ~1.5 billion ly).
1955	Le Roux discovers the 3 K microwave background radiation.
1963	Maarten Schmidt measures the redshift of the first quasar, 3C 273. The greatest quasar redshift found that year was in 3C 347 (redshift = 54%, v_{max} = 41% of light speed).
1964	Arno Penzias and Robert Wilson rediscover the 3 K microwave background radiation.
1958–70	Further revisions of the extragalactic distance scale quadruple the size of the universe (redshift = 220%, d_{max} ~15 billion ly, v_{max} = 82% of light speed).
1970s	The "red limit" is encountered. Quasars having redshifts greater than 3.0 (300% red shift, or v ≥ 88% of light speed) are found to be progressively less abundant.
1978	Paul LaViolette discovers that subquantum kinetics predicts the existence of a tired-light effect for photons traveling through intergalactic space.
1985	LaViolette publishes his theory of subquantum kinetics and also demonstrates that the tired-light model fits observational data better than expanding-universe models.
1986	A quasar with redshift of 4.0 is discovered (d_{max} = 18 billion ly, v_{max} = 92% of light speed).
1990s	Observations with the Hubble Space Telescope begin, providing further confirmation that the universe is not expanding and that the red-shift is due to a tired-light effect, as the ancient cosmology predicts.
2001	Hubble Space Telescope observations establish a $13^1/_2$ billion year age for the big bang universe. A quasar with redshift of 6.3 is discovered (d_{max} = 13 billion ly, v_{max} = 96% of light speed) and a galaxy candidate with a possible redshift of 12 is sighted.

14

SMASHING THE CRYSTALLINE SPHERE

Being has no coming-into-being and no destruction, for it is whole of limb, without motion, and without end. And it never Was, nor Will Be, because it Is now, a Whole all together, One, continuous; for what creation of it will you look for? . . . Nor shall I allow you to speak or think of it as springing from Not-Being for it is neither expressible nor thinkable that What-Is-Not Is. Also what necessity impelled it, if it did spring from Nothing, to be produced later or earlier? . . . The decision on these matters depends on the following: IT IS, or IT IS NOT. It is therefore decided—as is inevitable—[that one must] ignore the one way as unthinkable and inexpressible and take the other as the way of Being and Reality.

PARMENIDES

THE DISMAL PARADIGM

Big bang theorists offer no explanation for the source of energy needed to power the Big Bang. They cannot appeal to a preceding causal event, since they maintain that existence itself came into being simultaneously with the appearance of this primordial energy impulse. In effect, they claim that the cosmic "egg" came into being without any chicken to lay it.

At a 1981 cosmology conference at the Vatican, Pope John Paul II expressed the views of the Catholic Church that a divine creator is ultimately needed to explain the origin of the universe. Addressing the mathematicians and physicists attending the meeting, he stated:

> Any scientific hypothesis on the origin of the world, such as that of a primeval atom from which the whole of the physical world derived, leaves open the problem concerning the beginning of the Universe. Science cannot by itself resolve such a question: what is needed is that human knowledge that rises above physics and

astrophysics and which is called metaphysics; it needs above all the knowledge that comes from the revelation of God.[1]

When the pope says that "science" alone cannot resolve the question of the world's origin, he is referring to present-day positivist science, which recognizes the existence of only tangible, measurable things. By appealing to metaphysics and divine inner sight, he is suggesting that an inner awareness tells us that there is more to the world than just the physical, that subtle phenomena in dimensions not directly probed by physicists might play a key role in the origin of things. His perspective essentially agrees with the worldview offered by the ancient ether physics. Many modern cosmologists, however, believe it is unnecessary to posit a beyond or a divine agent to explain the origin of the universe. For example, the British theoretician Stephen Hawking claims to have developed a mathematical device that enables the big bang theory to avoid the question of how its initial energy pulse emerged. He views the physical universe as completely self-contained, with neither a beginning nor an end, and hence having no moment of creation. He asks, "What place, then, for a creator?"[2] In support of his view, the science writer John Gribbin concludes his book on modern cosmology with the following attack on religion and metaphysics:

> Perhaps it would be more accurate to say that Hawking has already indicated an end, not to physics, but to *metaphysics*. It is now possible to give a good scientific answer to the question "Where do we come from?" without invoking either God or special boundary conditions for the Universe at the moment of creation. As of the Vatican conference of 1981, it is the metaphysicians who are out of a job. Everything else shrinks into insignificance alongside such a claim, and the end of the road for metaphysics certainly seems to be a good place to end this book.[3]

This attack on metaphysics and mysticism is quite typical of the view of the logical positivist philosophy that still maintains a strong grip on the modern physical sciences. The big bang positivist cosmology may be summarized as follows: (a) all that exists is the tangible physical universe; (b) its origin and workings can be entirely described by physical theory; (c) there is no need to invoke the presence of God or a metaphysical beyond in accounting for the existence of this elegant machine; and (d) unless God can be physically measured in some way, he does not exist.

Big bang cosmologists conceive the universe to be contained within a finite sphere, an expanding bubble of space-time that has attained a radius of some 14 billion light-years. They claim that nothing lies beyond this bubble of physical manifestation, no existence of any kind. Just as theologians did in medieval times, today's cosmologists have confined the heavens within an Aristotelian "crystalline sphere."

When modern cosmologists are asked to comment on the present nature of the universe and on its ultimate fate, they project a dismal outlook. Basing their view on the second law of thermodynamics, they state that the universe was at its greatest state of order at the moment of the Big Bang, and that ever since, with the continual expansion of space, things have tended toward greater disorder. They see the universe in its final stages as primarily dark, a giant black hole whose gloom arises from its vast population of smaller black holes, avaricious beasts that continuously gobble up matter and snuff out light beams that approach too close. The cosmos is seen to tend toward a decay that has supposedly been going on for more than 10 billion years and is expected to continue for many billions of years to come.

By comparison, the event of creation, the emergence of light and order into physical manifestation, is said to have occurred in the briefest possible moment, an instant lasting less than 10^{-43} seconds. Modern cosmology on the one hand elevates the principle of decay to a supreme status, referring to it as the "second law," but on the other hand relegates the phenomenon of cosmic creation to the obscure realm of chance and uncertainty as a highly improbable event. It is not surprising that death and darkness should ultimately rule over light in this spiritually impoverished worldview.

BIG BANG HOT AIR

The theory of inflation developed in 1981 by Alan Guth attempts to overcome this energy-creation problem. It suggests that most of the universe's matter was not created all at once at the time of the initial explosion, but that it appeared over a span of time lasting just over 10^{-32} seconds. In this initial moment, space-time purportedly "inflated" from a point 1 trillion trillion trillion times smaller than a proton into a volume several hundred million light-years in diameter. This meant that space initially expanded at the phenomenal rate of 10^{48} times the velocity of light and then, for some reason, braked to its present light-velocity speed limit.

This inflation assumption is supposed to greatly reduce the total amount of energy needed to spawn the universe, thereby minimizing the extent to which the big bang theory violates the law of energy conservation (the first law of thermodynamics) by which physicists regularly abide. It also is supposed to eliminate the "flatness problem," which required the Big Bang to be fine-tuned to within 10^{-50} of the critical density. But the inflation theory is not without its problems. It predicts that many expanding bubbles of space-time arise during inflation, each having its own laws of physics, and that unusual interactions take place between those bubbles. On the contrary, the laws of physics are seen to be relatively invariant throughout the observable universe.

Moreover, all inflation theories make one testable prediction, namely, that protons have a finite lifetime of approximately 10^{30} years, after which they spontaneously decay into energy. Thus far, experimental results have not supported this claim. If the inflation prediction were true, a three-ton vat of water containing approximately 10^{30} protons would give rise to at least one proton decay per year. For years, scientists watched a vat the size of a swimming pool, but their detectors registered no decays.[4] Furthermore, inflation models do not offer a complete theory of creation, since they fail to explain how the original infinitesimally small bubble of uninflated space-time and its associated energy impulse first came into being.

Inflation theory also does not solve the so-called *field singularity problem* that has long plagued the big bang theory. This difficulty arises at the Big Bang's very beginning at time $t = 0$. At this zero time, the volume of the primordial space-time bubble is vanishingly small, and therefore the energy density of its contained fields must be infinitely large. There is no way around the singularity problem within the framework of general relativity; its field equations absolutely require that the universe emerged from such a state of infinitely concentrated energy.[5]

Some physicists have attempted to avoid the consequences of this singularity paradox by making the ad hoc assumption that the laws of physics did not apply in the first 10^{-43} seconds of the universe's existence. Hawking has suggested that for times earlier than $t = 10^{-43}$ seconds, the dimensions of space-time were "smeared out" because of *quantum uncertainty*, thereby replacing the singularity point at time $t = 0$ by a space-time sphere whose surface has no beginning and no end. Hawking's solution does not explain the origin of the primordial energy quantum; it merely blurs space-time so that there is no appropriate place where the question of origin might be asked.

The problem with Hawking's approach lies in his freehanded use of the quantum uncertainty principle. The notion of quantum uncertainty has its roots in the uncertainty principle put forth in the 1920s by Werner Heisenberg. This principle asserts the impossibility of simultaneously measuring with absolute precision both the position and the momentum of a particle, or the energy and time duration of an energy-releasing event. The physics community soon became divided on how this principle should be interpreted. Albert Einstein championed the more conservative, common-sense view that the uncertainty principle simply reflects a restriction of an observer's ability to sense phenomena at the quantum level. Since a scientist's measurement probes, even in their most refined form, are quite gross in comparison with the tiny structures of the microphysical realm, an observer would strongly affect what he or she was trying to observe. The Danish physicist Niels Bohr, who spoke for the logical positivists, believed otherwise: that there was no reality apart from observation. He regarded Heisenberg's uncertainty relation as a statement that the quantum world itself was inherently fuzzy, indeterminate, and incomprehensible to human thought, and that it became physically realistic and definite only when graced by human observation. Furthermore, Bohr regarded nature to be *acausally indeterminate*, meaning that he rejected the possibility that an underlying observable substrate such as an ether might be responsible for causing observable quantum phenomena. Swayed by Bohr's philosophy, the physics community adopted this positivist interpretation of the uncertainty relation, which came to be known as the "Copenhagen interpretation," in recognition of Niels Bohr of Copenhagen. It has since come to be the central creed of modern quantum mechanics.

Hawking's attempt to solve the big bang theory's energy-singularity problem begins with the assumption that there is such a thing as relativistic space-time, something that laboratory experiments have recently disproven. He then assumes that Bohr's interpretation of the uncertainty relation, along with its rejection of a metaphysical beyond, is a correct description of the microphysical realm. Finally, he assumes that this acausal uncertainty principle can be extended to describe the behavior of space-time in a primordial epoch devoid of both physical quanta and human experimenters. Thus he disposes of the two elements essential to Heisenberg's uncertainty relation (quanta and observers) and takes only the uncertainty itself together with its positivist view of nature, a view that denies the existence of the unseen. Further on he concludes that no precursor is needed to explain the origin of the Big Bang. But this is merely a restatement of the assumptions that he began with.

Fortunately we need not accept this materialistic paradigm, nor imprison ourselves within its crystalline walls. For there is a better way of seeing things. It is the outlook embodied in the futuristic science of creation handed down to us from a time in the distant past, a worldview that finds its modern expression in the cosmology of subquantum kinetics.

QUANTUM NONLOCALITY

In 1935 Einstein collaborated with Boris Podolsky and Nathan Rosen to demonstrate the absurdity of the Copenhagen interpretation by presenting a troubling thought experiment.[6] This came to be known as the Einstein-Podolsky-Rosen (EPR) experiment. It is best illustrated in the following version, suggested by the British physicist David Bohm.

A subatomic particle, such as a proton, may be thought of as a spinning top generating a small magnetic field, or "intrinsic magnetic moment," oriented along its spin axis. If one of two adjacent protons has its north magnetic pole aimed up, the other will necessarily have its north pole aimed down, since the two particles are in a state of mutual interaction. If these two protons suddenly fly apart in opposite directions and, after traveling a considerable distance, are each measured by separate devices that determine their spin orientations, the two devices would be expected to register opposite spin orientations assuming that the two protons had not been disturbed during their flight. If device 1 finds that one proton has an upward-pointing spin axis, device 2 should find that the other proton has a downward-pointing spin axis.

The probabilistic Copenhagen interpretation holds that the spin directions of the two protons should be indeterminate prior to being measured, and that the spin axes should initially point in all directions with equal probability and become uniquely defined in a real physical sense only at the time of measurement. If one proton's spin axis suddenly points up as a result of being measured by device 1, information about the outcome of that measurement would have to be conveyed through space (by some unknown means) to inform the other proton so that it could properly orient its spin axis in the opposite direction before being measured by device 2. If the two measuring devices are separated from one another by a sufficiently large distance and the two measurements conducted sufficiently close together in time, information about the spin orientation of the first particle would have to travel to the second particle at a speed far exceeding the velocity of light. Thus the Copenhagen interpretation was in gross conflict with the basic tenet of relativity theory, which holds that information can be transmitted no faster than the speed of light.

Einstein, Podolsky, and Rosen hoped that physicists would admit the absurdity of the notion that the properties of subatomic particles have no logical prior cause and arise in random fashion just at the moment of observation. They advocated the more commonsense view that properties such as a proton's spin are determined in a real sense prior to the moment of observation. The two protons, then, would have oppositely oriented spins from the very moment of their separation, whether or not a measurement is made. The act of observation would simply *inform* the observer of the proton's preexisting spin orientation.

The *principle of local causes* is basic to Einstein's philosophy. This states that whatever happens in a given region depends on events taking place in that region and not on the decision of an experimenter situated at some distant location. Hence any quantum structure, such as an electromagnetic wave or subatomic particle, has real properties that are at all times determined by hidden interactions occurring in its immediate vicinity that are beyond the reach of experiment. Such "hidden interactions" would also include field effects created by events taking place at some distant location and communicated to that location at the speed of light. Physicists often refer to theories advocating this point of view as *local hidden-variables theories.*

In 1964 Irish physicist John Bell showed mathematically that if the behavior of the microphysical world is correctly described in terms of local hidden variables, then an observer carrying out an EPR type of experiment should find certain maximum probabilities of registering coincidences among the measurements made with the two spin-measuring devices. Quantum mechanics, on the other hand, predicted slightly higher coincidence probabilities for certain orientations of the spin-measuring devices.

Most tests of the EPR experiment measured the polarization directions, rather than the spin directions, of coupled pairs of photons. The statistical results of these experiments favored the odd predictions of quantum mechanics rather than the more commonsense predictions of Einstein and the hidden-variables theorists. Ironically, the argument that Einstein and his colleagues had devised to demonstrate the absurdity of Bohr's perplexing philosophy has now been turned on its head and used to show that the subatomic world actually does behave in ways more peculiar than had previously been thought.

While these experimental results match the predictions of quantum mechanical acausality, they do not necessarily force us to reject causality in favor of this positivist philosophy. One may hold on to causality simply by rejecting the premise of locality, or in other words, by dispensing with the relativistic assumption that effects are communicated no faster than the speed of light. In such a case, the results of the EPR experiment can be explained by admitting that the two protons, or two photons, actually do behave as a correlated system even when spatially separated and that they exchange information about their respective spin or polarization orientations at superluminal speeds. Physicists term this state of affairs *causal nonlocality.*[7]

From the standpoint of the ancient ether physics, indeterminate behavior originates at the subquantum level, rather than at the quantum level. Indeed, with the untold numbers of etherons present in every cubic centimeter of space reacting and diffusing in complex ways, it is essentially impossible to predict what the ether's state will be from one moment to the next. The transmuting ether is inherently indeterminate. The law of averages is of no help in computing the emergence of its quantum structures (concentration patterns), because at any moment a concentration fluctuation might arise and throw off any expected outcome. This indeterminacy is the chaos that, according to ancient myth, gave birth to physical form. Although it incorporates indeterminacy, the ancient physics does not do away with causality, since the ether dynamics are conceived to be the cause of physical phenomena. It advocates what physicists today would call *indeterminate causality,* as distinguished from the *acausal indeterminacy,* of modern quantum theory.

Where does the ancient physics (and subquantum kinetics) stand in regard to the EPR experiment results? Although it is incompatible with the acausal positivist view, this ether physics does adapt well to causal nonlocality. As mentioned in chapter 4, the ancient physics describes an organic, Whiteheadian ether whose various volume increments take account of one another to yield a unified whole. This nonlocal integration is accomplished by means of diffusing etherons, which quite possibly could travel faster than the speed of light. This reaction-diffusion type ether thus automatically incorporates nonlocality at a very basic subquantum level.

The ancient ether physics lends itself to a conservative, as opposed to a liberal, construction of nonlocality. In other words, it refrains from unduly generalizing the EPR experiment results to infer nonlocal causal connections where there might actually be none (see table 14.1). The danger of adopting a more liberal interpretation

TABLE 14.1
Quantum Causal Nonlocality

CONSERVATIVE INTERPRETATION	LIBERAL INTERPRETATION
1. Restricts nonlocality to properties such as polarization and spin that have been specifically checked by EPR experiments.	1. Extends nonlocality to other quantum properties such as mass and charge.
2. Assumes that nonlocal connections exist only among correlated quanta, that is, among waves or particles that had the opportunity of being positioned very close together.	2. Presumes that such connections operate between all particles regardless of whether they had previously been proximally located.
3. Nonlocal connections extend only over a limited distance, after which point quanta begin to lose their ability to behave in a correlated fashion.	3. Nonlocal connections extend throughout the universe.
4. Nonlocal connections are mediated at some finite speed.	4. Nonlocality operates instantaneously.

of nonlocality is that it leads to a deterministic, monistic worldview. Many contemporary physicists, though, have tended toward a liberal interpretation, including David Bohm, who proposed that the entire universe is intricately integrated by a forest of nonlocal causal connections operating instantaneously at the quantum level.[8] By virtue of these quantum interconnections, each part of the universe would contain within it every other part, a condition Bohm refers to as *quantum interconnectedness* or *unbroken wholeness*.

In the early twentieth century, physical science was criticized for its extreme reductionist position, which assumed that a system's behavior could be understood by studying its parts in isolation and ignoring their collective interactions. Now, with the emergence of the liberal interpretation of quantum nonlocality, we find that the pendulum has swung to the other extreme. Bohm's new physics advocates that only the whole is causally effective, and the parts are completely determined by the state of the entire physical universe. Bohm's theory proposes that the whole universe operates as one gigantic clockwork mechanism determined from "above." As such, it ascribes Godlike macrodeterministic properties to the quantum material realm.

This new brand of "systems" thinking is at variance with the tenets of general system theory. Systems theorists recognize that nature's evolution is controlled neither by the parts of a system nor by the system as a whole. The behavior of the parts affects the functioning of the whole, and the whole likewise affects the behavior of the parts. Systems theorists realize that nature is most fully understood only by reflecting upon the interrelation of its various hierarchic levels—subsystem, system, and supersystem—and how they work together in harmony. Such a multilevel perspective is lost in Bohm's approach.

Bohm does not go to the extreme of completely denying the presence of an underlying order. He postulates the existence of a "nonmanifest" dynamic realm that he calls the "implicate order," whose activity is supposed to give rise to the observable universe, which he terms the "explicate order." Although his concept of explicate order is similar to the subquantum kinetics concept of explicit order, his implicate order differs substantially from the implicit order concept of subquantum kinetics. Bohm states that the implicate order is an entity whose nature is beyond comprehension, composed of a complex interweaving of Einsteinian spacetimes enfolded together with various subtle energies of an unknown nature. The implicit order of the ancient transmuting ether, on the other hand, is held to be potentially comprehensible, at least in part. In fact, by properly discerning this underlying order, it should be possible to account for the form and behavior of the subatomic realm. Whereas the ancients took a fundamentally organic approach, Bohm's approach is fundamentally mechanistic. In the ancient creation myths there is the recurring theme of the individualistic hero who, by challenging authority, ultimately establishes a new order. Such creative self-determination is prohibited in Bohm's worldview, as individuality is entirely absorbed into the whole.

THE FORCE FIELD FIASCO

Although it claims to be a theory of creation, the big bang theory offers no insights into the underlying makeup of the material particles and fields that the primordial explosion supposedly brought into being. In an attempt to fill this theoretical void, cosmologists turn to high-energy physics to discern the secrets of matter's composition. But no clear answers have been forthcoming.

In 1964 the physicist Murray Gell-Mann proposed that nuclear particles such as protons and neutrons are composed of hypothetical entities called "quarks" that are cemented together by bonding particles called "gluons." Gell-Mann originally spoke of three quarks denoted by the flavors "up," "down," and "strange." With the discovery of more massive subatomic particles, this scheme became inadequate. To save this model, physicists increased the number of proposed quarks and suggested that each quark comes in three different colors, "red," "green," and "blue," which brought the total number to nine. Again, new particle discoveries rendered this scheme obsolete and so a new flavor called "charm" was added to the previous three, bringing the total to twelve. Today this set is eighteen. Including the eighteen antiquarks that complement these, the total comes to thirty-six. With the eight gluons, the figure comes to forty-four. With the discovery of the massive U particles

(U⁺, U⁻, and U°), even forty-four fundamental units are not enough to explain the entire particle zoo.[9]

Many physicists have become disenchanted with the quark theory, finding its large variety of quarks rather unattractive. Moreover, in all the years since it was first proposed, no one has ever detected a quark. Increasingly powerful particle accelerators were built that allowed protons to be smashed together at increasingly higher speeds, but no quarks were found. Unlike etherons, which are inherently unobservable, quarks are supposed to be physical entities and in principle observable. Each quark is theorized to carry an electric charge measuring either one-third or two-thirds of an electron charge. Yet no such charges have been found.

In addition, quark theory fails to properly predict the spin properties of subatomic particles. The theory posits that a proton's spin arises from the sum total of the spins of its three component quarks, two spinning in one direction and one in the other. With this arrangement, one quark spin would remain uncanceled and thus be available to express itself as the proton's spin. If this model were valid, protons should interact approximately 25 percent more frequently if their spins were aligned in the same direction (parallel) as opposed to being oppositely aligned (antiparallel).* But one group of physicists conducting high-energy collision experiments with spin-aligned protons has discovered that colliding protons instead interact up to five times more frequently if their spins are parallel aligned.[10] This is over an order of magnitude greater than the quark model prediction. They also found that, in the case of the far more interactive parallel-spin collisions, protons tended to deflect in the same rotational sense as the spin rotation, counterclockwise, when the proton spins were going counterclockwise. Again, this finding could not be explained by the quark theory.

Eric Lerner has pointed out that these proton collision experiments favor a more classical notion that considers spin to be associated with the proton as a discrete entity, not with any quarklike subcomponents.[11] He suggests that an appropriate model for spin would be to suppose that protons are some form of vortex. Experiments with plasmoids indicate that this is a feasible model, for plasma vortices interact far more strongly when they are spinning in the same direction. Interestingly, subquantum kinetics and the ancient ether physics propose precisely such a vortex model of subatomic particle spin.

Another aspect of physical reality on which the big bang theory remains silent is the question of force fields or potential fields. What is their nature and how do they attract particles? Current physics recognizes four types of forces:

---

*In the case of two colliding protons with parallel-aligned spins, there would be a 5:4 chance that collisions would take place among parallel-aligned quarks as opposed to antiparallel-aligned quarks. If parallel alignments produced stronger interactions, such interactions should have been observed approximately 25 percent more frequently.

1. Gravitation, a long-range force
2. Electric and magnetic forces, a medium-range force
3. The weak force, a short-range force
4. The strong force (nuclear), a short-range force

The first two are the most commonly experienced of all, since they are encountered in our everyday life. The last two are active over a much shorter range and hence not accessible to our direct experience. The weak force manifests itself in certain kinds of particle collisions and radioactive decay processes, an example being the decay of a neutron into a proton, electron, and neutrino. The strong force operates in the atomic nucleus to bind together its neutrons and protons.

Contemporary physics explains gravitation on the basis of Einstein's general theory of relativity, which proposes that massive bodies exert their pull by warping the metric of space, but it does not explain how mass warps space. The famous Serbian-American scientist-inventor Nikola Tesla strongly disapproved of Einstein's spatial-warping assumption. In 1932 Tesla stated:

> I hold that space cannot be curved, for the simple reason that it can have no properties. . . . Of properties we can only speak when dealing with matter filling the space. To say that in the presence of large bodies space becomes curved is equivalent to stating that something can act upon nothing. I, for one, refuse to subscribe to such a view.[12]

Research conducted in 1991 by a group of Cornell University scientists suggests that Tesla's doubts about relativity may have been justified.[13] These researchers discovered something that upset previous beliefs about the plausibility of Einstein's space-time warping equations. Their computer simulations showed that if a very large oblong mass were allowed to collapse upon itself, it would produce a spindle-shaped gravitational singularity of infinite energy—a black hole—whose extremities would extend outside the black hole's central region of invisibility. Such a "naked" singularity would radiate infinite quantities of energy into surrounding space: an absurd result that is fatal for general relativity.

Moreover, there is no way to verify unequivocally whether or not matter does warp space, since all of the standard tests for this effect may be accounted for by alternate theories. For example, in 1898, before the development of general relativity, Paul Gerber derived a formula for the angular advance of the long axis of Mercury's elliptical orbit in classical physics terms by assuming that gravity is transmitted by a Newtonian potential that propagates with the velocity of light.[14] Also, the gravitational bending of starlight passing near the limb of the Sun, another standard test for general relativity, is just as easily explained as a refraction effect.[15] For example, subquantum kinetics predicts that the velocity and wavelength of a light wave should vary in proportion to gravitational field intensity, and therefore that the Sun's gravitational field should refract passing light rays just as is observed.[16]

As an explanation for the remaining three forces, physicists suggest that such interactions arise as a result of the exchange of subatomic particles between the interacting bodies. The electrostatic and magnetic forces are supposedly mediated by the exchange of photons (electromagnetic quanta), the strong force by *pions* (pi mesons), and the weak force by some as yet unknown heavy subatomic particle. Yet experimental supporting evidence for all of these particle exchange theories is completely lacking. In fact, no one has ever observed such particles being exchanged.

Proponents of these theories respond by explaining that the exchanged components are emitted and absorbed so quickly that they remain invisible to experimental detection. They refer to these as "virtual" particles, particles that are only imminently in existence. According to this quantum electrodynamic view, a real particle, such as an electron or proton, would be continually emitting and reabsorbing virtual particles, with the emitted virtual particles forming an invisible cloud, or "hairdo," around the parent particle. In emitting a virtual particle, the parent particle would receive a recoil impulse, and upon reabsorption, the proponents claim, it would receive an opposite recoil canceling out the emission impulse, resulting in no net movement. Movement, they say, occurs only when the emitted virtual particles are absorbed by a neighboring real particle.

Such particle-exchange field theories fall within a branch of physics known as *quantum field theory*. But all is not well for quantum field theory. Although the theory offers a clear conceptual model of electrostatic repulsion, its description of particle attraction, such as electrostatic attraction or nuclear binding, is not so clear-cut. To explain attraction, quantum field theory departs from the familiar collision concepts of Newtonian mechanics and proposes that upon emitting a virtual particle, the real particle moves in the same direction as the virtual particle, rather than recoiling in the opposite direction.

Another area of concern centers on the question of energy conservation. The proposal that virtual particles are spontaneously created upon emission by the parent particle and destroyed upon reabsorption blatantly violates the law of energy conservation. Physicists justify their mistreatment of this long-cherished fundamental law by hypothesizing that the created particles exist for a period of time shorter than the time uncertainty predicted by the Heisenberg relation.

The energy summation of all virtual particles momentarily existing around a particle also poses a problem since quantum field theory predicts that a given virtual particle cloud should have an infinitely large mass. For example, in creating each virtual photon the electron is assumed to "borrow" energy from the uncertainty relation "bank." Although this energy is assumed to be "paid back" shortly thereafter to avoid violating the law of energy conservation, there would always be a net reservoir of virtual photon energy in the electron's photon cloud. When the individual packets of energy associated with each virtual photon are added up, however, the energy (and mass) of the electron together with its virtual particle cloud is found to be infinite. This blatantly contradicts experimental observation that shows the electron as having a finite rest mass, whose energy equivalent amounts to 510,000 electron volts.*

Physicists have attempted to resolve this difficulty through a process called *renormalization,* or the ad hoc assumption that the "bare" electron core has an infinite negative mass of just the right amount to cancel out most of the infinite positive mass of the electron's photon cloud and leave a net positive residual equivalent to the electron's observed mass. Further complicating matters, this ad hoc finely tuned renormalization procedure must be repeated for each force field that a particle is presumed to generate, since the clouds of virtual pions, gluons, and other particles required to mediate strong and weak forces and quark binding would similarly produce singularities. Many physicists find this technique unsatisfactory, since there is no reason to justify the infinite negative mass assumption other than the purported need to patch up the failing quantum electrodynamics theory. Paul Dirac, a famous British physicist known for his 1928 prediction of the existence of antimatter, played an important role in developing the foundations of quantum electrodynamics. But throughout his life he expressed his displeasure with the business of renormalization. He felt it was no more than a poorly executed attempt to sweep under the carpet problems of a theory that was inherently flawed.

Contemporary field theory resembles a patchwork quilt. On the one hand, it explains gravitational force in terms of spatial warping, and, on the other hand, it explains nature's other forces in terms of the exchange of evanescent particles. This wide diversity of mechanisms more likely reflects the current diversity of thinking rather than an actual diversity of force field mechanisms in nature itself. It is not surprising that efforts to unify the four forces under a single theory have not met with much success. Just as modern cosmology does not provide a satisfactory explanation for the origin of the universe, modern physics likewise fails to explain the underlying makeup of the physical world.

Another unsolved problem concerns neither subatomic particles nor their force fields, but space itself. According to quantum field theory, pairs of virtual particles can spontaneously pop into existence right out of "empty" space and then disappear again (figure 14.1). Each pair consists of a particle and its corresponding antiparticle, such as a virtual proton and virtual antiproton. These transient matter-antimatter pairs are termed *quantum fluctuations,* or alternatively, *vacuum fluc-*

---

*The rest energy of the electron is calculated by multiplying its rest mass by the square of the speed of light, i.e., $E = mc^2$.

tuations. Space is thought to be full of them. Although the vacuum fluctuation concept bears strong similarities to the ether fluctuations described in ancient mythology and in subquantum kinetics, there are several distinct differences. Ether fluctuations are much smaller in size, perhaps by as much as a millionfold; they arise as statistical variations in the concentration of the subquantum medium rather than from empty space; and they arise as individual pulses (of either plus or minus polarity), not as matched plus-minus pairs.

Virtual particles pose another major problem for modern physics: they drastically distort the geometry of space. Upon summing up the mass/energies of all the virtual particles that are theorized to reside throughout space, one finds that each cubic centimeter of space should have a prohibitively large density, ranging from 10^{12} to 10^{89} times the density of water.[17] Expressed in energy terms, each cubic centimeter of space would contain anywhere from 10^{36} to 10^{113} ergs. For comparison, at the low end of this range, one cubic centimeter of space would contain as much energy as would be released if each person on Earth were to detonate a one-thousand-megaton hydrogen bomb.

It has become popular to use the term "zero-point energy" to describe this hypothesized vacuum energy, the term "zero-point" signifying that this energy would be present even in the case where the temperature of space is at absolute zero with all electromagnetic energy absent. However, the zero-point energy concept applies just as well to the ether physics of subquantum kinetics where the term may be used to describe the spontaneous subquantum energy fluctuations that spontaneously arise and subside in the ether, even in regions that are empty of matter and energy, except in this case fluctuations that are large enough to be comparable in size to a virtual particle would be extremely rare. When they did emerge, they would stand a very good chance of nucleating a real subatomic particle. As a consequence, subquantum kinetics would predict a much lower zero-point energy density for space.

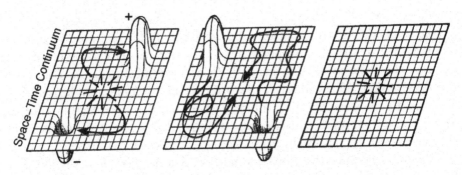

Figure 14.1. An idealized representation of how virtual particle fluctuations would spontaneously appear, briefly interact, and disappear in the quantum mechanical vacuum. Adapted from L. Abbott, "The mystery of the cosmological constant," pp. 106–13.

The high zero-point energy density predicted by quantum field theory, however, raises serious problems for conventional physics, particularly with general relativity, where high energy densities translate into high mass densities for space. A density of 10^{25} grams per cubic centimeter at the modest end of the predicted range would significantly distort space-time geometry over distances of just a few centimeters. The aberration would be so great that it would be impossible to draw two parallel lines on a sheet of paper that wouldn't cross. Moreover, the angles of a triangle would no longer add up to 180°. Light from objects on the far side of the room would be redshifted into the infrared spectral region or beyond, making them invisible to the eye.

If space were curved by even the slightest amount, evidence of this would have shown up in astronomical surveys. When the data are checked, however, no evidence of curvature is found. Observations of the density of galaxies found at distant locations of the universe indicate that space is Euclidean out to the farthest limits of observation. These data show that the mass/energy density of the universe is no greater than 10^{-29} grams per cubic centimeter, a mass equivalent to about ten protons in each cubic meter of space. Compared with the density estimates from quantum mechanics, the predictions of quantum theory and observational astronomy differ by more than 41 orders of magnitude, and possibly by as much as 118 orders of magnitude. As physicist Larry Abbott notes, "few theoretical estimates in the history of physics made on the basis of what seemed to be reasonable assumptions have ever been so inaccurate."[18]

If it were unopposed, the attractive gravitational field produced by such large zero-point energy-mass densities would be so great that the big bang fireball should have collapsed on itself the instant that it was born. To extricate themselves from this embarrassing situation, cosmologists would need to introduce a repulsive cosmological gravitational force term called the *cosmological constant*. Einstein was the first to propose this term. He introduced it into his general relativity field equations to ensure that they would predict a stationary space-time continuum. Later, having become convinced that the universe was really expanding, he eliminated this constant from his equations. But it refuses to be thrown out. It has now returned to haunt cosmology. Astronomers recently reintroduced it in order to make the big bang theory fit the new redshift-distance data obtained from observations of distant galactic supernovae. But to counteract the immense attractive forces predicted by quantum field theory, the cosmological constant would have to be absurdly large, far surpassing the modest values required to fine-tune the expanding-universe theory.

The cosmological constant mystery indicates that modern physics is seriously flawed. At present no solution seems to be in sight, leaving physics in a state of massive contradiction. Quantum theory and general relativity constitute the two main pillars supporting the edifice of modern physics and astronomy. To understand the microcosm and to account for short-range forces holding together atomic nuclei, atoms, and molecules, physicists rely on quantum field theory. On the other hand, to understand the macrocosm and to account for long-range gravitational attraction, physicists rely on general relativity and its associated big bang cosmology. But when these two major branches of physics are brought together, they mutually annihilate, like two virtual particles of matter and antimatter.

15

ENERGY IN THE UNIVERSE

THE CREATION AND
ANNIHILATION OF ENERGY

Although the ancient creation myths we have examined explain allegorically how physical form first emerged in the universe, they say little about matter except to imply that the particles that emerge have a wavelike form and, possibly, spin characteristics. Little is said about the nature and behavior of radiant energy, or about the energetic properties of celestial bodies, other than to suggest that the core of the galaxy is the prime center of matter and energy creation in the Milky Way. Nevertheless, by exploring the ramifications of subquantum kinetics' Model G, we can learn something about these aspects of the ancient ether physics.

A careful examination of Model G reveals that its X and Y ethers not only can configure themselves as stationary reaction-diffusion waves (material subatomic particles), but also as propagating reaction-diffusion waves, physically evident as waves of radiant energy (photons). As pointed out in chapter 3, reaction-diffusion waves differ from mechanical waves in that the former are produced indigenously through the underlying activity of a thermodynamically open reaction-diffusion medium, whereas the latter arise through external mechanical disturbance of an inert mechanical medium. As a result of this fundamental difference, subquantum kinetics and the ancient ether physics make predictions about photon behavior that are substantially different from those arising out of conventional physics. For example, modern electromagnetic theory, which developed out of the nineteenth-century frictionless mechanical ether concept, requires that a photon perfectly conserve its energy and maintain its energy constant during the course of its flight through space.

Subquantum kinetics, on the other hand, is not so restrictive in this regard. The Model G reaction-diffusion equations indicate that photons do not necessarily maintain constant energies. Depending on the state of the ether in their vicinity, they may either gradually gain or gradually lose energy, albeit at an extremely slow rate. In regions of space where the ether reactions are supercritical, a photon's

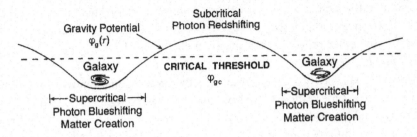

Figure 15.1. Galaxies would form supercritical G-wells in subcritical space in the manner shown above. Photons would experience energy creation or energy annihilation in the respective regions.

energy would gradually increase over time, while in regions where the reactions are subcritical, its energy would gradually decrease over time. Since the ether's critical-ity is controlled by the value of the G ether concentration, the rate at which this energy change occurs, and the sense in which it occurs (increase or decrease), will depend on the value of the ambient gravitational potential. In any case, these energy changes take place at such an exceedingly slow rate that the resulting gain or loss in the energy would pass undetected in laboratory experiments, even when using the most sensitive instruments available.

Such a physics opens up new ways of understanding astronomical and cosmolog-ical phenomena. Light photons traversing the vast stretches of space between galax-ies, where the gravity potential is high and the ether is subcritical, would experience tired-light redshifting (see figure 15.1).* Their wavelengths would increase over time, and their frequency would gradually shift toward the red end of the light spectrum. On the other hand, photons traveling in the vicinity of galaxies where the gravity potential is most negative and the ether reactions most supercritical would tend to gain energy over time and gradually blueshift. Perfect energy conservation would be rare, occurring only in intergalactic regions of space at places where subcritical and supercritical regions happened to interface one another. At such locations, the ether reaction system would be operating precisely at its critical threshold.

This progressive increase or decrease of photon energy should not be viewed as energy coming from "nowhere" or going to "nowhere." For, as pointed out in ear-lier chapters, the physical universe is an epiphenomenon, a concentration "water-mark" that emerges in the transmuting ether. A subatomic particle or quantum of energy owes the continuation of its particular state of existence to the continuous

---

*This gravity potential–dependent cosmological redshifting effect predicted by subquantum kinetics should not be confused with what astrophysicists term the *gravitational* redshift, which occurs near the surface of a star and involves a rapid lengthening in the wavelength of photons as they emerge from the star's gravity well. This is an entirely different phenomenon that is also predicted by this open-systems physics.

operation of these etheric processes. When a photon's energy increases or decreases, there is no corresponding gain or loss in the amount of subquantum vital energy powering the ether processes, but merely a change in the way these processes are physically manifested.[1]

Since the expanses of subcritical space between individual galaxies are far larger in size than the supercritical G-wells within each galaxy, light coming from distant galaxies would be predominantly redshifted. A photon energy loss of just 7.4 percent for every billion years of its travel would entirely account for the observed increase of redshift with distance. This leads to a stationary-universe tired-light cosmology that, as we saw earlier, fits observational data better than the big bang theory. But what about the blueshifting prediction? Is that also compatible with astronomical observation?

Light coming from stars within our own galaxy would be blueshifted as it journeys to us through our galaxy's supercritical region, although the degree of this spectral shifting would be far smaller than the Doppler shifts produced by the stars' relative motions. Most of the supercritical blueshifting would instead occur inside a star before a photon had left the star's surface. Since a star generates a relatively deep gravity well and supercritical region, photons would blueshift at a much faster rate inside a star than they would on their journey through interstellar space. Moreover, they would spend a considerable length of time in these fertile birthplaces. Since photons are repeatedly absorbed and reemitted as a result of frequent encounters with gas nuclei in a star's interior, they can take some ten thousand years or more to reach a star's surface, a distance that in open space would be covered in only a few seconds. Energy quanta would thus experience most of their energy gain while inside their stellar birthplaces. Stars, in effect, are the wombs where this self-generated *genic energy* comes into being.[2]

The individual genic-energy contributions from the enormous numbers of photons jostling around inside a star would yield a tremendous energy output. Taking Earth as an example, if each photon making up Earth's internal geothermal energy reserve were to increase its energy content at a rate of just 7 billionths of a percent per year (or 2×10^{-18} per second), the excess energy generated would be sufficient to supply all of Earth's geothermal energy flux. By analogy, Earth's internal energy reservoir would produce genic energy in much the same way that money in a savings account produces interest; Earth's energy savings account, however, would produce interest over a billion times more slowly than most bank accounts.

Doctrinaire physicists who maintain that photons strictly conserve their energy would find the notion of spontaneous photon blueshifting and genic-energy production just as unacceptable as the idea of spontaneous photon redshifting. But given the extent to which scientists currently know the energy conservation "law" to be true, the energy creation and destruction processes predicted by subquantum kinetics are not in violation of any physical observation. The precision attainable with the most up-to-date instruments is such that if photons were violating the energy conservation law by increasing their energies at the rate of 0.1 parts per million per second (10^{-7}/sec), their transgression would pass unnoticed, being just below the threshold of detection. By comparison, the energy change rates that subquantum kinetics predicts for an Earth-based laboratory are 100 billion times smaller than this detection threshold.

Physicists' current belief that a photon's energy and wavelength should stay constant over time has its roots in the nineteenth-century mechanical ether theory that assumed the mechanical ether was perfectly elastic and frictionless and capable of transmitting wave energy great distances without any energy losses. Whatever amount of energy was initially impressed upon the ether to create the wave was expected to be faithfully conveyed as the wave moved forward in space and time. Although physicists later abandoned the mechanical ether concept, they retained its underlying energy conservation assumption in their field equations. As a result, the first law of thermodynamics, which states that energy can neither be created nor destroyed, continued to be interpreted in its strictest sense. In view of the lack of laboratory evidence substantiating this extreme position, however, this "law" might be better termed the first *approximation* of thermodynamics.

This assumption that energy is strictly conserved is a great drawback to progress in the field of astronomy, where photon energy change rates far below the limits of laboratory measurement can be of enormous consequence. Our stand on energy conservation influences how we interpret a wide variety of energy generation phenomena, ranging from the gradual release of heat from the interior of our own planet to the tremendous outpouring of energy in a supernova explosion. It is also fundamental to explaining the origin of the universe.

GENIC ENERGY IN STARS AND PLANETS

The genic-energy prediction accounts for the radiant output from planets and low-mass stars and, in particular, it accounts for the mathematical relation that exists between the masses of such celestial bodies and their luminosities.[3] It predicts that the genic-energy luminosity of a celestial body should increase according to the 2.7 power of mass $(L_g \propto M^{2.7})$.*

To see how this relation compares with actual data on the masses and luminosities

---

*The heat reservoir within a planet or star produces genic energy at a rate that depends on the amount of stored heat and on the prevailing rate of heat amplification. Since a celestial body's heat reserve is a function of its mass times its internal temperature, and since internal temperature itself increases with increasing mass, more massive bodies will necessarily produce genic energy at greater rates. Also, since more massive bodies have deeper supercritical G-wells, photons within them will amplify their energy at a faster rate, and this too will increase the body's output of genic energy.

of planets and stars, it is best to use the logarithmic graph shown in figure 15.2, which allows us to plot bodies that span a wide range of masses and luminosities. The graph plots the masses and luminosities of *red dwarf* stars, stars that range from approximately 8 to 45 percent of the Sun's mass. Regression analysis shows that the luminosity of such stars increases according to the 2.75 ± 0.15 power of their mass, in very close agreement with the genic-energy prediction.

The mass-luminosity (*M-L*) coordinates for the jovian planets—Jupiter, Saturn, Uranus, and Neptune—also fall quite close to the genic-energy *M-L* relation, again confirming the predictions of the genic-energy hypothesis.* Jupiter, the most massive of the four, has been found to radiate the greatest flux of internally generated heat, with Saturn, Uranus, and Neptune radiating lesser amounts in accordance with their smaller masses. The large intrinsic heat fluxes coming from the interiors of these planets make their surface temperatures considerably hotter than would be expected based on the insolation these planets normally receive from the Sun. Saturn, for example, radiates almost as much heat from its interior as it receives from the Sun.

Of the four planets, Jupiter's anomalous surface temperature was the first to be discovered. In an attempt to explain it, astronomers suggested that Jupiter's hydrogen/helium core might be storing a very large quantity of primordial heat left over from billions of years in the past when the planet formed from a collapsing gas cloud. But this theory did not adequately account for the excess heat flux coming from the other giant planets. As a result, other mechanisms had to be devised that were specifically tailored for each planet. Moreover, none of these theories explains why the intrinsic luminosities for these planets lie along the mass-luminosity relation for red dwarf stars.

Excluding the possibility that this good fit is a coincidence, the conformance of both planets and red dwarf stars to the same *M-L* relation strongly suggests that both are powered by the same kind of energy generation mechanism. Nuclear fusion, commonly considered the energy source for stars, must be ruled out as the power source for planets, since their temperatures and densities are far too low to trigger nuclear reactions. For example, Jupiter, the hottest of the giant planets, has a core temperature of approximately 10,000 to 20,000° Centigrade, which is only a fraction of the million-degree temperatures needed to initiate nuclear fusion.

By the same token, energy mechanisms that have been suggested to explain the

---

*The planetary luminosities plotted in figure 15.2 represent just the internal energy outputs of the respective planets, excluding the energy contribution from solar radiation. If the *M* and *L* coordinates for the Earth were plotted here, the point would lie substantially above this trend line. In this low-mass range, however, the genic-energy relation actually bends to a lower slope, since the background gravity potential field produced by the Galaxy as a whole swamps the planet's own gravity potential contribution, thereby raising its genic-energy output above the value it would otherwise have. When this factor is taken into account, along with the differences in composition and core temperature, the genic-energy relation actually predicts 75 percent of the Earth's heat flux.

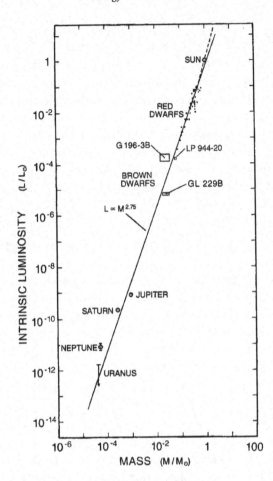

Figure 15.2. The mass-luminosity relation for planets and low-mass stars. The solid line charts the M-L regression line for red dwarf stars. The axes chart mass and luminosity as fractions of solar mass ($M_\odot$) and solar luminosity ($L_\odot$). The boxes denote the M-L coordinate range for the brown dwarfs GL 229B and G-196-3B.

heat coming from the planets, such as stored primordial heat, can be ruled out, since the energy output from these mechanisms is far too meager to produce the tremendous outpouring of energy that comes from red dwarf stars. For example, some of the more massive red dwarfs positioned at the upper end of the *M-L* relation radiate over 100 million times as much energy as Jupiter. This dilemma is easily resolved if celestial bodies continuously generate genic energy. Such a photon energy-amplification mechanism, operating within all masses, could explain why planets and red dwarfs alike share the same *M-L* relation. In such a case, there would be no clear qualitative distinction between planets and red dwarf stars, just one of degree.

The smaller planets and moons of the solar system also continuously radiate heat from their interiors. But the fluxes are so small that they do not produce a change in the planet's overall surface temperature sufficiently large to be measured from space; they can be accurately determined only by penetrating thermal probes into the orb's surface soil. So far, such measurements have only been made for the Earth and the Moon.

It has long been known that the Earth evolves heat from its interior, more commonly termed geothermal energy. In places where it most prominently surfaces, the geothermal flux produces hot springs and volcanic hot spots and occasionally is released suddenly in volcanic eruptions. Approximately 30 to 60 percent of this energy flux can be attributed to the decay of radioactive elements naturally present in the Earth's crust. The remaining 40 to 70 percent is thought to come from the Earth's core and to be responsible for driving convective processes that generate the Earth's magnetic field.

A variety of mechanisms have been proposed to account for the source of the Earth's core heat, such as the gradual release of heat trapped since primordial times, heat released from the gradual solidification of the Earth's molten core, or the radioactive decay of potassium-40. Since we have no way of directly knowing what is actually happening in the Earth's core, however, all of these suggestions are speculative. Interestingly, the genic-energy relation predicts that approximately 73 percent of the entire geothermal output is of genic origin, a value that accounts for all of the Earth's core heat flux.[4] In the case of the Moon, genic energy is able to account for approximately 40 percent of the Moon's measured heat output, leaving radioactive decay to account for the remaining 60 percent.

As for the other moons and small planets of the solar system, spacecraft observations have provided qualitative evidence that Neptune's largest moon, Triton, is producing a substantial amount of internal heat. Owing to its great distance from the Sun, Triton's frigid surface is just 350 above absolute zero (-238° Centigrade), or about 28° below the freezing point for nitrogen. Yet during the 1989 Voyager 2 fly-by, astronomers detected a geyser of liquid nitrogen spouting from the planet's solid nitrogen surface, indicating that the temperature of this part of Triton's surface has somehow increased by at least 28° Centigrade. Scientists have been at a loss to explain the source of this heat. Whereas the volcanic activity of Jupiter's moon Io can be entirely attributed to heat produced by Jupiter's tidal effects, no comparable mechanism is available for Triton. So genic energy provides a solution to this energy-creation mystery.

The genic-energy prediction also shows that if red dwarf stars are powered entirely by genic energy, as the data in figure 15.2 seem to indicate, astrophysicists have been wrong in assuming that nuclear burning can take place in such low-mass stars. It is generally agreed that a star must be greater than a certain mass if it is to develop temperatures and densities in its core sufficiently high to ignite nuclear reactions. The best theoretical estimates had placed this critical mass for nuclear ignition at approximately 8 percent of the Sun's mass (approximately 0.05 percent of the Sun's luminosity). These estimates appear to be in error, for the mass-luminosity data instead suggest that nuclear ignition occurs in stars more massive than 0.45 solar masses, or alternatively, more luminous than 0.07 solar luminosities. At this critical value, the mass exponent of the *M-L* relation abruptly increases from 2.75 to approximately 4.0. In other words, at these higher masses, stellar luminosity increases much faster with increasing mass, as indicated by the dashed line in figure 15.2. This upturn in the *M-L* relation is most likely a result of the onset of nuclear burning. When nuclear fusion turns on, the star's luminosity becomes elevated above the amount expected solely on the basis of the genic-energy *M-L* relation. Since the output of nuclear fusion plus genic energy increases faster with mass than the output of genic energy alone, the *M-L* relation proceeds upward at a steeper slope.

Consequently, stars such as red dwarfs that lie along the lower branch of the *M-L* relation would be powered exclusively by genic energy, while stars like the Sun that lie along the upper branch would be powered by a combination of both genic and nuclear energy. When the red dwarf *M-L* trend line shown in figure 15.2 is extrapolated upward using a nonlinear regression fit to recent data, it predicts that about 15 percent of the Sun's energy should be of genic origin and about 85 percent should be of nuclear origin. Although stellar astrophysicists claim that their nuclear fusion models should adequately account for the Sun's total energy output, they admit that the uncertainty range of their models is such that an 85 percent to 90 percent fusion contribution would still make an adequate fit to recent solar neutrino data.

The ignition of fusion reactions at the critical 0.07 solar luminosity threshold should be apparent in the shape of the stellar *luminosity function,* a statistical graph that plots the number of stars found in each of a consecutive series of luminosity increments. Such a profile, constructed from observations of stars in our galaxy, in fact, shows a pronounced inflection at this critical luminosity value (figure 15.3). Stars with luminosities greater than this 7 percent value form a small hump that stands out from the main part of the distribution. The added energy output resulting from fusion ignition increases a star's luminosity, shifting it leftward in the luminosity function profile and thus producing the observed population hump.

Stars powered exclusively by genic energy, composing the lower branch of the *M-L* relation, make up the main hump of the luminosity function, while stars powered by both genic and nuclear fusion energy, composing the upper branch of the *M-L* relation, make up the small added hump. The predominance of stars belonging to the lower branch suggests that genic energy serves as the primary source of energy for over 90 percent of the stars in the universe.

Presently, the gap in the *M-L* relation between Jupiter and the bottom of the red dwarf sequence arises because stars of such low luminosity, the so-called *brown*

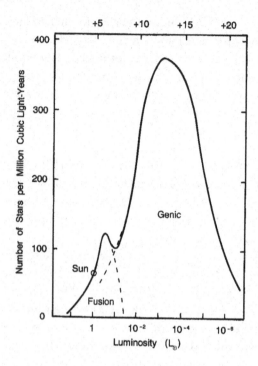

Figure 15.3. The luminosity function for stars in our galaxy. The profile charts the prevalence of stars in each of a sequence of star-luminosity categories. The inflection is a result of the onset of nuclear burning.

dwarfs, are difficult to see. Nevertheless, in 1995 and later in 1998 and 1999, astronomers successfully determined the masses and luminosities of three brown dwarfs, GL 229B found in the constellation of Gliese, G 196-3B in Ursa Major, and LP 944-20 in Hydra. As seen in figure 15.2, the mass-luminosity error boxes for these dwarfs fall along the planetary-stellar *M-L* trend line. For conventional astrophysics, the discovery that brown dwarfs conform to the same *M-L* trend line as red dwarfs comes as a total surprise since brown dwarfs are unable to power themselves by thermonuclear fusion due to their low mass. On the other hand, this finding strongly supports the genic-energy hypothesis. It confirms predictions that I previously published specifically stating that brown dwarfs should be found to conform to this relation.

Spacecraft observations may now have made a direct confirmation of the genic-energy effect. In 1985 and 1994 I had predicted that maser signals transponded between spacecraft should be found to exhibit a small blueshift amounting to about one part in 10^{18} per second.[5] In 1998 and later in 2002 astronomers reported detecting an "anomalous" blueshift in the transponded maser signals they had been using to track the Pioneer 10 and 11 spacecraft.[6] It turns out that the blueshifting rate they found comes very close to the value subquantum kinetics had predicted.[7]

STELLAR PULSATIONS AND EXPLOSIONS

Many astrophysical phenomena that are not easily explained by current physical theory are readily accounted for on the basis of Model G's prediction that stars spontaneously produce genic energy. Pulsating stars, for example, are a class of stars that expand and contract in a regular fashion, cyclically varying their luminosity and temperature. The phenomenon occurs in giant and supergiant stars such as Cepheids, long-period variables, and RR Lyrae stars. It also occurs in medium-sized Beta Cephei stars as well as in extremely dense white dwarf stars called ZZ Ceti stars.

Conventional astronomy has no theory capable of explaining this general phenomenon. Stellar pulsation produces essentially no radial movement at the star's center, so there is no way for the motion of the star's envelope to affect fusion reactions in the core and thereby cause the star's energy output to pulsate. As an alternative explanation, astronomers finally proposed that the phenomenon might be caused by cyclical changes in the opacity of a star's photosphere, its outer luminous surface. They suggested that as the star contracts and heats up, the gas in its photosphere ionizes and becomes more opaque to the star's radiation. As a result, this outgoing radiation couples more efficiently to the photosphere and the resulting pressure causes the star to expand. As the star expands and cools, the degree of gas ionization decreases, resulting in an increase in photospheric transparency. With the star's radiation exerting less outward pressure on the photosphere, the star contracts and the cycle repeats. But this mechanism does not explain why Beta Cephei or ZZ Ceti stars pulsate, for the photospheres of those stars remain fully ionized throughout the pulsation cycle.

Subquantum kinetics, though, offers a very simple solution. Unlike nuclear fusion, genic-energy production *is* affected by a star's radial pulsation. Genic energy is produced throughout the star, not just at its core. As a star contracts and becomes hotter, its rate of genic-energy production necessarily increases throughout, which increases the star's luminosity; the increased outward pressure eventually brings the star's contraction to a halt and initiates its expansion. As it expands, the star becomes cooler, and its rate of genic-energy production decreases. This reduces the radiation pressure, which in turn induces the star to contract. This simple mechanism has the advantage that it works equally well for all types of pulsating stars.

Standard physics provides no clue as to where the supernova gets its energy. It is conventionally believed that a supernova occurs at the end of a star's lifetime, at a time when fusion reactions have exhausted almost all of the star's fuel. As the nuclear fires in its core flicker out, the star contracts and heats up. This fires up high-temperature, high-output fusion reactions that consume the last remnants of its fuel. The energy released from these reactions causes the star to expand severalhundred-fold, transforming it into a cool red supergiant. When these fusion reactions finally expire and no outgoing radiation is left to support its mass, the star is assumed to collapse inward upon itself. Then somehow, when the star has compressed itself to very small dimensions, this inward collapse is supposed to miraculously reverse and direct this same energy outward as a supernova explosion.

But this sudden reversal is at the crux of the problem. Nuclear fusion cannot supply the energy for this outward explosion, since as we have said, the star's nuclear reactions already have been totally exhausted. Moreover, the kinetic energy stored in the star's collapsing mass is not the answer, since that energy would have been absorbed by the core-squeezing process. It is also not known how the star would manage to reverse its collapse. Supernova explosions rocket the star's mass many light-years into space at velocities approaching 3 percent of the speed of light. Such feats require far greater amounts of energy than could possibly be available from its gravitational potential energy.

Given all of these inadequacies, it is not surprising that in 1987 when astronomers finally did observe a supernova relatively close by, in the Magellanic Cloud, they found a flaw in their collapse theory. The progenitor star was not a cool red supergiant, as expected, but a hot blue supergiant, called Sanduleak -69 202.[8] Since energy production was at a peak in this star, the expected collapse should never have occurred. The supernova phenomenon is easily explained by the idea that stars are partly powered by genic energy. Here, an increase in a star's internal temperature increases its rate of genic-energy production, which in turn leads to a further increase in temperature, and so on. Under circumstances where this energy is produced faster than it is able to escape, as in certain blue supergiant stars, energy production soars and ultimately explodes the star. This explosion demonstrated to modern astronomers something that the ancients knew long ago: in supercritical space, energy can be spontaneously created.

Like supernova explosions, galactic core explosions also pose great difficulty for current astronomical theory. Their prodigious energy outputs are often attributed to black holes swallowing large amounts of matter and spitting back a fraction of the swallowed tidbits as energy. But matter is almost always seen to be moving away from galactic cores, not toward them. Moreover, some galactic core explosions are so luminous that black holes cannot attract matter fast enough to produce the required energy outputs. One example is quasar 3C 279, an exploding galactic core that emits most of its radiant energy in the form of gamma rays. Its core radiates 10 million times as much energy as our entire galaxy, an outpouring so intense that an entirely new type of physical process is needed to account for it.

Hubble Space Telescope observations of fifteen distant quasars, disclosed in January 1995, dealt a damaging blow to the black hole theory. Astronomers found that, contrary to their expectations, eleven of the fifteen quasars were devoid of any surrounding material and hence had no matter available to fall into any hypothesized black holes. Yet these quasar galaxies were somehow producing intense radio emissions, indicating a huge energy output. John Bahcall, the Princeton University astronomer who headed the investigation, considered the findings "a giant leap backward in our understanding of quasars." He stated, "It is absolutely clear that we need to rethink how quasars shine."[9]

Again, genic energy holds the answer to the galactic core explosion mystery. The massive object at the center of each galaxy is an extremely dense, highly supercritical "mother star" spontaneously generating genic energy and matter at a prodigious rate. Nothing is consumed to produce this energy; as in stars, planets, novae, and supernovae, the energy springs into the physical world out of the underlying transmuting ether. At periodic intervals the core mother star flares up and enters a highly luminous mode. The core's high-energy output—unlike that of a supernova explosion, which transpires within a few days—is sustained over hundreds or thousands of years. As the galactic core expels matter from its immediate vicinity, the depth of its gravity well is reduced, and it becomes less supercritical. Its genic-energy production rate subsides, and it enters a temporary quiescent period.

Table 15.1 reviews some of the main points of the cosmology of subquantum kinetics. In overview, ancient creation myths, astrology, the Tarot, and the I Ching encode the basis for an astrophysics capable of solving many of the problems that trouble modern science today. It is interesting that by going back to the past we come face to face with a science of the future.

TESLA ENERGY WAVES

Model G's reaction-diffusion energy waves are actually composed of two traveling ether waves, one of X and one of Y, with the X wave lagging behind the Y wave by one-fourth of a wavelength. They are not in a reciprocal relation, as they would be when forming a particle's electrostatic energy field. Thus the X and Y waves together compose a traveling electric potential wave.

While this energy wave model reproduces all the optical and electromagnetic wave effects familiar to physicists, it is substantially different from the electromagnetic wave model that James Clerk Maxwell proposed during the nineteenth century and that is still in use today. Laboratory experiments performed by Panagiotis Pappas on the motion of the Pi-frame electrodynamic pendulum,[10] by Pappas and T. Vaughan on the Z-shaped microwave antenna,[11] and by Peter Graneau on the Ampère bridge and railgun experiments[12] confirm that Maxwell was incorrect in hypothesizing the existence of a *displacement current,* a quantity that plays a crucial role in his electric

TABLE 15.1

A Summary of Subquantum Kinetics Cosmology

1. The transmuting ether existed prior to the appearance of matter.

2. The transmuting ether is not characterized by expansion or contraction, although its etheron constituents are always in motion, diffusing throughout Euclidean space.

3. Our experience of time arises as a result of the continuous flux of etherons transmuting through our physical universe.

4. There was no "Big Bang." Creation began with the spontaneous materialization in space of individual subatomic particles. Particle materialization has been continuing at an exponentially increasing rate as matter creates more matter. Creation proceeds at its most rapid rate within massive bodies, and this rate depends on the value of the ambient gravitational potential field. The highest materialization rate occurs in the core of a galaxy.

5. Photons traveling through intergalactic space gradually lose energy over time and as a result become redshifted (the cosmological redshift effect).

6. Photons traveling through supercritical regions of space, such as within galaxies, gradually gain energy over time and as a result become blueshifted. Such spontaneously generated energy, genic energy, is produced in the interiors of all celestial masses. There are no such things as "black holes."

7. Under certain conditions genic-energy production within massive stars or galactic cores can enter an explosive energy generation mode. The resulting energy outburst can show up as a nova, supernova, or galactic core explosion. A dwarf elliptical galaxy gradually evolves into a spiral galaxy as gas is periodically expelled from its core by recurrent explosions.

and magnetic field equations.[13] Although Maxwell's equations accurately model the electromagnetic wave phenomenon, they are based on a fiction—the displacement current.

Nikola Tesla disagreed with Maxwell's theory that energy waves were transverse vibrations. Early in the twentieth century, Tesla proposed that light waves instead travel forward by longitudinally compressing and rarefying the ether, in much the same way that sound waves compress and rarefy the air medium. The X-Y electric potential waves of Model G in some ways resemble Tesla's ether sound waves in that they too involve alternating regions of high and low ether concentration.

Model G's energy waves can produce the same observable effects as Maxwell's transverse waves. For example, electric charges oscillating from side to side in the dipole antenna of a radio transmitter would produce X-Y waves having *transverse* electric potential gradients with a distinct direction of polarization. These transverse gradients would induce transverse forces on electric charges in a distant radio receiver antenna, causing them to oscillate. This is the same effect that Maxwell claimed his waves would have. Such energy waves are usually termed Hertzian waves, after Heinrich Hertz, the German physicist who discovered radio wave transmission in 1888. It is this kind of wave that physicists and engineers speak of when they refer to electromagnetic radiation, a category that includes microwaves, infrared radiation, light waves, ultraviolet rays, X rays, and gamma rays.

Model G also predicts the emission of another kind of energy wave, one that is non-Hertzian. Such waves would be radiated by a monopole antenna rather than a dipole antenna, a single sphere rather than a linear wire. If alternately charged and discharged, such an antenna should radiate pure longitudinal electric potential waves. Unlike Hertzian waves, these so-called *Tesla waves* or *scalar waves* would carry no transverse potential gradients and hence would be unpolarized, so conventional radio receivers would be unable to pick them up. Certain devices such as the Bendini detector, however, have reportedly received them. Whereas classical electromagnetic wave theory fails to predict the existence of Tesla waves or account for their behavior, Model G readily explains the phenomenon. Compared with conventional physics, subquantum kinetics provides the basis for a more encompassing theory of energy wave behavior.

A NEW THEORY OF FORCE

The ether physics of ancient mythology leads to a unified theory of force that avoids the problems that plague conventional force field theory. According to subquantum kinetics, force intensity should scale according to the steepness (or gradient) of the energy potential field. Although classical field theory adopts the same gradient relation, the two theories regard force and potential in fundamentally different ways.

An electric potential field exerts its force on a charged particle by stressing or distorting the particle's etheron concentrations, altering the concentrations by a greater amount on the side facing the field source. The particle responds to this stress or force by readjusting its wave pattern. In so doing, it accelerates and continues to accelerate as long as it is in a potential gradient. Depending on the polarity of the electric field and target particle, whether X or Y (Castor or Pollux) is dominant, the particle will move either up or down the field gradient—like polarities repelling and unlike polarities attracting.

In effect, the Model G ether exerts its force through the operation of Le Chatlier's principle: if a system in dynamic equilibrium is subjected to a stress, then the system will change so as to relieve that stress. Although originally developed to explain the equilibration of chemical systems, this principle explains equally well how concentration gradients in a reaction-diffusion ether induce force and motion. A gravity potential gradient (*G*-field) would induce gravitational force the way an electric potential gradient induces electrostatic force. In the case of an electrostatic or gravitational potential field that decreases inversely with increasing distance from the center of a charged particle, Model G predicts that if a unit charge or unit mass is placed in such a field, it should feel a force whose magnitude decreases according to the inverse square of its distance from the particle, consistent with experimental observation.*

Subquantum kinetics and classical physics view force and potential energy in two very different ways. Classical field theory assumes that force fields physically exist. This conception may be traced to the nineteenth-century idea that force fields are stresses impressed on the mechanical ether. Although physicists later rejected the notion of an ether, they continued to regard forces as physically real entities, representing them as mathematical vector quantities. They regarded energy potential as a more abstract property, giving it a secondary status to force. It was defined in terms of force and distance as the ability of a unit charge or mass to induce a body to move by exerting a force on it over a measured distance. The situation is just the reverse in subquantum kinetics, where energy potential fields (etheron concentration gradients) are the cause of motion. Force, on the other hand, is merely an effect produced by a potential field gradient, a perturbation induced on a particle at a given instant. Force fields in subquantum kinetics are virtual quantities; they have no independent existence.

It is understandable why classical physicists favored force over energy potential in constructing their field theories. Force is something that is directly experienced through the physical senses and is something easily detectable by simple instruments. Science progressed a step closer to the field-potential view in the early part of the twentieth century when the Austrian physicist Erwin Schroedinger formulated the theory of quantum mechanics. Quantum mechanics, like subquantum kinetics, regards energy potentials as the real physical actors and forces as derived effects. Quantum mechanics, however, is merely an abstract mathematics for representing these field interactions without explaining how they occur. Subquantum kinetics goes several steps further in that it explains how energy potentials arise and generate forces on distant objects.

We have seen from the above how the X-, Y-, and G-fields generated by material particles exert electrostatic, magnetic, and gravitational forces on distant bodies. In addition, these dissipative-structure particles can also exert short-range forces on one another when their respective wave patterns interlock. For example, when two

---

*While the radial etheron flux also varies according to the inverse square of the distance from the particle's center, it is the particle's potential field, not its etheron flux, that is directly responsible for causing force. This force mechanism may be more apparent with the aid of a computer. By running a computer simulation of Model G, it should be possible to generate a three-dimensional image of a subatomic particle and observe its motion in response to a superimposed concentration gradient. Simulations of this sort should help eliminate much of the long-standing mystery behind the phenomenon of force field action at a distance.

particles of similar wavelength and matter polarity as a proton and neutron are separated by just one-half wavelength, the high Y/low X core of one particle becomes trapped in the low Y/high X sheath that surrounds its partner's core (figure 15.4). While the repulsive action of their cores prevents the particles from approaching closer together, the attractive force exerted on each particle by the outer wall of its partner's sheath prevents the particles from separating. In this way, a proton and neutron could combine to form a deuterium nucleus (heavy hydrogen). Or two protons could mutually interlock with a neutron to form a helium-3 nucleus, as their nuclear binding is strong enough to overcome the mutual electrostatic repulsion of the protons. By interlocking in this manner, increasing numbers of protons and neutrons combine in various ways to form the range of atomic nuclei that make up the periodic chart of elements.

The explanation given above does not require the ad hoc introduction of a special "nuclear force field" to mediate this binding. Nuclear binding instead emerges as a natural consequence of the wavelike configuration of a particle's electric field, which in turn is merely a manifestation of the implicit order inherent in the underlying ether reactions. In other words, these mutually bonding subatomic structures and their energy potential fields emerge as a direct consequence of the basic ether transmutation equations. From this standpoint subquantum kinetics may well be the unified field theory for which physicists have long been searching.

GRAVITY AND ANTIGRAVITY

In their attempts to unify field theory, physicists have sought a way in which gravitational and electrical forces might somehow be connected. As we have seen in the last chapter, present-day theories attempting to explain gravitational and electrodynamic phenomena, general relativity and quantum electrodynamics, are fundamentally incompatible with one another in that quantum electrodynamics predicts absurdly large curvatures for the geometry of space. Classical field theory also does not offer much hope for unification since Newton's theory of gravitation and Maxwell's theory of electrodynamics are isolated from one another. Nevertheless, subquantum kinetics' Model G offers a way to achieve this long-sought goal. Its equations provide a natural connection between electricity and gravitation.

Although the Model G ether reactions proceed primarily in the forward direction, very limited reverse reactions can also occur. In fact, the *reverse* reaction $G \leftarrow X$, which transforms X-ons back into G-ons, turns out to be quite important from the standpoint of gravitation. Although it is extremely feeble in comparison to Model G's forward reactions, it provides an explanation for the generation of the gravity field of a subatomic particle. Without this reverse reaction, G would remain uniformly distributed throughout space at its steady-state value, whether or not a material particle was present. This reverse reaction, in essence, is what links the

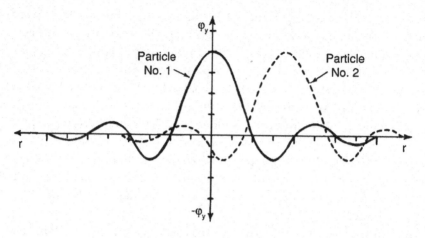

Figure 15.4. A proton and neutron of similar polarity shown mutually interlocked to form an atomic nucleus.

electric and gravitational fields. It indicates that a positive or negative X-field (electric potential) should induce a positive or negative G-field (gravity potential), or that electric charge should generate gravitational mass.

Since electric charge comes in two polarities, Model G predicts, accordingly, that a particle should have either a positive or a negative gravitational mass (G-well or G-hill). The G-on production rate surplus or deficit at the center of the particle represents the particle's gravitational mass, since it is this production rate discrepancy that generates the particle's gravity field.* Subatomic particles, which have a G-on production rate deficit and associated gravity well, are assigned a positive gravitational mass polarity. Those having a G-on production rate surplus and associated gravity hill are assigned a negative mass polarity. Electrically neutral matter, made up of equal numbers of plus and minus charges, would have a net positive gravitational mass, since Model G predicts that the gravitational mass of a proton should be slightly greater than that of an electron.[14] The electron's G-hill only partially cancels out the proton's G-well, and as a result, an electrically neutral atom is left with

---

*More specifically, we may call this the particle's active gravitational mass, since it actively generates its gravity field. This is distinguished from the particle's passive gravitational mass, which quantifies the particle's tendency to respond to an externally imposed gravity field. In addition to these two gravitational mass quantities, there is also inertial mass, the quantity that represents a particle's tendency to resist motion. The terms *active mass* and *passive mass* were developed by cosmologist Hermann Bondi ("Negative mass in general relativity," *Reviews of Modern Physics* 29 [1957]: 432–28). In classical physics, all three mass quantities are represented by the same mass symbol, *m*, which can sometimes lead to confusion. Similar active-passive categories may also be used to describe electrostatic charge to distinguish a particle's electrostatic field-generating capabilities from its field-responding capabilities.

a residual matter-attracting G-well. Particles producing G-wells would attract neutral matter, whereas particles producing G-hills would repel neutral matter.*

If it is indeed true that positively charged particles have positive mass and negatively charged particles negative mass, it may be possible to artificially generate a gravitational potential field simply by placing a high-voltage charge across the plates of an electrical capacitor. Since positively charged particles are net G-on consumers and negatively charged particles are net G-on producers, this electric potential difference would establish a gravitational potential gradient between the capacitor's positive and negative plates. This gradient, in turn, would exert a force on the material situated between the plates, tugging the capacitor toward its positively charged side.[15] If the capacitor is turned so that its positively charged side faces up, this force would partially oppose the pull of gravity.

Such an "electrogravitic" effect was first demonstrated in high-voltage experiments conducted during the early 1920s by the American physicist and inventor Thomas Townsend Brown.[16] Brown got best results when he used capacitors made with massive high-k dielectrics that were both massive and able to store large quantities of electrical charge. In his early experiments he demonstrated that he could induce a 1 percent loss of weight by charging capacitors with electric potentials of the order of 100,000 volts. He took out many patents on the concept and by 1958 perfected a disk-shaped apparatus that, when energized with between 50,000 and 250,000 volts, exhibited a substantial lifting force.[17] In one set of tests in France he demonstrated that this electrogravitic force could be sustained even in a vacuum of one-billionth of an atmosphere, thereby demonstrating that the effect was not a result of electrostatic ion propulsion. Brown also staged a demonstration in which he supplied 150,000 volts to a pair of three-foot-diameter electrostatic disks. The results were so impressive that the subject immediately became classified.[18]

A recently declassified aviation industry intelligence report indicates that as early as 1955 the U.S. government was planning a classified research program to develop a Mach 3 disk-shaped combat plane based on Brown's electrogravitic effect.[19] Both this report and magazine articles from the late 1950s name several major aerospace companies that either were conducting or had plans to conduct electrogravitics research.[20] Inspired by articles about Brown's work, a number of independent researchers, working in their spare time in makeshift high-voltage laboratory facilities, have also carried out experiments that successfully duplicated some of

---

*To fit the standard view of gravity, a neutron should have a positive gravitational mass comparable to that of a proton-electron pair; that is, it should produce a G-well whose depth is comparable to the residual G-well produced by an electrically neutral hydrogen atom. Since Model G predicts that electrostatic and gravitational fields are coupled, however, this would imply that a neutron should exhibit a very small positive electric charge, perhaps a billion times smaller than that of a proton. The presence of such a charge might be detected by conducting sufficiently sensitive measurements.

Brown's results. Most of their work, however, has remained unpublished. One professional and well-documented study was carried out in 1991 by physicist Robert Talley of Veritay Technology, Inc., with funding from the U.S. Small Business Innovation Research Program under the sponsorship of the Air Force Systems Command Propulsion Directorate at Edwards Air Force Base.[21] Talley conducted vacuum chamber experiments similar to those Brown carried out in France, but using much lower voltages (5 to 19 kilovolts rather than the 100 to 200 kilovolts). He concluded that three of his test runs registered the presence of an "anomalous propulsive force" that could not be attributed to normal electrostatic effects, and he suggested that future research be conducted to further investigate this anomalous force.

When the great variety of evidence is considered—patents; research papers by Brown and others; numerous demonstrations of electrogravitic rigs to top military officials and aerospace company executives; declassified documents describing secret government funding of antigravity research; rumors of actual test flights of antigravity vessels—one begins to suspect that electrogravitics may not be mere science fiction. If Brown's findings are valid, they could produce a revolution in space propulsion and energy generation technologies. Positive evidence for an electrogravitic effect would also be highly significant from the standpoint of physical theory. For example, it would pose a serious challenge to general relativity, which has long claimed that matter, regardless of its electrical charge, produces only an attractive gravitational force. Antigravity, the ability to artificially engineer the gravity field, is an idea totally foreign to Einstein's theory.

Brown's findings also contradict the grand unified theory of particle physics, which claims that gravity is unified with electromagnetism and other forces only at very high energies (10^{28} electron volts), beyond the reach of any manmade particle accelerator. Even the ill-conceived "superconducting supercollider," unwittingly financed by the U.S. government, would have fallen short by fourteen orders of magnitude in producing the particle energies required to test this theory. Those billions of dollars would have been far more prudently spent on investigating the astounding antigravity implications of electrogravitics.

In summary, Model G of subquantum kinetics truly forms the basis for a unified field theory. Its set of five simple transmutation relations, whose basic features are described in ancient myths, are able to model the nucleation of individual subatomic particles and the ultimate generation of an entire physical cosmos. In one elegant sweep, this one reaction system produces fields capable of exerting electrostatic, magnetic, gravitational, and nuclear forces, achieving a unitary synthesis surpassing anything devised by today's relativity theorists (see table 15.2). By accounting in a consistent fashion for both the very small and the very large, the ether cosmology of ancient times has done what modern physics has failed to do.

TABLE 15.2

A Comparison of Conventional Physics to Subquantum Kinetics

A. Philosophy and Metaphysics

CONVENTIONAL PHYSICS	SUBQUANTUM KINETICS
1. Conventional physics is based on positivism. Does not recognize the existence of an unobservable spiritual realm. Contradicts mystical teachings.	1. Subquantum kinetics (SQK) is based on inference from general systems principles. Admits the possibility of an unseen realm. Harmonizes with mystical teachings.
2. Is the only science that hangs on to the mechanical, closed-system model.	2. Adopts the open-system model similar to that eventually adopted in other sciences.
3. Is made up of a conglomeration of theories that are sometimes incompatible with one another.	3. Consists of a unitary theory.

B. Quantum Theory and Field Theory

CONVENTIONAL PHYSICS	SUBQUANTUM KINETICS
1. Special and general relativity are disproven by the Sagnac, Silvertooth, and Ampère force law experiments.	1. The ether concept is supported by the Sagnac, Silvertooth, and Ampère force law experiments.
2. Special relativity suffers from the twin clock paradox and the light-source-velocity paradox.	2. This problem does not arise in SQK.
3. Classical field theory is plagued by the field-particle dualism.	3. This problem does not arise in SQK.
4. Classical field theory suffers from the infinite energy absurdity.	4. This problem does not arise in SQK.
5. The wave packet model used in quantum mechanics to model a subatomic particle has the problem of gradually spreading out and dissipating. It cannot model a particle at rest.	5. This problem does not arise in SQK. Field patterns composing a subatomic particle remain structurally coherent over time, even when the particle . is stationary.
6. Is plagued by the nonintuitive wave-particle dualism concept.	6. This problem does not arise in SQK. Subatomic particles naturally incorporate wave aspects in their structure.
7. Does not explain what charge and mass are, or how they generate electric and gravitational fields.	7. SQK explains what they are and how they generate such fields.

B. Quantum Theory and Field Theory (cont'd)

CONVENTIONAL PHYSICS	SUBQUANTUM KINETICS
8. Fails to account for experiments showing that gravitational fields may be electrostatically induced. Considers antigravity an impossibility.	8. SQK explains such experiments. Allows the possibility of antigravity propulsion.
9. Maxwellian electrodynamics fails to explain the induction of Tesla waves, nonpolarized longitudinal energy waves.	9. SQK explains the induction of both transversely polarized Hertzian waves and Tesla waves.
10. Advocates the nonintuitive notion that natural events are inherently indeterminate.	10. Retains the commonsense notion of causality. Nonlocal superluminal interac-explain the EPR experiment results.
11. Accepts the Copenhagen interpretation of the uncertainty relation.	11. Maintains that the uncertainty relation is merely a statement about the limitations inherent to quantum level observation.
12. Quantum electrodynamics and general relativity, two pillars of conventional field theory, contradict one another, a problem known as the cosmological constant conundrum.	12. This problem does not arise in SQK. All fields (electrostatic, magnetic, gravitational, nuclear) are encompassed within a single internally consistent theory.
13. General relativity is fatally flawed in that it allows the the formation of cosmologically disruptive naked singularities.	13. This problem does not arise in SQK.

C. Cosmology and Astrophysics

CONVENTIONAL COSMOLOGY AND ASTROPHYSICS	SUBQUANTUM KINETICS
1. The big bang theory proposes the counterintuitive notion that the universe emerged out of a state of nonexistence.	1. This problem does not arise in SQK. Proposes that physical form emerged from a preexisting ether substrate.
2. Fails to explain how subatomic particles originate. It merely postulates that they form out of the vacuum.	2. SQK explains in detail how subatomic particles arise from subquantum fluctuations in the ether.
3. Fails to explain why our universe is made up of matter as opposed to antimatter.	3. SQK predicts a matter/antimatter bias to particle materialization.

C. Cosmology and Astrophysics (cont'd)

CONVENTIONAL COSMOLOGY AND ASTROPHYSICS	SUBQUANTUM KINETICS
4. Introduces the ad hoc assumption that the universe is expanding, to account for the cosmological redshift phenomenon.	4. SQK naturally predicts a tired-light cosmological redshift effect without introducing any ad hoc assumptions.
5. The expanding-universe model fails to make a good fit to astronomical data on four cosmology tests.	5. The tired-light static-universe cosmology makes a superior fit to astronomical data on all four cosmology tests.
6. Conventional cosmology fails to explain observations of galaxies with redshifts greater than 4.0.	6. SQK accounts for the existence of galaxies at redshifts many times higher than 4.0.
7. Conventional physics fails to explain why the jovian planets and brown dwarfs fall along the M-L relation for low mass stars, attributes this to chance	7. SQK explains this conformance by predicting that planets, brown dwarfs, and low mass stars are similarly powered by genic energy.
8. Fails to explain the source of the excess heat coming from the Earth's core.	8. SQK attributes the excess heat from the Earth's core to genic energy.
9. Fails to account for the inflection (at 0.45 solar masses) in the stellar luminosity function and for the accompanying upward bend in the stellar mass-luminosity relation.	9. SQK explains this inflection and upward bend as arising from the onset of fusion energy production and the formation of a radiative core at the star's center.
10. Fails to adequately explain the phenomenon of stellar pulsation.	10. The SQK genic-energy prediction explains this phenomenon.
11. Fails to explain the source of energy powering supernova explosions and why supernovae arise from blue giant stars.	11. The SQK genic-energy prediction explains this phenomenon.
12. Fails to account for the energy source powering galactic core explosions.	12. The SQK genic-energy prediction explains this phenomenon.
13. General relativity predicts that massive, highly luminous galactic cores should exist as matter-consuming black holes, but this prediction fails to conform with observation.	13. SQK predicts that black holes should not form, that galactic core should instead consist of very dense stellar bodies that continually create matter and energy. This prediction is supported by observation.

16

BACK TO THE FUTURE

AS THE END OF THE SECOND MILLENNIUM APPROACHES, long-cherished theories about our universe are crumbling before us. The crystalline sphere that has long imprisoned us is shattering, revealing a boundless expanse of space and an infinity of time. Equipped with increasingly sophisticated electronic sensors and advanced control systems for atmospheric aberration correction, our telescopes reveal a new cosmic landscape. Galaxies no longer rush away from us at incredible speeds, but instead float gently in the waters of the cosmos, like so many glittering lilies on a vast lake. In this new view of things, the mythical Big Bang has become an imaginative thought of a past era. Creation, which once was thought to occur in the briefest of possible moments, now proceeds over countless billions of years, its very earliest stirrings extending back possibly even trillions of years. We recognize that the matter composing the Milky Way and the other galaxies around us arose through a process of continuous creation, each galaxy's family tree stemming from a single self-created subatomic particle lying within its matter-expelling core.

It is impressive that creation myths and secret lore of ancient times describe a cosmology that finds validation in sophisticated astronomical observations secured only in the past few decades. Armed with a vast array of modern technology, ranging from space-borne gamma-ray telescopes to supercomputers and laser interferometers, our civilization is just now arriving at a level of understanding about the physical world comparable to that of these early mythmakers.

As we review the concepts presented in these ancient metaphorical "texts," we are confronted with advanced ideas at every step, indicating that the originators of these works had an intimate understanding of how open reaction systems are able to produce ordered patterns. This ancient science portrays modern scientific concepts such as entropy, order through fluctuation, circular causality, positive feedback, critical mass, spontaneous symmetry breaking, bifurcation, matter/antimatter creation asymmetry, wave pattern self-stabilization, stable periodic states, and sequential quantum jumps to successive steady states. In the creation myth of Atlantis, these ancient theoreticians even present a two-dimensional diagram of a wavelike dissipative structure, something that modern scientists have discovered only with the help of sophisticated computers.

Astrology, one of the boldest presentations of this ancient physics, writes these principles of cosmic creation across the sky in imaginative pictures 15 to 45° of arc in width, visible only at night when inquisitive minds idly peer into the vast reaches of space. The placement of these signs in the heavens indicates considerable scientific aptitude. Was it just good fortune that Castor was chosen as the lead star of Gemini, the constellation that depicts the principle of duality? Or did these stellar artists know that of all the stars in the zodiac, Castor is truly the king of binaries, something modern astronomers have discovered only through years of observation with powerful telescopes?

We must also wonder about their decision to place Scorpio and Sagittarius on either side of the Galactic center, two signs that specifically designate the emergence of matter and energy into physical manifestation. This core region is known to be the most energetic site in the Galaxy. It is also impressive that the "pointers" of these two zodiac signs accurately designate the location of the Galactic center, a region that is hidden from direct view by dense clouds of interstellar dust. Around 15,800 years ago, when Sagittarius's arrow pointer was aligned with its target, the heart of the Scorpion, its trajectory passed within 0.7 full moon diameters of the Galactic center. A designation more accurate than this did not come until after the development and construction of large radio telescopes.

The placement of the Virgo constellation is also noteworthy. Can it be just chance that Virgo, which symbolizes the virgin birth concept, happens to be pointing to the center of the largest and densest cloud of galaxies in the heavens and indicating the equator of the local galaxy supercluster as she scatters her "cosmic seeds"? To purposely make such significant alignments a civilization would not only have required access to a powerful telescope, but also would have needed to know that these unusual luminous objects that Virgo contains in such great abundance are not mere stars but entire galaxies of stars. Modern astronomy did not acknowledge the existence of other galaxies until the beginning of the twentieth century.

Moreover, the physics conveyed by these ancient myths and esoteric lore is itself something to stir wonder. Subquantum kinetics, the modern counterpart of this ancient cosmic science, avoids many of the problems plaguing conventional physics and astronomy, and its cosmology fits observational data far better than the big bang theory. Yet subquantum kinetics is still at a relatively early stage of development. There is still much work to be done to explore its various ramifications. It will probably be many years before physicists and cosmologists, who until now have been thinking in mechanical and relativistic terms, will dare venture into this new and unfamiliar territory. Perhaps an entirely new generation of scientists must be born before subquantum kinetics will be adopted as the new physics of the twenty-first century.

Meanwhile the pressing question still remains: How did such advanced scientific concepts come into the hands of ancient civilizations? This physics may be traced back to the dawning of civilization in Egypt and Mesopotamia. Vestiges of it seen in ancient Indian, Chinese, and Polynesian mythology probably have equally ancient roots. This science may have come to these ancient civilizations from some much more ancient time, just as ancient writings attest. If this knowledge originated from well before the time of early Egypt, then our traditional concepts of prehistoric humans may need to be drastically changed. Could there be some truth to the legends stating that there once existed a noble prehistoric civilization of whose existence we have no written record because it was destroyed by a sudden global catastrophe?[1]

Perhaps this ancient science was developed by a civilization that left behind few physical traces of its existence, and that concentrated on developing the inner faculties of the human mind rather than on building an industrial economy. It is said that the human unconscious, the wellspring of creative ideas, is far wiser than the conscious mind, and that each of us knows more than we might think. Could a few very gifted individuals living in a relatively nontechnical prehistoric culture have devised this creation metaphysics by relying solely on their intuitive faculties? Some clairvoyants claim the ability to view directly the structure of subatomic matter. Could psychics in ancient times have developed similar powers and through a series of such observations pieced together an entire cosmogenic science? If so, how could they be sure of the accuracy of their insights without checking them against solid experimental observations every step of the way?

The existence of certain artifacts requires a reevaluation of the technological sophistication of ancient cultures. For example, the complex gear mechanism retrieved from a first-century B.C.E. Aegean shipwreck has been determined to be part of a mechanical computing device used to predict the positions of the Sun, Moon, and planets.[2] The two-thousand-year-old electric storage batteries unearthed from an Iraqi village south of Baghdad are another surprise from the past.[3] When these six-inch-high clay jars are filled with an acid solution, their protruding copper and iron electrodes generate electric potentials of up to two volts.

Strange artifacts of unknown origin have been found that date much further back in time. One example is the metallic bell-shaped vase dynamited from solid pudding stone in Dorchester, Massachusetts, in 1851.[4] The vessel is composed of an alloy of brass, zinc, iron, and lead and is decorated with beautiful floral designs made of pure silver inlay. Although not an instrumental device, its elegant craftsmanship and seeming prehistoric age give us reason to question traditional concepts about the history of our race.

The ruins of immense structures found at various sites around the world provide additional evidence that a highly advanced civilization once existed on our planet. One example is the ancient stone fortress of Baalbek, situated forty miles east of Beirut in the mountains of Lebanon. Local tradition considers the fortress to be antediluvian.[5] It consists of an enormous dressed stone platform that rises fifty feet above the neighboring ground level and is surrounded on three sides by a wall composed of huge rectangular blocks of granite. The blocks are fitted together with such precision that a sheet of paper cannot be passed between them. On the western side, this 35-foot-high cyclopean wall is approximately 193 feet in length. It is composed of six blocks of stone, each 33 feet long and 14 feet high, overlain by three enormous stones, each measuring approximately 65 feet in length, $14^{1}/_{2}$ feet in height, and 12 feet in thickness and weighing about 750 tons. How the blocks were moved over the half-mile distance from their quarry to the fortress still eludes archaeologists and engineers. By one estimate it would have taken the combined efforts of forty thousand men to move the larger of these stones. Curiously, there are no traces of a roadbed leading from the quarry.

This edifice is not an exception. Similar cyclopean structures are found in the Peruvian Andes. The pre-Incan fortress at Ollantaytambo has walls constructed from tight-fitting stone blocks ranging in weight from 150 to 250 tons.[6] These were somehow transported from a quarry located on another mountaintop seven miles away, the descent from which was impeded by a river canyon with one-thousand-foot vertical rock walls. This had to be accomplished at an altitude of two miles above sea level, where atmospheric pressure is less than 60 percent of normal.

The ancient city at Tiahuanaco, located at an altitude of thirteen thousand feet on the shores of Lake Titicaca, is also a puzzle for archaeologists. The site is of unknown age, although local tradition claims it is antediluvian. The walls of its temple complex are constructed from stones weighing sixty tons each. Blocks weighing one hundred to two hundred tons each were used in constructing the 440-by-390-foot stone platform that forms the foundation for the "Temple of the Sun." These immense stones were somehow transported from quarry sites over thirty miles away.[7] Clearly, the hostile alpine climate and oxygen-deficient atmosphere pose problems for anyone attempting to explain the construction of these edifices simply in terms of muscular effort. Like the ancient megalithic structures at Baalbek, the means by which these Andean edifices were constructed still remains a mystery. Perhaps these enduring constructions are the visible remains of a prehistoric civilization that possessed technologies far in advance of anything usually attributed to preindustrial societies.

Yet another possibility is that this knowledge came to Earth from some other star system. Each year there are reports of numerous sightings of alien spacecraft, despite official government denials and efforts by certain intelligence agencies to suppress knowledge of these visitations.[8] Some researchers have suggested that extraterrestrials have been visiting our solar system off and on for many millennia.[9] If so, perhaps an advanced race visited our planet at some time in the prehistoric past and shared with our ancestors a summary of their knowledge of physics and cosmology. Seeing that the people of that time did not have the background to fully understand what they were being told, these visitors could have encoded the basics of this science in allegorical myths and lore and instructed their hosts to transmit these from generation to generation in the hope that at some future date humankind would be ready to comprehend their contents. Perhaps it is no accident that this science forms the basis for a new theory of gravity that might one day enable us to build craft capable of traveling to the stars.

Finally, what about the catastrophe connection? Did an outburst of cosmic rays from the core of our galaxy wash over our planet thousands of years ago, and did that event lead to the near-extinction of the human race? Was the real purpose behind encoding this ancient physics to help future generations understand the cause of a global cataclysm and to warn humanity that a similar event could recur at some future date? The prospect of impending annihilation might very well have encouraged a prehistoric civilization to develop a scientific understanding of their cosmos. Or, seeing that the disaster had caught Earth's citizens by surprise, could galactic visitors have shared some of their knowledge with the survivors with the hope that future generations might avoid the same fate? Such an ET connection could explain why certain pulsating radio beacons located thousands of light-years from our sun appear to be conveying a galactic core explosion warning message identical to that encoded in our zodiac.[10] If anything, the discovery of advanced science in ancient myths and lore should give us reason to pause and take a broader view of modern achievements, to realize that we may not be the first scientifically advanced culture to inhabit this planet.

NOTES

CHAPTER 1 A LOST SCIENCE REDISCOVERED

1. Plato, *Timaeus*, in *The Collected Dialogues of Plato*, ed. E. Hamilton and H. Cairns, trans. B. Jowett (Princeton, N.J.: Princeton University Press, 1961), section 23.
2. R. A. Schwaller de Lubicz, *Sacred Science: The King of Pharaonic Theocracy*, trans. André Vandenbroeck and Goldian Vandenbroeck (Rochester, Vt.: Inner Traditions, 1961, 1982), pp. 86–87.
3. Ibid.
4. Ibid., p. 117.
5. J. A. West, *The Serpent in the Sky: The High Wisdom of Ancient Egypt* (Wheaton, Ill.: Quest, 1993), p. 1.
6. A. Zaikin and A. N. Zhabotinskii, "Concentration wave propagation in two-dimensional liquid-phase self-oscillating system," *Nature* 225 (1970): 535–37.
7. P. A. LaViolette, "A reaction-diffusion model of space-time" (paper presented at a workshop on nonlinear chemical systems chaired by I. Prigogine, Austin, Texas, March 1980); "An introduction to subquantum kinetics," 3 parts, *International Journal of General Systems, Special Issue on Systems Thinking in Physics* 11 (1985): 281–345; *Subquantum Kinetics: A Systems Approach to Physics and Cosmology*, 2nd edition (Schenectady, N.Y.: Starlane Publications, 1994, 2003); www.etheric.com.
8. I. Prigogine, G. Nicolis, and A. Babloyantz, "Thermodynamics of evolution," *Physics Today* 25, no. 11 (1972): 23–28; 25, no. 12 (1972): 38–44.
9. C. G. Jung, *The Archetypes and the Collective Unconscious*, trans. R. F. C. Hull (New York: Pantheon Books, 1959).

CHAPTER 2 PROCESS AND ORDER

1. R. T. Rundle Clark, *Myth and Symbol in Ancient Egypt* (New York: Thames & Hudson, 1959), p. 142.
2. Ibid., p. 115.
3. L. von Bertalanffy, General System Theory: Foundations, Development, Applications (New York: Braziller, 1968), pp. 145–46; B. Hess, "Fliessgleichgewicht der Zellen," *Dt. Med. Wschr* 88 (1963): 668–76.
4. Bertalanffy, *General System Theory*, p. 147.
5. I. Prigogine and I. Stengers, *Order out of Chaos: Man's New Dialogue with Nature* (New York: Bantam Books, 1984), pp. 286–87.
6. Rundle Clark, *Myth and Symbol*, p. 111.
7. Ibid., pp. 114–15.
8. R. A. Schwaller de Lubicz, *Sacred Science* (Rochester, Vt.: Inner Traditions, 1961, 1982), pp. 146–47.
9. L. von Bertalanffy, *Theoretische Biologic* (Ann Arbor, Mich.: J. W. Edwards, 1948); R. Delay, "Introduction à la thermodynamique des systèmes ouvertes," *Academic Royale de Belgique, Bulletin classe de sciences* 53 (1929).
10. Rundle Clark, *Myth and Symbol*, pp. 107–108, 121–22.
11. Ibid., pp. 124–25.

12. W. Buckley, *Sociology and Modern Systems Theory* (Englewood Cliffs, N.J.: Prentice-Hall, 1967); F. E. Emery, ed., *Systems Thinking* (Baltimore: Penguin Books, 1969).
13. W. Gray, "Understanding creative thought processes: An early formulation of the emotional-cognitive structure theory," *Man-Environment Systems* 9 (1979): 3–14; P. A. LaViolette, "Thoughts about thoughts about thoughts: The emotional-perceptive cycle theory," *Man-Environment Systems* 9 (1979): 15–47; "The thermodynamics of the 'aha' experience," *Proceedings of the 24th Annual Meeting of the Society for General Systems Research* (January 1980), pp. 460–72, reprinted in *General Systems Theory and the Psychological Sciences*, vol. 1, ed. W. Gray, J. Fidler, and J. Battista (Seaside, Calif.: Intersystems Press, 1982), pp. 275–87.
14. P. Davies, *The Runaway Universe* (New York: Penguin Books, 1980), p. 197.

CHAPTER 3 THE NEW ALCHEMY

1. L. von Bertalanffy, *Modern Theories of Development* (New York: Harper Torchbook, 1961).
2. L. von Bertalanffy, *General System Theory: Foundations. Development, Applications* (New York: Braziller, 1968), pp. 55–56, 83–84; L. von Bertalanffy, *Robots, Men and Minds* (New York: Braziller, 1967); L. von Bertalanffy, *A Systems View of Man*, ed. P. LaViolette (Boulder, Colo.: Westview Press, 1981); M. Davidson, *Uncommon Sense: The Life and Thought of Ludwig von Bertalanffy, Father of General Systems Theory* (Los Angeles: Tarcher, 1983); E. Laszlo, *The Systems View of the World* (New York: Braziller, 1972); E. Laszlo, ed., *The Relevance of General Systems Theory* (New York: Braziller, 1972).
3. P. Glansdorff and I. Prigogine, *Thermodynamic Theory of Structure, Stability, and Fluctuations* (New York: Wiley, 1971); R. Lefever, "Dissipative structures in chemical systems," *Journal of Chemical Physics* 49 (1968): 4977–78; R. Lefever and G. Nicolis, "Chemical instabilities and sustained oscillations," *Journal of Theoretical Biology* 30 (1971): 267–84; I. Prigogine, G. Nicolis, and A. Babloyantz, "Thermodynamics of evolution," *Physics Today* 25, no. 11 (1972): 23–28; 25, no. 12 (1972): 38–44; G. Nicolis and I. Prigogine, *Self-Organization in Nonequilibrium Systems* (New York: Wiley-Interscience, 1977); I. Prigogine and I. Stengers, *Order out of Chaos* (New York: Bantam Books, 1984); J. F. G. Auchmuty and G. Nicolis, "Bifurcation analysis of nonlinear reaction-diffusion equations: I. Evolution equations and the steady state solutions," *Bulletin of Mathematical Biology* 37 (1975): 323–65.
4. Auchmuty and Nicolis, "Bifurcation analysis of nonlinear reaction-diffusion equations," 323–65.
5. R. A. Schwaller de Lubicz, *Sacred Science* (Rochester, Vt.: Inner Traditions, 1961, 1982), pp. 7, 47.
6. Prigogine, Nicolis, and Babloyantz, "Thermodynamics of evolution," pp. 23–28, 38–44.

CHAPTER 4 THE TRANSMUTING ETHER

1. A. N. Whitehead, *Science and the Modern World* (New York: Free Press, 1925); *Process and Reality* (New York: Macmillan, 1929).
2. P. A. LaViolette, "A reaction-diffusion model of space-time" (paper presented at a workshop on nonlinear chemical systems chaired by I. Prigogine, Austin, Texas, March 1980); "An introduction to subquantum kinetics," 3 parts, *International Journal of General Systems. Special Issue on Systems Thinking in Physics* 11 (1985): 281–345; *Subquantum Kinetics: A Systems Approach to Physics and Cosmology*, 2nd edition (Schenectady, N.Y.: Starlane Publications, 1994, 2003).
3. D. A. Hyland, *The Origins of Philosophy* (New York: Putnam, 1973), p. 163, Heraclitian Fragment No. 30.
4. W. Y. Evans-Wentz, *Tibetan Yoga and Secret Doctrines* (London: Oxford University Press, 1958), p. 17.
5. H. Zimmer, *Myths and Symbols in Indian Art and Civilization* (Princeton, N.J.: Princeton University Press, 1946, 1972), p. 155.
6. A. K. Coomaraswamy, *The Dance of Shiva* (New York: Noonday Press, 1957), p. 71.
7. H. Bergson, *An Introduction to Metaphysics*, trans. T. E. Hulme (New York: Bobbs-Merrill, 1903, 1949), pp. 24–25.
8. I. Prigogine and I. Stengers, *Order out of Chaos* (New York: Bantam Books, 1984), p. 285.

CHAPTER 5 COSMOGENESIS

1. W. Gray, *General System Formation Precursor Theory* (self-published, 1977; available from 58 Pine Crest Road, Newton Centre, Mass., 02159); "Understanding creative thought processes: An early formulation of the emotional-cognitive structure theory," *Man-Environment Systems* 9 (1979): 3–14; "The evolution of emotional-cognitive and system precursor theory," in *Living Groups: Group Psychotherapy and General System Theory,* ed. James E. Durkin (New York: Brunner/Mazel, 1981), pp. 199–215.

2. I. Prigogine and I. Stengers, *Order out of Chaos* (New York: Bantam Books, 1984), pp. 160–70; J. F. G. Auchmuty and G. Nicolis, "Bifurcation analysis of nonlinear reaction-diffusion equations: I. Reaction systems," *Bulletin of Mathematical Biology* 37 (1975): 323–65; M. Herschkowitz-Kaufman, "Bifurcation analysis of nonlinear reaction-diffusion equations: II. Steady-state solutions and comparison with numerical simulations," *Bulletin of Mathematical Biology* 37 (1975): 589–635.

3. Prigogine and Stengers, *Order out of Chaos,* p. 230.

4. P. A. LaViolette, "An introduction to subquantum kinetics: III. The cosmology of subquantum kinetics," *International Journal of General Systems, Special Issue on Systems Thinking in Physics* 11 (1985): 329–45.

5. Ibid.; P. A. LaViolette, *Subquantum Kinetics: The Alchemy of Creation* (Schenectady, N. Y.: Starlane Publications, 1994), p. 118.

6. S. P. Driver, R. A. Windhorst, and R. E. Griffiths, "The contribution of late-type/irregulars to the faint galaxy counts from Hubble Space Telescope medium deep survey images," *Astrophysical Journal* 453 (1995): 48–64; S. P. Driver, et al. "Morphological number counts and redshift distributions to I < 26 from the Hubble Deep Field: Implications for the evolution of ellipticals, spirals, and irregulars," *Astrophysical Journal* 496 (1998): L93–L96.

7. Meher Baba, *The Everything and the Nothing* (Sydney: Meher House, 1963), p. 87.

8. P. Yogananda, *Autobiography of a Yogi* (Los Angeles: Self-Realization Fellowship, 1946, 1985), p. 361.

9. H. Zimmer, *Myths and Symbols in Indian Art and Civilization* (Princeton, N.J.: Princeton University Press, 1946, 1972), p. 25.

10. R. T. Rundle Clark, *Myth and Symbol in Ancient Egypt* (New York: Thames & Hudson, 1959), pp. 139–40.

CHAPTER 6 THE EGYPTIAN CREATION MYTHS

1. R. A. Schwaller de Lubicz, *Sacred Science* (Rochester, Vt.: Inner Traditions, 1961, 1982), p. 162.

2. R. A. Schwaller de Lubicz, *The Egyptian Miracle: An Introduction to the Wisdom of the Temple* (Rochester, Vt.: Inner Traditions, 1963, 1985), pp. 101–105.

3. Ibid.

4. R. T. Rundle Clark, *Myth and Symbol in Ancient Egypt* (New York: Thames & Hudson, 1959), p. 35.

5. Ibid., p. 39.

6. Ibid., pp. 42–43.

7. Ibid.

8. A. Piankoff, *Egyptian Religious Texts and Representations,* vol. 3, *Mythological Papyrii* (New York: Pantheon Books, 1957), p. 49.

9. Plutarch, *Moralia,* trans. Frank C. Babbitt (Cambridge, Mass.: Harvard University Press, 1936), pp. 35–49; Rundle Clark, *Myth and Symbol,* pp. 103–12.

10. Rundle Clark, *Myth and Symbol,* p. 50.

11. Ibid., p. 51.

12. Schwaller de Lubicz, *Sacred Science,* p. 146.

13. Rundle Clark, *Myth and Symbol,* pp. 235–38.

14. Ibid.

15. Ibid.

16. Ibid., p. 67.

17. H. Zimmer, *Myths and Symbols in Indian Art and Civilization* (Princeton, N.J.: Princeton University Press, 1946, 1972), p. 37, citing *Matsya Purana,* vol. 167, pp. 13–25.

18. H. Zimmer, *The Art of Indian Asia*, vol. 1, ed. J. Campbell (Princeton, N.J.: Princeton University Press, 1960), p. 165.

19. *Brahmavaivarta Purana*, vol. 2. Gurumandal Series no. 14 (Calcutta: Gurumandal, 1954–55), pp. 836–43; Zimmer, *Myths and Symbols*, pp. 3–11.

20. P. A. LaViolette, "The thermodynamics of the 'aha' experience," *Proceedings of the 24th Annual Meeting of the Society for General Systems Research* (January 1980), pp. 460–72, reprinted in *General Systems Theory and the Psychological Sciences*, vol. 1, ed. W. Gray, J. Fidler, and J. Battista (Seaside, Calif.: Intersystems Press, 1982), pp. 275–87.

CHAPTER 7 THE EGYPTIAN MYSTERIES

1. P. Christian, "The Mysteries of the Pyramids," in *The History and Practice of Magic*, vol. 1, book 2, trans. J. Kirkup and J. Shaw, ed. R. Nichols (New York: Citadel Press, 1870, 1963), p. 19.

2. R. T. Rundle Clark, *Myth and Symbol in Ancient Egypt* (New York: Thames & Hudson, 1959), pp. 62, 172.

3. P. D. Ouspensky, *A New Model of the Universe* (New York: Vintage Books, 1931, 1971), p. 196.

4. Rundle Clark, *Myth and Symbol*, p. 157.

5. Christian, "Mysteries of the Pyramids," p. 19.

6. Rundle Clark, *Myth and Symbol*, pp. 162–63.

7. Iamblichus, *An Egyptian Initiation*, trans. P. Christian, trans. from the French by Genevieve Stebbins Astley (1901; Denver: Edward Bloom, 1965); see also P. Christian, *The History and Practice of Magic*, vol. 1, trans. J. Kirkup and J. Shaw, ed. R. Nichols (New York: Citadel Press, 1870, 1963), pp. 89–112; C. C. Zain, *The Sacred Tarot* (Los Angeles: The Church of Light, 1936).

8. C. McIntosh, *Eliphas Lévi and the French Occult Revival* (London: Rider, 1972), p. 129.

9. *Egyptian Mysteries: An Account of an Initiation* (York Beach, Me.: Weiser, 1988).

10. *Ammianus Marcellinus*, vol. 2, book 22, trans. J. C. Rolfe (Cambridge, Mass.: Harvard University Press, 1935), p. 295.

11. J. A. West, *Serpent in the Sky: The High Wisdom of Ancient Egypt* (Wheaton, Ill.: Quest, 1993), p. 228.

12. Ibid., p. 195.

13. Ouspensky, *A New Model*, p. 320.

14. R. Bauval and A. Gilbert, *The Orion Mystery* (New York: Crown, 1994), pp. 117–19.

15. R. A. Schwaller de Lubicz, *Sacred Science* (Rochester, Vt.: Inner Traditions, 1982).

16. West, *Serpent in the Sky*, pp. 184–231.

17. Rundle Clark, *Myth and Symbol*, p. 145.

18. Ibid., p. 147.

19. Ibid., p. 148.

20. P. A. LaViolette, "An introduction to subquantum kinetics: II. An open systems description of particles and fields," *International Journal of General Systems* 11 (1985): 307–12.

21. Rundle Clark, *Myth and Symbol*, p. 166.

22. Christian, "Mysteries of the Pyramids," p. 114.

23. J. A. West, *The Traveler's Key to Ancient Egypt* (New York: Knopf, 1985), pp. 282–97.

24. Ibid., pp. 168–69.

25. Ibid., p. 167.

CHAPTER 8 THE TAROT

1. R. Tilley, *Playing Cards* (London: Octopus Books, 1973), p. 28.

2. R. Cavendish, *The Tarot* (New York: Harper & Row, 1975), p. 26.

3. Papus, *The Tarot of the Bohemians: The Most Ancient Book in the World*, trans. P. Morton (Hollywood, Calif.: Wilshire Book Co., 1889, 1973), pp. 341–42.

4. Iamblichus, *An Egyptian Initiation*, trans. P. Christian, trans. from the French by Genevieve Stebbins Astley (1901; Denver: Edward Bloom, 1965), p. 12; see also P. Christian, *The History and Practice of Magic*, vol. 1, trans. J. Kirkup and J. Shaw, ed. R. Nichols (New York: Citadel Press, 1870, 1963), pp. 89–112.

5. Cavendish, *The Tarot*, pp. 11, 18, 22.
6. Papus, *Tarot of the Bohemians*, p. 111.
7. R. T. Rundle Clark, *Myth and Symbol in Ancient Egypt* (New York: Thames & Hudson, 1959), p. 51.
8. W. Y. Evans-Wentz, *Tibetan Yoga and Secret Doctrines* (London: Oxford University Press, 1958), pp. 161–62.
9. Iamblichus, *An Egyptian Initiation*, p. 17.
10. Evans-Wentz, *Tibetan Yoga*, pp. 161–62.
11. H. Jennings, *The Rosicrucians: Their Rites and Mysteries* (New York: Arno Press, 1976), p. 156.
12. Papus, *Tarot of the Bohemians*, p. 124.
13. J. Campbell, *The Mythic Image* (Princeton, N.J.: Princeton University Press, 1974), pp. 340–41.
14. Christian, *History and Practice* of Magic, p. 144.
15. Papus, *Tarot of the Bohemians*, pp. 19–23.

CHAPTER 9 THE THERMODYNAMICS OF ASTROLOGY

1. D. C. Doane and K. Keyes, *How to Read Tarot Cards* (New York: Funk & Wagnalls, 1968), pp. 17, 24.
2. P. Christian, "The Mysteries of the Pyramids," in *The History and Practice of Magic*, vol. 1, book 2, trans. J. Kirkup and J. Shaw, ed. R. Nichols (New York: Citadel Press, 1963), pp. 115–16.
3. J. A. West and J. G. Toonder, *The Case for Astrology* (Baltimore: Penguin Books, 1973), p. 26.
4. R. A. Schwaller de Lubicz, *Sacred Science* (Rochester, Vt.: Inner Traditions, 1961, 1982), pp. 177–79.
5. O. E. Scott, *Stars in Myth and Fact* (Caldwell, Idaho: Caxton Printers, 1947), p. 67.
6. A. Sachs, "Babylonian horoscopes," *Journal of Cuneiform Studies* 6 (1952): 49–75.
7. P. A. LaViolette, "Astrological symbolism and its depiction of the theory of self-organizing, dissipative systems" (unpublished paper, 1976).
8. G. H. Mees, *The Book of Stars* (Dordrecht, Netherlands: Kluwer, 1954), p. 233.
9. E. Meyer, *Chronologie Egyptienne*, trans. Moret, Annales du Musée Guimet (Paris, 1912), pp. 51–55; Schwaller de Lubicz, *Sacred Science*, Appendix VIII; J. A. West, *Serpent in the Sky: The High Wisdom of Ancient Egypt* (Wheaton, Ill.: Quest, 1993), pp. 95–99.
10. West and Toonder, *Case for Astrology*, p. 52.
11. Scott, *Stars in Myth and Fact*, pp. 160–61.
12. R. Burnham, Jr., *Burnham's Celestial Handbook: An Observer's Guide to the Universe beyond the Solar System*, vol. 2 (New York: Dover, 1978), pp. 912–15.
13. Ibid., p. 1057.
14. E. A. W. Budge, *The Gods of the Egyptians*, vol. 1 (New York: Dover, 1969), p. 473.
15. A. Piankoff, *The Tomb of Ramses VI*, vol. 1 (New York: Pantheon Books, 1954), p. 158.
16. Scott, *Stars in Myth and Fact*, p. 288.
17. B. D. Ruben and J. Y. Kim, *General Systems Theory and Human Communication* (Rochelle Park, N.J.: Hayden, 1975), p. 211.
18. E. Laszlo, *Introduction to Systems Philosophy: Toward a New Paradigm of Contemporary Thought* (New York: Harper & Row, 1972); I. Prigogine, P. M. Allen, and R. Herman, "Long-term trends and the evolution of complexity," in *Goals in a Global Community*, vol. 1, *Studies on the Conceptual Foundations*, ed. E. Laszlo and J. Bierman (New York: Pergamon Press, 1977), p. 39.
19. R. Brown, *Researches into the Origin of the Primitive Constellations of the Greeks, Phoenicians, and Babylonians* (London: Williams & Norgate, 1899), pp. 291–92.
20. Burnham, *Burnham's Celestial Handbook*, pp. 915–18.
21. J. Jeans, *Astronomy and Cosmogony* (London: Cambridge University Press, 1928), p. 352; V. A. Ambartsumian, in *Report on Eleventh Solvay Conference* (Brussels: Institute Internationale de Physique Solvay, 1958); W. H. McCrea, "Continual creation," *Monthly Notices of the Royal Astronomical Society* 128 (1964): 335–44; J. Gribbin, *White Holes: Cosmic Gushers in the Universe* (New York: Delta, 1977).

CHAPTER 10 SUBATOMIC ATLANTIS

1. Plato, *Critias* sections 115e–116a, in *The Collected Dialogues of Plato,* ed. E. Hamilton and H. Cairns, trans. A. E. Taylor (Princeton, N.J.: Princeton University Press, 1961).
2. Ibid.
3. J. Kelly, "Nuclear charge and magnetization densities from Sachs form factors," *Physical Review C* 66, no. 6 (2002), id: 065203; P. A. LaViolette, *Subquantum Kinetics: A Systems Approach to Physics and Cosmology,* 2nd ed. (Schenectady, N.Y.: Starlane Publications, 1994, 2003), ch. 4.
4. O. Muck, *The Secret of Atlantis* (New York: Pocket Books, 1976, 1978), p. 9.
5. Plato, *Critias,* section 114d.
6. Ibid., section 115d.
7. R. P. Feynman, R. B. Leighton, and M. Sands, *The Feynman Lectures on Physics,* vol. 2 (Reading, Mass.: Addison-Wesley, 1964), pp. 12.6–12.7.
8. Ibid., pp. 12.12–12.13.
9. A. Einstein, "On the generalized theory of gravitation," *Scientific American* 182, no. 4 (1950): 14.
10. Ibid., p. 15.
11. A. Besant and C. W. Leadbeater, *Occult Chemistry* (Adyar, India: Theosophical Publishing House, 1951), p. 13; Besant and Leadbeater's emphasis.
12. Ibid., p. 14.
13. Ibid.
14. E. Lerner, *The Big Bang Never Happened* (New York: Vintage Books, 1992), pp. 243–45.
15. Besant and Leadbeater, *Occult Chemistry,* p. 14.
16. K. I. Kellerman et al., "The small radio source at the Galactic center," *Astrophysical Journal* 214 (1977): L61–L62.
17. A. M. Ghez et al., "The accelerations of stars orbiting the Milky Way's central black hole," *Nature* 407 (2000): 349–351.
18. E. E. Becklin, I. Gatley, and M. W. Werner, "Far-infrared observations of Sagittarius A: The luminosity and dust density in the central parsec of the galaxy," *Astrophysical Journal* 258 (1982): 135–42.
19. K. Y. Lo et al., "On the size of the Galactic centre compact radio source: Diameter <2OAU," *Nature* 315 (1985): 124–26.
20. P. A. LaViolette, "Cosmic ray volleys from the Galactic center and their impact on the earth environment," *Earth, Moon, and Planets* 38 (1987): 241–86; "Galactic explosions, cosmic dust invasions, and climatic change," diss. Portland State University, 1983.
21. C. H. Townes et al., "The centre of the galaxy," *Nature* 301 (1983): 661–66; J. Barnes, L. Hernquist, and F. Schweizer, "Colliding galaxies," *Scientific American* 265, no. 2 (1991): 40–47.
22. G. R. Burbidge, T. W. Jones, and S. L. O'Dell, "Physics of compact nonthermal sources: III. Energetic considerations," *Astrophysical Journal* 193 (1974): 43–54.
23. P. A. LaViolette, *Earth Under Fire* (Schenectady, N.Y.: Starlane Publications, 1997).
24. Plato, *Critias,* section 121c.

CHAPTER 11 MYTHS FROM THE ANCIENT EAST AND MEDITERRANEAN

1. S. N. Kramer, *Sumerian Mythology* (Philadelphia: American Philosophical Society, 1944), p. 40.
2. The Babylonian creation myth is recounted in A. Heidel, *The Babylonian Genesis* (Chicago: University of Chicago Press, 1942); J. Gray, *Near Eastern Mythology* (New York: Hamlyn, 1969); and F. G. Bratton, *Myths and Legends of the Ancient Near East* (New York: Crowell, 1973).
3. Heidel, *Babylonian Genesis,* p. 10.
4. P. Feyerabend, *Against Method* (London: Verso, 1975), p. 49.
5. Hesiod's story is retold in E. Hamilton, *Mythology: Timeless Tales of Gods and Heroes* (New York: Mentor, 1940), chapter 3 and R. Graves, *The Greek Myths, vol. 1* (Edinburgh: Penguin Books, 1955), chapters 6 and 7.
6. Hesiod, *Theogony,* in E. Hamilton, *Mythology: Timeless Tales of Gods and Heroes,* p. 64.

7. Plato, *Cratylus*, section 402, in *The Dialogues of Plato*, vol. 1, trans. B. Jowett (New York: Scribner, Armstrong, 1874).
8. This story is recounted in Graves, *Greek Myths*, vol. 1, pp. 69–70.
9. Fung Yu-lan, *A Short History of Chinese Philosophy* (New York: Macmillan, 1959), p. 169.
10. Lao Tze, *Tao Te Ching*, trans. James Legge (New York: Dover, 1997), chapters 25, 34.
11. Fung Yu-lan, *Short History*, pp. 193, 279.
12. Ibid., pp. 269–70.
13. Ibid.

CHAPTER 12   ETHER OR VACUUM?

1. R. McCormmach, "H. A. Lorentz and the electromagnetic view of nature," *Isis* 61 (1970): 459–97.
2. H. Ives, "Revisions of the Lorentz transformations," *Proceedings of the American Philosophical Society* 95 (1951): 125–31.
3. H. A. Lorentz, *The Theory of Electrons*, 2nd ed. (New York: Dover, 1909, 1952).
4. G. Mie, "Grundlagen einer Theorie der Materie," *Annals of Physics* 37 (1912): 511–34.
5. Fung Yu-lan, *A Short History of Chinese Philosophy* (New York: Macmillan, 1959), pp. 695–97.
6. H. Ives, "The aberration of clocks and the clock paradox," *Journal of the Optical Society of America* 27 (1937): 305–9; "The clock paradox in relativity theory," *Nature* 168 (1951): 246; D. Turner and R. Hazelett, *The Einstein Myth and the Ives Papers: A Counter-Revolution in Physics* (Old Greenwich, Conn.: Devin-Adair, 1979), pp. 34–75.
7. Ives, "Revisions," pp. 125–31.
8. G. Sagnac, "The luminiferous ether demonstrated by the effect of the relative motion of the ether in an interferometer in uniform rotation," *Comptes rendus de l'Academie des Sciences* (Paris) 157 (1913): 708–10, 1410–13. (For a translation, see Turner and Hazelett, *The Einstein Myth*, pp. 247–50.)
9. D. Z. Anderson, "Optical gyroscopes," *Scientific American* (April 1986): 94–99.
10. H. Ives, "Light signals sent around a closed path," *Journal of the Optical Society of America* 28 (1938): 296–99.
11. Ives, "Revisions," pp. 125–31.
12. Ibid.; H. Ives, "Derivation of the Lorentz transformations," *Philosophical Magazine* 36 (1945): 392–403; "Lorentz-type transformations as derived from performable rod and clock operations," *Journal of the Optical Society of America* 39 (1949): 757–61; "Extrapolation from the Michelson-Morley experiment," *Journal of the Optical Society of America* 40 (1950): 185–91.
13. H. Erlichson, "The rod contraction-clock retardation ether theory and the special theory of relativity," *American Journal of Physics* 41 (1973): 1068–77.
14. E. W. Silvertooth, "Experimental detection of the ether," *Speculations in Science and Technology* 10 (1987): 3–7; "Motion through the ether," *Electronics and Wireless World* (May 1989): 437–38.
15. E. W. Silvertooth and C. K. Whitney, "A new Michelson-Morley experiment," *Physics Essays* 5 (1992): 82–89.
16. G. F. Smoot et al., "Preliminary results from the COBE differential microwave radiometers: Large angular scale isotropy of the cosmic microwave background," *Astrophysical Journal* 371 (1991): L1–L5.
17. P. T. Pappas, "The original Ampère force and Biot-Savart and Lorentz forces," *Il nuovo cimento* 76B (1983): 189–97; "On Ampère electrodynamics and relativity," *Physics Essays* 3 (1990): 117–21; "The non-equivalence of the Ampère and Lorentz/Grassmann force laws and longitudinal contact interactions," *Physics Essays* 3 (1990): 15–23; P. T. Pappas and T. Vaughan, "Forces on a stigma antenna," *Physics Essays* 3 (1990): 211–16.
18. P. Graneau, "Electromagnetic jet-propulsion in the direction of current flow," *Nature* 295 (1982): 311–12; "Amperian recoil and the efficiency of railguns," *Journal of Applied Physics* (1987): 3006–9; "First indication of Ampère tension in solid electric conductors," *Physics Letters* 97A (1983): 253–55; P. Graneau and P. N. Graneau, "Electrodynamic explosions in liquids," *Applied Physics Letters* 46 (1985): 468–70.

CHAPTER 13 THE TWENTIETH-CENTURY CREATION MYTHOS

1. T. Wright, *An Original Theory of the Universe* (London: Macdonald, 1750).
2. E. Lerner, *The Big Bang Never Happened* (New York: Vintage Books, 1992), p. 156.
3. F. Zwicky, "On the red shift of spectral lines through interstellar space," *Proceedings of the National Academy of Sciences* 15 (1929): 773–79.
4. W. von Nernst, *The Structure of the Universe in Light of Our Research* (Berlin: Springer, 1921), p. 40, trans. R. Monti in *SeaGreen* 4 (1986): 32–36.
5. W. von Nernst, "Additional test of the assumption of a stationary state in the universe," *Zeitschrift für Physik* 106 (1938): 633–61.
6. E. Hubble and R. C. Tolman, "Two methods of investigating the nature of the nebular red-shift," *Astrophysical Journal* 82 (1935): 302–37.
7. E. Hubble, "Effects of red shifts on the distribution of nebulae," *Astrophysical Journal* 84 (1936): 517.
8. Ibid., p. 554.
9. Nernst, "Additional test," pp. 639–40.
10. R. B. Tully, "Origin of the Hubble constant controversy," *Nature* 334 (1988): 209–12.
11. G. Burbidge, "Why only one big bang?" *Scientific American* 266, no. 2 (1992): 120.
12. W. Freedman, et al., "Final results from the Hubble Space Telescope key project to measure the Hubble constant." *Astrophysical Journal* 553 (2001): 47–72.
13. C. L. Carilli et al., "A 250 GHz survey of high-redshift quasars from the Sloan Digital Sky Survey." *Astrophysical Journal* 555 (2001): 625–32.
14. M. Dickinson et al., "The unusual infrared object HDF-N J123656.3+621322," *Astrophysical Journal* 531 (2000): 624–34.
15. Lerner, *The Big Bang Never Happened*, p. 220.
16. M. Valtonen and G. Byrd, "Redshift asymmetries in systems of galaxies and the missing mass," *Astrophysical Journal* 303 (1986): 523–34; Lerner, *The Big Bang Never Happened*, pp. 35–39.
17. T. J. Broadhurst, R. S. Ellis, D. C. Koo, and A. S. Szalay, "Large-scale distribution of galaxies at the galactic poles," *Nature* 343 (1990): 726–28.
18. D. Lindley, "Cold dark matter makes an exit," *Nature* 349 (1991): 14.
19. Broadhurst et al., "Large-scale distribution of galaxies," pp. 726–28.
20. P. A. LaViolette, "Is the universe really expanding?" *Astrophysical Journal* 301 (1986): 544–53; reprinted in the first edition of *Subquantum Kinetics: The Alchemy of Creation* (Schenectady, N.Y.: Starlane Publications, 1994).
21. "New study questions expanding universe," *Astronomy* (August 1986): 64; A. Ubachukwu and L. Onuora, "Radio source orientation and the cosmological interpretation of the angular size-redshift relation for double-lobed quasars" *Astrophysics and Space Science* 209 (1993):169–80; A. Sandage, "Observational tests of world models," *Annual Reviews of Astronomy and Astrophysics* 26 (1988): 591; V. K. Kapahi, "The angular size-redshift relation as a cosmological tool," *Observational Cosmology*; Proc. 124th IAU Symposium, Beijing, Aug. 25–30, 1986. A. Hewitt et al., eds. (Dordrecht: D. Reidel Publishing, 1987), pp. 251–66; J.-C. Pecker and J.-P. Vigier, "A possible tired-light mechanism." *Observational Cosmology*; Proc. 124th IAU Symposium, Beijing, Aug. 25–30, 1986. A. Hewitt et al., eds. (Dordrecht: D. Reidel Publishing, 1987), p. 507; J.-C. Pecker, "How to describe physical reality?" Nobel Symposium, 1986.
22. L. M. Lubin and A. Sandage, "The Tolman surface brightness test for the reality of the expansion. IV. A measurement of the Tolman signal and the luminosity evolution of early-type galaxies." *Astronomical Journal* 122 (2001): 1084–1103; G. Goldhaber, et al., "Timescale stretch parameterization of type I-a supernova B-band light curves." *Astrophysical Journal* 558 (2001): 359–68.
23. P. A. LaViolette, *Subquantum Kinetics: A Systems Approach to Physics and Cosmology,* 2nd ed. (Schenectady, N.Y.: Starlane Publications, 1994, 2003).
24. A. Le Floch and F. Bretenaker, "Early cosmic background," *Nature* 352 (1991): 198.
25. E. Regener, "Der Energiestrom der Ultrastrahlung," *Zeitschrift für Physik* 80 (1933): 666–69.
26. W. von Nernst, "Additional test," pp. 639–40; trans. R. Monti in *SeaGreen* 4 (1986): 32–36.

27. I. Peterson, "Cosmic evidence of a smooth beginning," *Science News* 137 (1990): 36.
28. E. Lerner, "Radio absorption by the intergalactic medium," *Astrophysical Journal* 361 (1990): 63–68.
29. E. Lerner, "The cosmologists' new clothes," *Sky and Telescope* (February 1992): 124; (August 1992): 126.

CHAPTER 14 SMASHING THE CRYSTALLINE SPHERE

1. Pope John Paul II quoted in J. Gribbin, *In Search of the Big Bang* (New York: Bantam Books, 1986), pp. 387–88.
2. S. W. Hawking, *A Brief History of Time: From the Big Bang to Black Holes* (New York: Bantam Books, 1988), p. 141.
3. Gribbin, *In Search of the Big Bang*, p. 392; Gribbin's emphasis.
4. E. Lerner, *The Big Bang Never Happened* (New York: Vintage Books, 1992), pp. 159–60.
5. Gribbin, *In Search of the Big Bang*, p. 381.
6. A. Einstein, B. Podolsky, and N. Rosen, "Can a quantum-mechanical description of physical reality be considered complete?" *Physical Review* 47 (1935): 777.
7. H. Stapp, "Are superluminal connections necessary?" *Il nuovo cimento* 40B (1977): 191-205; G. Zukov, *The Dancing Wu Li Masters: An Overview of the New Physics* (New York: Bantam Books, 1979), pp. 295-305.
8. D. Bohm and B. Hiley, "On the intuitive understanding of nonlocality as implied by quantum theory," *Foundations of Physics* 5 (1975): 93–109; D. Bohm, *Wholeness and the Implicate Order* (London: Routledge & Kegan Paul, 1979).
9. D. E. Thomsen, "Unconventional physics: Quite crazy after all these years," *Science News* (26 July 1986): 55.
10. A. D. Krisch, "Collisions between spinning protons," *Scientific American* 257, no. 2 (1987): 42–50.
11. Lerner, *The Big Bang Never Happened*, pp. 347, 370–71.
12. *New York Times*, 10 July 1932, p. 19, col. 1.
13. S. L. Shapiro and S. A. Teukolsky, "Formation of naked singularities: Violation of cosmic censorship," *Physical Review Letters* 66 (1991): 994–97.
14. P. Gerber, *Zeitschrift für Mathematik und Physik* 43 (1898) : 93–104.
15. H. C. Hayden, "Light speed as a function of gravitational potential," *Galilean Electrodynamics* 1 (1990): 15–17.
16. P. A. LaViolette, "An introduction to subquantum kinetics, II," *International Journal of General Systems* 11(1985): 324–25.
17. "Cosmological constant conundrum," *Sky and Telescope* (August 1989): 132–33.
18. L. Abbott, "The mystery of the cosmological constant," *Scientific American* 258, no. 5 (1988): 106–13.

CHAPTER 15 ENERGY IN THE UNIVERSE

1. P. A. LaViolette, "The matter of creation," *Physics Essays* 7, no. 1 (1994): 1–8.
2. P. A. LaViolette, "An introduction to subquantum kinetics: III. The cosmology of subquantum kinetics," *International Journal of General Systems, Special Issue on Systems Thinking in Physics* 11 (1985): 329–45; "A Tesla wave physics for a free energy universe," in *Proceedings of the 1990 International Tesla Society Conference* (Colorado Springs, Colo., 1991).
3. Ibid.; P. A. LaViolette, "Subquantum kinetics: Exploring the crack in the first law," in *Proceedings of the Twenty-Sixth Annual Intersociety Energy Conversion Engineering Conference* (Boston, 1991); "The planetary-stellar mass-luminosity relation: Possible evidence of energy nonconservation?" *Physics Essays* 5 (1992): 536–42; erratum: *Physics Essays* 6, no. 4 (1993): 616.
4. LaViolette, "The planetary-stellar mass-luminosity relation" pp. 536–42,
5. P. A. LaViolette, "An Introduction to Subquantum Kinetics," *International Journal of General Systems* 11 (1985): 340; P. A. LaViolette, *Subquantum Kinetics: The Alchemy of Creation*, (Schenectady, N.Y: Starlane Publications, 1994), p. 135.
6. J. D. Anderson et al., "Indication, from Pioneer 10/11, Gallileo, and Ulysses Data of an Apparent Anomalous, Weak, Long-Range Acceleration." *Physical Review Letters* 81 (1998): 2858–61; J. D.

Anderson, et al., "Study of the anomalous acceleration of Pioneer 10 and 11." *Physical Review D* 65 (2002), article no. 082004.

7. P. A. LaViolette, *Subquantum Kinetics: A Systems Approach to Physics and Cosmology,* 2nd ed. (Schenectady, N.Y.: Starlane Publications, 1994, 2003), chapter 9; P. A. LaViolette,. "The Pioneer maser signal anomaly: Possible confirmation of spontaneous photon blueshifting," 2003, submitted for review.

8. M. M. Waldrop, "Sanduleak -69 202: Guilty as charged," *Science* 236 (May 1987): 523.

9. Paul Recer, "Quasar data has scientists re-examining their theories," *The Daily Gazette,* Schenectady, N.Y., January 12, 1995.

10. P. T. Pappas, "The original Ampère force and Biot-Savart and Lorentz forces," *Il nuovo cimento* 76B (1983): 189–97; "The non-equivalence of the Ampère and Lorentz/Grassmann force laws and longitudinal contact interactions," *Physics Essays* 3 (1990): 15–23.

11. P. T. Pappas and T. Vaughan, "Forces on a stigma antenna," *Physics Essays* 3 (1990): 211–16.

12. P. Graneau, "Electromagnetic jet-propulsion in the direction of current flow," *Nature* 295 (1982): 311–12; "Amperian recoil and the efficiency of railguns," *Journal of Applied Physics* 62 (1987): 3006–9.

13. LaViolette, "A Tesla wave physics"; P. A. LaViolette, "An open system approach to electromagnetic wave propagation," in *Proceedings of the Thirty-Fourth Annual Meeting of the International Society for the Systems Sciences,* vol. 2 (Portland, Ore., 1990), pp. 1119–26; P. T. Pappas, "Energy creation in electrical sparks and discharges," in *Proceedings of the Twenty-Sixth Annual Intersociety Energy Conversion Engineering Conference* (Boston, 1991).

14. P. A. LaViolette, "An introduction to subquantum kinetics, II," *International Journal of General Systems* 11 (1985): 309–10; *Subquantum Kinetics: A Systems Approach,* chapter 3.

15. P. A. LaViolette, "A theory of electrogravitics," *Electric Spacecraft Journal* 8 (1992): 33–36; *Subquantum Kinetics,* chapter 9.

16. T. T. Brown, "How I control gravity," *Science and Invention* (August 1929); G. Burridge, "Another step toward anti-gravity," *American Mercury* 86, no. 6 (1958): 77–82; Aviation Studies (International) Ltd., *Electrogravitics Systems: An Examination of Electrostatic Motion, Dynamic Counterbary, and Barycentric Control,* Report GRG 013/56, February 1956 (Library of Congress no. 3,1401,00034,5879), reprinted in *Electrogravitics Systems: Reports on a New Propulsion Methodology,* ed. T. Valone, (Washington, D.C.: Integrity Research Institute, 1994).

17. T. T. Brown, letter to R. Schaffranke, 5 April 1973. In Rho Sigma, *Ether Technology: A Rational Approach to Gravity Control* (Lakemont, Ga.: self-published, 1977), pp. 46–49.

18. Intel, "Towards flight without stress or strain . . . or weight," *Interavia* (Switzerland) 11, no. 5 (1956): 373–74.

19. Aviation Studies, *Electrogravitics Systems;* Aviation Studies (International) Ltd., *The Gravitics Situation* (December 1956).

20. Intel, "Towards flight," pp. 373–74; "Electrogravitics: Science or daydream?" *Product Engineering* (30 December 1957): 12; "How to 'fall' into space," *Business Week* (8 February 1958): 51–53; C. Carew, "The key to travel in space," *Canadian Aviation* 32, no. 6 (1959): 27–32; P. A. LaViolette, "Electrogravitics: An energy-efficient means of spacecraft propulsion," submission to the 1990 NASA Space Exploration Outreach Program; reprinted in *Explore* 3, no. 1 (1991): 76–79; "Electrogravitics: Back to the future," *Electric Spacecraft Journal* 4 (1992): 23–28.

21. R. L. Talley, *Twenty-First-Century Propulsion Concept,* U.S. Air Force report No. PL-TR-91-3009 (Washington, D.C.: National Technical Information Service, 1991).

CHAPTER 16 BACK TO THE FUTURE

1. Plato, *Timaeus,* in *The Collected Dialogues of Plato,* ed. E. Hamilton and H. Cairns, trans. B. Jowett (Princeton, N.J.: Princeton University Press, 1961), section 23.

2. B. Steiger, *Worlds before Our Own* (New York: Berkeley Books, 1978), pp. 120–21.

3. R. Noorbergen, *Secrets of the Lost Races* (New York: Barnes & Noble, 1977), pp. 48–50.

4. Steiger, *Worlds before Our Own,* pp. 120–21; *Scientific American* 7 (June 1851): 298.

5. M. M. Alouf, *History of Baalbek* (Beirut: American Press, 1938), pp. 38–39, 103–4, 111–12.

6. Noorbergen, *Secrets of the Lost Races,* pp. 194–96.

7. Ibid.

8. S. M. Greer, *Disclosure* (Crozet, Va.: Crossing Point, 2001); T. Good, *Above Top Secret: The World-wide UFO Cover-Up* (New York: Morrow, 1988).

9. R. C. Hoagland, *The Monuments of Mars: A City on the Edge of Forever* (Berkeley, Calif.: North Atlantic Books, 1987, 1992); R. L. Thompson, *Alien Identities: Ancient Insights into Modern UFO Phenomena* (San Diego, Calif.: Govardhan Hill, 1993).

10. P. A. LaViolette, *The Talk of the Galaxy: An ET Message for Us?* (Schenectady, N.Y.: Starlane Publications, 2000).

GLOSSARY

Autocatalysis. An ordering of reaction-kinetic processes in which the output from a reaction serves as an input to that same reaction.

Autogenesis. The circumstance in which a system generates conditions that favor its continued existence.

Bifurcation. The division of a mathematical solution into two branches.

Blueshift. The shortening of the wavelength of starlight photons (increase in frequency) due to the progressive amplification of their energy during propagation in a supercritical region of space, or alternatively, due to Doppler compression of the wavelength as a result of the star's approaching motion.

Brusselator. A nonlinear two-variable reaction system specified by four kinetic equations.

Cosmological constant. A mathematical device used by Einstein to give space-time an inbuilt tendency to expand.

Closed system. A system considered to be isolated from its environment.

Critical fluctuation. A deviation of a system variable from its steady-state value that is large enough to allow a spontaneous increase in size.

Critical threshold. A condition in which a nonlinear system borders between subcritical and supercritical behavior.

Cross-catalysis. An ordering of reaction-kinetic processes in which the output from one reaction serves as an input to a second reaction and the output from the second reaction serves as an input to the first.

Dissipative structure. An ordered pattern whose form is maintained by the operation of underlying entropy-increasing processes; characteristically observed in open systems.

Entropy. A quantity measuring a system's state of disorder.

Ether. The unobservable subtle substance that is hypothesized to fill all of space and to compose all physical form.

Etheron. An etheric component.

Explicit order. The outwardly visible physically observable order of a system.

First law of thermodynamics. The law that states that energy (or its equivalent in mass) can be neither created nor destroyed.

G-well. A gravitational potential well.

General system theory (GST). A scientific discipline that attempts to discern general systems laws of nature. A holistic way of thinking based on an awareness of the behavior of systems in general.

Genic energy. The energy produced in a supercritical region of space as a result of the progressive amplification of photon energy.

Hubble constant (H_0). The ratio between a galaxy's hypothesized recessional velocity and its distance, measured to be about thirty kilometers per second per million light-years.

Implicit Order. The underlying functional order responsible for generating a system's observable explicit order.

Logical positivism. The view that the only valid knowledge is that which can be empirically tested.

Model G. A nonlinear reaction system, similar to the Brusselator, specified by five kinetic equations.

Nonequilibrium thermodynamics. A branch of thermodynamics concerned with describing the behavior of open systems.

Open system. A system that continuously exchanges matter/energy with its environment; includes all living systems.

Parthenogenesis. The process of virgin birth, in which the ordered form emerges spontaneously from a preexisting system or medium without requiring external causation.

Reaction-diffusion system. A system whose constituents mutually react with one another and also move through space by means of diffusion.

Reaction-diffusion wave. A wave produced by a reaction-diffusion system.

Redshift. The lengthening of the wavelength of starlight photons (decrease in frequency) due to the progressive attenuation of their energy through a tired-light effect, or alternatively as a result of the star's receding motion.

Second law of thermodynamics. The law that states that a closed system will naturally evolve toward a condition of greater disorder.

Spin. An internal property of elementary particles related to, but not identical to, the everyday concept of spin.

Spontaneous symmetry breaking. The termination of a condition of concentration uniformity by the spontaneous onset of a periodic pattern that can adopt either of two polarities.

Steady state. A state in which the system variables (for example, ether concentrations) do not vary with time.

Subcritical. A nonlinear system state characterized by a tendency for its variables to return progressively to their steady-state values.

Subquantum kinetics. A physics methodology that postulates a reaction-diffusion ether as a substrate for physical form.

Supercritical. A nonlinear system state characterized by a tendency for its variables to depart progressively from their steady-state values.

System. An entity maintained by the mutual interaction of its parts.

Tired-light effect. An explanation of the cosmological redshift effect in which a photon progressively loses energy and increases its wavelength as it travels through intergalactic space.

Virtual particle. In quantum mechanics, a hypothetical particle that can never be directly detected, but whose existence is theorized to have measurable effects.

Wave-particle dualism. The concept in quantum mechanics that there is no distinction between waves and particles; particles may sometimes behave like waves, and waves like particles.

BIBLIOGRAPHY

Abbott, E. A. *Flatland: A Romance of Many Dimensions.* New York: Dover, 1952.

Abbott, L. "The mystery of the cosmological constant." *Scientific American* 258, no. 5 (1988): 106–13.

Alouf, M. M. *History of Baalbek.* Beirut: American Press, 1938.

Ambartsumian, V. A. In *Report on Eleventh Solvay Conference.* Brussels: Institute Internationale de Physique Solvay, 1958.

Ammianus Marcellinus. Translated by J. C. Rolfe. Cambridge, Mass.: Harvard University Press, 1935.

Anderson, D. Z. "Optical gyroscopes." *Scientific American* 254, no. 4 (1986): 94–99.

Auchmuty, J. F G., and G. Nicolis. "Bifurcation analysis of nonlinear reaction-diffusion equations: Reaction systems." *Bulletin of Mathematical Biology* 37 (1975): 323–65.

———. "Bifurcation analysis of nonlinear reaction-diffusion equations: Evolution equations and the steady state solutions." *Bulletin of Mathematical Biology* 37 (1975): 589–635.

Aviation Studies (International) Ltd. *Electrogravitics Systems: An Examination of Electrostatic Motion, Dynamic Counterbary, and Barycentric Control.* Report GRG 013/56, February 1956. Library of Congress no. 3,1401,00034,5879, call no. TL565.A9.

———. *The Gravitics Situation.* December 1956.

Barnes, J., L. Hernquist, and F. Schweizer. "Colliding galaxies." *Scientific American* 265, no. 2 (1991): 40–47.

Bauval, R., and A. Gilbert. *The Orion Mystery.* New York: Crown, 1994.

Becklin, E. E., I. Gatley, and M. W. Werner. "Far-infrared observations of Sagittarius A: The luminosity and dust density in the central parsec of the galaxy." *Astrophysical Journal* 258 (1982): 135–42.

Bergson, H. *An Introduction to Metaphysics.* Translated by T. E. Hulme. 1903. New York: Bobbs-Merrill, 1949.

Bertalanffy, L. von. *Modern Theories of Development.* 1928. New York: Harper Torchbook, 1961.

———. *Theoretische Biologie.* 1932. Ann Arbor, Mich.: J. W. Edwards, 1948.

———. *Robots, Men, and Minds.* New York: Braziller, 1967.

———. *General System Theory: Foundations, Development, Applications.* New York: Braziller, 1968.

———. *A Systems View of Man.* Edited by P. LaViolette. Boulder, Colo.: Westview Press, 1981.

Besant, A., and C. W. Leadbeater. *Occult Chemistry.* Adyar, India: Theosophical Publishing House, 1951.

Bohm, D. *Wholeness and the Implicate Order.* London: Routledge & Kegan Paul, 1979.

Bohm, D., and B. Hiley. "On the intuitive understanding of nonlocality as implied by quantum theory." *Foundations of Physics* 5 (1975): 93–109.

Bondi, H. "Negative mass in general relativity." *Reviews of Modern Physics* 29 (1957): 423–28.

Brahmavaivarta Purana. Vol. 2, series no. 14. Calcutta: Gurumandal, 1954–55.

Bratton, F. G. *Myths and Legends of the Ancient Near East.* New York: Crowell, 1973.

Broadhurst, T. J., R. S. Ellis, D. C. Koo, and A. S. Szalay. "Large-scale distribution of galaxies at the galactic poles." *Nature* 343 (1990): 726–28.

Brown, R. *Researches into the Origin of the Primitive Constellations of the Greeks, Phoenicians, and Babylonians.* London: Williams & Norgate, 1899.

Brown, T. T. "How I control gravity." *Science and Invention,* August 1929.

————. Personal communication, 9 February 1982.

Buckley, W. *Sociology and Modern Systems Theory.* Englewood Cliffs, N.J.: Prentice-Hall, 1967.

Budge, E. A. W. *The Gods of the Egyptians.* Vol. 1. New York: Dover, 1969.

Burbidge, G. R. "Why only one big bang?" *Scientific American* 266, no. 2 (1992): 120.

Burbidge, G. R., T. W. Jones, and S. L. O'Dell. "Physics of compact nonthermal sources: III. Energetic considerations." *Astrophysical Journal* 193 (1974): 43–54.

Burnham, R., Jr. *Burnham's Celestial Handbook: An Observer's Guide to the Universe beyond the Solar System.* Vol. 2. New York: Dover, 1978.

Burridge, C. "Another step toward anti-gravity." *American Mercury* 86, no. 6 (1958): 77–82.

Campbell, J. *The Mythic Image.* Princeton, N.J.: Princeton University Press, 1974.

Capra, F. *The Tao of Physics.* New York: Bantam Books, 1975.

Carew, C. "The key to travel in space." *Canadian Aviation* 32, no. 6 (1959): 27–32.

Carilli, C. L., et al. "A 250 GHz survey of high-redshift quasars from the Sloan Digital Sky Survey." *Astrophysical Journal* 555 (2001): 625–32.

Case, P. F. *The Tarot: A Key to the Wisdom of the Ages.* Richmond, Va.: Macoy, 1947.

Cavendish, R. *The Tarot.* New York: Harper & Row, 1975.

Christian, P. "The mysteries of the pyramids." In *The History and Practice of Magic,* vol. 1, book 2. Translated by J. Kirkup and J. Shaw. Edited by R. Nichols. 1870. New York: Citadel Press, 1963.

Coomaraswamy, A. K. *The Dance of Shiva.* New York: Noonday Press, 1957.

"Cosmological constant conundrum." *Sky and Telescope* (August 1989): 132–33.

Davidson, M. *Uncommon Sense: The Life and Thought of Ludwig von Bertalanffy, Father of General Systems Theory.* Los Angeles: Tarcher, 1983.

Davies, P. *The Runaway Universe.* New York: Penguin Books, 1980.

Defay, R. "Introduction à la thermodynamique des systèmes ouvertes." *Academie Royale de Belgique, Bulletin classe de sciences* 53 (1929).

Dickinson, M., et al. "The unusual infrared object HDF-N J123656.3+621322." *Astrophysical Journal* 531 (2000): 624–34.

Doane, D. C., and K. Keyes. *How to Read Tarot Cards.* New York: Funk & Wagnalls, 1968.

Egyptian Mysteries: An Account of an Initiation. York Beach, Me: Weiser, 1988.

Einstein, A. "On the generalized theory of gravitation." *Scientific American* 182, no. 4 (1950): 1–17.

Einstein, A., B. Podolsky, and N. Rosen. "Can a quantum-mechanical description of physical reality be considered complete?" *Physical Review* 47 (1935): 777.

"Electrogravitics: Science or daydream?" *Product Engineering* (30 December 1957): 12.

Erlichson, H. "The rod contraction: Clock retardation ether theory and the special theory of relativity." *American Journal of Physics* 41(1973): 1068–77.

Evans-Wentz, W. Y. *The Tibetan Book of the Dead.* London: Oxford University Press, 1957.

————. *Tibetan Yoga and Secret Doctrines.* London: Oxford University Press, 1958.

"Farthest galaxy is cosmic question." *Science News* 133 (1988): 262–63.

Feyerabend, P. *Against Method.* London: Verso, 1975.

Feynman, R. P., R. B. Leighton, and M. Sands. *The Feynman Lectures on Physics.* Vol. 2. Reading, Mass.: Addison-Wesley, 1964.

Field, R. J., and R. M. Noyes. "Oscillations in chemical systems: IV. Limit cycle behavior in a model of a real chemical reaction." *Journal of Chemical Physics* 60 (1974): 1877–84.

Freedman, W., et al. "Final results from the Hubble Space Telescope key project to measure the Hubble constant." *Astrophysical Journal* 553 (2001): 47–72.

Fung Yu-lan. *Chinese Philosophy.* Vol. 2. Princeton, N.J.: Princeton University Press, 1953.

————. *A Short History of Chinese Philosophy.* New York: Macmillan, 1959.

Geller, M. J., and J. P. Huchra. "Mapping the universe." *Science* 246 (1989): 897–903.

Gerber, P. *Zeitschrift für Mathematik und Physik* 43 (1898): 93–104.

Glansdorff, P., and I. Prigogine. *Thermodynamic Theory of Structure, Stability, and Fluctuations.* New York: Wiley, 1971.

Good, T. *Above Top Secret: The Worldwide UFO Cover-Up.* New York: Morrow, 1988.

Graneau, P. "Electromagnetic jet-propulsion in the direction of current flow." *Nature* 295 (1982): 311–12.

———. "First indication of ampere tension in solid electric conductors." *Physics Letters* 97A (1983): 253–55.

———. "Amperian recoil and the efficiency of railguns." *Journal of Applied Physics* 62 (1987): 3006–9.

Graneau, P., and P. N. Graneau. "Electrodynamic explosions in liquids." *Applied Physics Letters* 46 (1985): 468–70.

Graves, R. *The Greek Myths.* Vol. 1. Edinburgh: Penguin Books, 1955.

Gray, J. *Near Eastern Mythology.* New York: Hamlyn, 1969.

Gray, W. *General System Formation Precursor Theory.* Newton Centre, Mass.: self-published, 1977.

———. "Understanding creative thought processes: An early formulation of the emotional-cognitive structure theory." *Man-Environment Systems* 9 (1979): 3–14.

———. "The evolution of emotional-cognitive and system precursor theory." In *Living Groups: Group Psychotherapy and General System Theory,* edited by James E. Durkin, 199–215. New York: Brunner/Mazel, 1981.

Gribbin, J. *White Holes: Cosmic Gushers in the Universe.* New York: Delta, 1977.

———. *In Search of the Big Bang.* New York: Bantam Books, 1986.

Hamilton, E. *Mythology: Timeless Tales of Gods and Heroes.* New York: Mentor, 1940.

Hawking, S. W. *A Brief History of Time: From the Big Bang to Black Holes.* New York: Bantam Books, 1988.

Hayden, H. C. "Light speed as a function of gravitational potential." *Galilean Electrodynamics* 1(1990): 15–17.

Heidel, A. *The Babylonian Genesis.* Chicago: University of Chicago Press, 1942.

Herschkowitz-Kaufman, M. "Bifurcation analysis of nonlinear reaction-diffusion equations: II. Steady-state solutions and comparison with numerical simulations." *Bulletin of Mathematical Biology* 37 (1975): 589–635.

Hesiod. *Theogony.* In *Mythology: Timeless Tales of Gods and Heroes.* New York: Mentor, 1940.

Hickson, P., and P. J. Adams. "Evidence for cluster evolution from the Θ-z relation." *Astrophysical Journal* 234 (1979): L91–L95.

Hoagland, R. C. *The Monuments of Mars: A City on the Edge of Forever.* 1987. Berkeley, Calif.: North Atlantic Books, 1992.

"How to 'fall' into space." *Business Week* (8 February 1958): 51–53.

Hubble, E. "A relation between distance and radial velocity among extra-galactic nebulae." *Proceedings of the National Academy of Sciences* 15 (1929):168–73.

———. "Effects of red shifts on the distribution of nebulae." *Astrophysical Journal* 84 (1936): 517.

Hubble, E., and R. C. Tolman. "Two methods of investigating the nature of the nebular red-shift." *Astrophysical Journal* 82 (1935): 302–37.

Huffaker, C. B. "Experimental studies on predation: Dispersion factors and predator-prey oscillations." *Hilgardia* 27 (1958): 343.

Hyland, D. A. *The Origins of Philosophy.* New York: Putnam, 1973.

Iamblichus. *De mysteriis aegyptiorum.* Oxford folio, 1678. Rare book collection, New York Public Library.

———. *An Egyptian Initiation.* Translated by P. Christian, translated from the French by Genevieve Stebbins Astley. 1901. Denver: Edward Bloom, 1965.

———. *Iamblichus on the Mysteries of the Egyptians, Chaldeans, and Assyrians.* Translated by T. Taylor. 1821. London: Stuart & Watkins, 1968.

Intel. "Towards flight without stress or strain . . . or weight." *Interavia* (Switzerland) 11, no. 5 (1956): 373–74.

Ives, H. "The aberration of clocks and the clock paradox." *Journal of the Optical Society of America* 27 (1937): 305–9.

———. "Light signals sent around a closed path." *Journal of the Optical Society of America* 28 (1938): 296–99.

———. "Derivation of the Lorentz transformations." *Philosophical Magazine* 36 (1945): 392–403.

———. "Lorentz-type transformations as derived from performable rod and clock operations." *Journal of the Optical Society of America* 39 (1949): 757–61.

———. "Extrapolation from the Michelson-Morley experiment." *Journal of the Optical Society of America* 40 (1950): 185–91.

———. "Revisions of the Lorentz transformations." *Proceedings of the American Philosophical Society* 95 (1951): 125–31.

———. "The clock paradox in relativity theory." *Nature* 168 (1951): 246.

Jeans, J. *Astronomy and Cosmogony.* London: Cambridge University Press, 1928.

Jennings, H. *The Rosicrucians: Their Rites and Mysteries.* New York: Arno Press, 1976.

Kaplan, S. R. *The Encyclopedia of Tarot.* Vol. 2. New York: U.S. Games Systems, 1986.

Kellermann, K. I. "Radio galaxies, quasars, and cosmology." *Astronomical Journal* 77 (1972): 531–42.

Kellermann, K. I., et al. "The small radio source at the Galactic center." *Astrophysical Journal* 214 (1977): L61–L62.

Kinsley, D. *The Goddesses' Mirror: Visions of the Divine from East and West.* Albany: State University of New York Press, 1989.

Kramer, S. N. *Sumerian Mythology.* Philadelphia: American Philosophical Society, 1944.

Krisch, A. D. "Collisions between spinning protons." *Scientific American* 257, no. 2 (1987): 42–50.

Laszlo, E. *Introduction to Systems Philosophy: Toward a New Paradigm of Contemporary Thought.* New York: Harper & Row, 1972.

———. *The Systems View of the World.* New York: Braziller, 1972.

———, ed. *The Relevance of General Systems Theory.* New York: Braziller, 1972.

LaViolette, P. A. "Thoughts about thoughts about thoughts: The emotional-perceptive cycle theory." *Man-Environment Systems* 9 (1979): 15–47.

———. "The thermodynamics of the 'aha' experience." In *Proceedings of the 24th Annual Meeting of the Society for General Systems Research* (January 1980): 460–72. Reprinted in *General Systems Theory and the Psychological Sciences,* vol. 1., edited by W. Gray, J. Fidler, and J. Battista, 275–87. Seaside, Calif.: Intersystems Press, 1982.

———. "A reaction-diffusion model of space-time." Paper presented at a workshop on nonlinear chemical systems chaired by I. Prigogine, Austin, Texas, March 1980.

———. "Galactic explosions, cosmic dust invasions, and climatic change." Ph.D. diss., Portland State University, 1983.

———. "An introduction to subquantum kinetics." *International Journal of General Systems* 11(1985): 281–345.

———. "Is the universe really expanding?" *Astrophysical Journal* 301 (1986): 544–53.

———. "Cosmic ray volleys from the Galactic center and their impact on the earth environment." *Earth, Moon, and Planets* 38 (1987): 241–86.

———. "An open system approach to electromagnetic wave propagation." *Proceedings of the 34th Annual Meeting of the International Society for the Systems Sciences* 2: 1119–26. Portland, Ore., 1990.

———. "A Tesla wave physics for a free energy universe." In *Proceedings of the 1990 International Tesla Symposium,* edited by S. R. Elswick, 5.15–.19. Colorado Springs, Colo.: International Tesla Society, 1991.

———. "Electrogravitics: An energy-efficient means of spacecraft propulsion." *Explore* 3, no. 1(1991): 76–79.

———. "Subquantum kinetics: Exploring the crack in the first law." In *Proceedings of the Twenty-sixth Annual Intersociety Energy Conversion Engineering Conference.* Boston, 1991.

———. "Electrogravitics: Back to the future." *Electric Spacecraft Journal* 4 (1992): 23–28.

———. "A theory of electrogravitics." *Electric Spacecraft Journal* 8 (1992): 33–36.

———. "The planetary-stellar mass-luminosity relation: Possible evidence of energy nonconservation?" *Physics Essays* 5 (1992): 536–42. Erratum: *Physics Essays* 6, no. 4 (1993): 616.

———. "The matter of creation." *Physics Essays* 7, no. 1 (1994): 1–8.

———. *Subquantum Kinetics: A Systems Approach to Physics and Cosmology.* Second edition. Schenectady, N.Y.: Starlane Publications, 1994, 2003.

———. *Earth Under Fire: Humanity's Survival of the Apocalypse.* Schenectady, N.Y.: Starlane Publications, 1997.

———. *The Talk of the Galaxy: An ET Message for Us?* Schenectady, N.Y.: Starlane Publications, 2000.

Le Floch, A., and F. Bretenaker. "Early cosmic background." *Nature* 352 (1991): 198.

Lefever, R. "Dissipative structures in chemical systems." *Journal of Chemical Physics* 49 (1968): 4977–78.

Lefever, R., and G. Nicolis. "Chemical instabilities and sustained oscillations." *Journal of Theoretical Biology* 30 (1971): 267–84.

Lerner, E. "Radio absorption by the intergalactic medium." *Astrophysical Journal* 361 (1990): 63–68.

———. *The Big Bang Never Happened.* New York: Vintage Books, 1992.

———. "The cosmologists' new clothes." *Sky and Telescope* (February 1992): 124; (August 1992): 126.

Lindley, D. "Cold dark matter makes an exit." *Nature* 349 (1991): 14.

Lo, K. Y., et al. "On the size of the Galactic centre compact radio source: Diameter <2OAU." *Nature* 315 (1985): 124–26.

Lorentz, H. A. *The Theory of Electrons.* 1909. New York: Dover, 1952.

McCormmach, R. "H. A. Lorentz and the electromagnetic view of nature." *Isis* 61(1970): 459–97.

McCrea, W. H. "Continual creation." *Monthly Notices of the Royal Astronomical Society* 128 (1964): 335–44.

McIntosh, C. *Eliphas Lévi and the French Occult Revival.* London: Rider, 1972.

Mees, G. H. *The Book of Stars.* Dordrecht, Netherlands: Kluwer, 1954.

Meher Baba. *The Everything and the Nothing.* Sydney: Meher House, 1963.

Meyer, E. *Chronologie égyptienne.* Translated by Moret. Paris: Annales du Musée Guimet, 1912.

Mie, G. "Grundlagen einer Theorie der Materie." *Annals of Physics* 37 (1912): 511–34.

Monti, R. "Albert Einstein and Walter Nernst: A comparative cosmology." *SeaGreen* 4 (1986): 32–36.

Moor, E. *The Hindu Pantheon.* 1810. Los Angeles, Calif.: The Philosophical Research Society, 1976.

Muck, O. *The Secret of Atlantis.* New York: Pocket Books, 1978.

Needham, J. *Science and Civilization in China.* Vol. 4. Cambridge: Cambridge University Press, 1956.

Nernst, W. von. *The Structure of the Universe in Light of Our Research.* Berlin: Springer, 1921.

———. "Additional test of the assumption of a stationary state in the universe." *Zeitschrift für Physik* 106 (1938): 633–61.

"New study questions expanding universe." *Astronomy* (August 1986): 64.

Nicolis, G. and I. Prigogine. *Self-Organization in Nonequilibrium Systems.* New York: Wiley Interscience, 1977.

Noorbergen, R. *Secrets of the Lost Races.* New York: Barnes & Noble, 1977.

Oken, A. *The Horoscope, the Road, and Its Travelers.* New York: Bantam Books, 1974.

Ouspensky, P. D. *A New Model of the Universe.* 1931. New York: Vintage Books, 1971.

Pappas, P. T. "The original Ampère force and Biot-Savart and Lorentz forces." *Il nuovo cimento* 76B (1983): 189–97.

———. "The non-equivalence of the Ampère and Lorentz/Grassmann force laws and longitudinal contact interactions." *Physics Essays* 3 (1990): 15–23.

———. "On Ampère electrodynamics and relativity." *Physics Essays* 3 (1990): 117–21.

———. "Energy creation in electrical sparks and discharges." In *Proceedings of the Twenty- sixth Annual Intersociety Energy Conversion Engineering Conference.* Boston, 1991.

Pappas, P. T., and T. Vaughan. "Forces on a stigma antenna." *Physics Essays* 3 (1990): 211–16.

Papus. *The Tarot of the Bohemians: The Most Ancient Book in the World.* Translated by P. Morton. 1889. Hollywood, Calif.: Wilshire, 1973.

Peterson, I. "Cosmic evidence of a smooth beginning." *Science News* 137 (1990): 36.

Piankoff, A. *The Tomb of Ramses VI.* Vol. 1. New York: Pantheon Books, 1954.

———. *Egyptian Religious Texts and Representations.* Vol. 3, *Mythological Papyrii.* New York: Pantheon Books, 1957.

Plato. *The Collected Dialogues of Plato.* Edited by E. Hamilton and H. Cairns. Translated by A. E. Taylor. Princeton, N.J.: Princeton University Press, 1961.

Plutarch. "Isis and Osiris." In *Moralia,* translated by E. Babbitt. Cambridge, Mass.: Harvard University Press, 1936.

Prigogine, I., P. M. Allen, and R. Herman. "Long-term trends and the evolution of complexity." In *Goals in a Global Community.* Vol. 1, *Studies on the Conceptual Foundations,* edited by E. Laszlo and J. Bierman. New York: Pergamon Press, 1977.

Prigogine, I., G. Nicolis, and A. Babloyantz. "Thermodynamics of evolution." *Physics Today* 25, no. 11 (1972): 23–28; 25, no. 12 (1972): 38–44.

Prigogine, I., and I. Stengers. *Order out of Chaos: Man's New Dialogue with Nature.* New York: Bantam Books, 1984.

Regener, E. "Der Energiestrom der Ultrastrahlung." *Zeitschrift für Physik* 80 (1933): 666–69.

Rho Sigma. *Ether Technology: A Rational Approach to Gravity Control.* Lakemont, Ga.: self-published, 1977.

Rieke, G. H., and M. J. Rieke. "Stellar velocities and the mass distribution in the Galactic center." *Astrophysical Journal* 330 (1988): L33–L37.

Roberts, N., et al. "Timing of the younger Dryas event in East Africa from lake-level changes." *Nature* 366 (1993): 146–48.

Ruben, B. D., and J. Y. Kim. *General Systems Theory and Human Communication.* Rochelle Park, N.J.: Hayden, 1975.

Rundle Clark, R. T. *Myth and Symbol in Ancient Egypt.* New York: Thames & Hudson, 1959.

Sachs, A. "Babylonian horoscopes." *Journal of Cuneiform Studies* 6 (1952): 49–75.

Sagnac, G. "The luminiferous ether demonstrated by the effect of the relative motion of the ether in an interferometer in uniform rotation." *Comptes rendus de l'Academie des Sciences* (Paris) 157 (1913): 708–10, 1410–13.

Sandburg, C. *Chicago Poems.* New York: Holt, Rinehart & Winston, 1916.

Schlegel, G. *Uranographie chinoise.* Vol. 2. Leyden, 1875.

Schwaller de Lubicz, R. A. *Sacred Science: The King of Pharaonic Theocracy.* 1961. Translated by André and Goldian Vandenbroeck. Rochester, Vt.: Inner Traditions, 1982.

———. *The Egyptian Miracle: An Introduction to the Wisdom of the Temple.* 1963. Translated by André and Goldian Vandenbroeck. Rochester, Vt.: Inner Traditions, 1985.

Scott, O. E. *Stars in Myth and Fact.* Caldwell, Idaho: Caxton, 1947.

Shapiro, S. L., and S. A. Teukolsky. "Formation of naked singularities: Violation of cosmic censorship." *Physical Review Letters* 66 (1991): 994–97.

Shapley, H., and A. Ames. "A survey of the external galaxies brighter than the thirteenth magnitude." *Annals of the Astronomical Observatory of Harvard College* 88, no. 2 (1932): 42–75.

Silvertooth, E. W. "Experimental detection of the ether." *Speculations in Science and Technology* 10 (1987): 3–7.

———. "Motion through the ether." *Electronics and Wireless World* (May 1989): 437–38.

Silvertooth, E. W., and C. K. Whitney. "A new Michelson-Morley experiment." *Physics Essays* 5 (1992): 82–89.

Smith, D. E. *History of Mathematics.* Vol. 1. 1923. New York: Dover, 1951.

Smoot, G. F., et al. "Preliminary results from the COBE differential microwave radiometers: Large angular scale isotropy of the cosmic microwave background." *Astrophysical Journal* 371 (1991): L1–L5.

Stapp, H. "Are superluminal connections necessary?" *Il nuovo cimento* 40B (1977): 191–205.

Steiger, B. *Worlds before Our Own.* New York: Berkley Books, 1978.

Talley, R. L. *Twenty-First Century Propulsion Concept.* U.S. Air Force report no. PL-TR-91-3009. Washington, D.C.: National Technical Information Service, 1991.

Thompson, R. L. *Alien Identities: Ancient Insights into Modern UFO Phenomena.* San Diego, Calif.: Govardhan Hill, 1993.

Thomsen, D. E. "Unconventional physics: Quite crazy after all these years." *Science News* (26 July 1986): 55.

Tilley, R. *Playing Cards.* London: Octopus Books, 1973.

Townes, C. H., et al. "The centre of the galaxy." *Nature* 301 (1983): 661–66.

Tully, R. B. "Origin of the Hubble constant controversy." *Nature* 334 (1988): 209–12.

Turner, D., and R. Hazelett. *The Einstein Myth and the Ives Papers: A Counter-Revolution in Physics.* Old Greenwich, Conn.: Devin-Adair, 1979.

Tyson, J. A., and J. F. Jarvis. "Automated faint galaxy counts and galactic evolution." In *Objects of High Redshift,* edited by G. O. Abell and P. J. E. Peebles. Paris: International Astronomical Union, 1980.

Valone, T., ed. *Electrogravitics Systems: Reports on a New Propulsion Methodology.* Washington, D.C.: Integrity Research Institute, 1994.

Valtonen, M., and G. Byrd. "Redshift asymmetries in systems of galaxies and the missing mass." *Astrophysical Journal* 303 (1986): 523–34.

Waldrop, M. M. "Sanduleak -69 202: Guilty as charged." *Science* 236 (May 1987): 523.

Walker, F. L. "A contradiction in the theory of universal expansion." *Apeiron* (Fall 1989): 1–6.

———. "State of the universe: A theoretical proof." Unpublished paper, 1992.

Walker, J. "Chemical systems that oscillate between one color and another." *Scientific American* 239, no. 1(1978): 152-58.

West, J. A. *The Traveler's Key to Ancient Egypt.* New York: Knopf, 1985.

———. *Serpent in the Sky: The High Wisdom of Ancient Egypt.* Wheaton, Ill.: Quest, 1993.

West, J. A., and J. G. Toonder. *The Case for Astrology.* Baltimore: Penguin Books, 1973.

Whitehead, A. N. *Science and the Modern World.* New York: Free Press, 1925.

———. *Process and Reality.* New York: Macmillan, 1929.

Winfree, A. T. "Spiral waves of chemical activity." *Science* 175 (1972): 634–36.

Wirth, O. "Essay upon the astronomical Tarot." In *The Tarot of the Bohemians,* by Papus, translated by P. Morton, 242–43. 1889. Hollywood, Calif.: Wilshire, 1973.

Wright, T. *An Original Theory of the Universe.* London: Macdonald, 1750.

Yogananda, P. *Autobiography of a Yogi.* 1946. Los Angeles: Self-Realization Fellowship, 1985.

Zaikin, A., and A. N. Zhabotinskii. "Concentration wave propagation in two-dimensional liquid-phase self-oscillating system." *Nature* 225 (1970): 535–37.

Zain, C. C. *The Sacred Tarot.* Los Angeles: The Church of Light, 1936.

Zhao, J., W. M. Goss, K. Y. Lo, and R. D. Ekers. "High-resolution VLA images of the Galactic center at 2-cm wavelength with large dynamic range." *Nature* 354(1991): 46–48.

Zimmer, H. *Myths and Symbols in Indian Art and Civilization.* 1946. Princeton, N.J.: Princeton University Press, 1972.

———. *The Art of Indian Asia.* Vol. 1. Edited by J. Campbell. Princeton, N.J.: Princeton University Press, 1960.

Zukov, G. *The Dancing Wu Li Masters: An Overview of the New Physics.* New York: Bantam Books, 1979.

Zwicky, F. "On the red shift of spectral lines through interstellar space." *Proceedings of the National Academy of Sciences* 15 (1929): 773–79.

INDEX

BOOKS OF RELATED INTEREST

Secrets of Antigravity Propulsion
Tesla, UFOs, and Classified Aerospace Technology
by Paul A. LaViolette, Ph.D.

Earth Under Fire
Humanity's Survival of the Ice Age
by Paul A. LaViolette, Ph.D.

Decoding the Message of the Pulsars
Intelligent Communication from the Galaxy
by Paul A. LaViolette, Ph.D.

Science and the Akashic Field
An Integral Theory of Everything
by Ervin Laszlo

The Akashic Experience
Science and the Cosmic Memory Field
by Ervin Laszlo

Transcending the Speed of Light
Consciousness, Quantum Physics, and the Fifth Dimension
by Marc Seifer, Ph.D.

Chaos, Creativity, and Cosmic Consciousness
*by Rupert Sheldrake, Terence McKenna,
and Ralph Abraham*

Precessional Time and the Evolution of Consciousness
How Stories Create the World
by Richard Heath

Inner Traditions • Bear & Company
P.O. Box 388
Rochester, VT 05767
1-800-246-8648
www.InnerTraditions.com

Or contact your local bookseller